Bioremediation of Petroleum Hydrocarbons in Cold Regions

This practical guide to bioremediation in cold regions is designed to aid environmental practitioners, industry, and regulators in the remediation of petroleum spills and contaminated sites in cold regions. Remediation design and technology used in temperate climates does not necessarily work in cold climates, and cleanup takes longer due to shorter treatment seasons, sub-freezing temperatures, ground freezing and thawing, and limited bioactivity. Environmental engineers and scientists from eight countries working in the polar regions combine their experiences and expertise with petroleum contamination in this book. It contains in-depth discussions on regulations, freezing and frozen ground, identification and adaptations of cold-tolerant bacteria, contaminant transport in cold soils and permafrost, temperature effects on biodegradation, analytical methods, treatability studies, and nutritional requirements for bioremediation. Emphasis is given to practical and effective bioremediation methods for application in cold regions using case studies and cost–benefit analyses. Emerging technologies are also discussed. This volume will be an important reference for students, researchers, and practitioners of environmental science and engineering.

DENNIS FILLER is an Assistant Professor and Fulbright Scholar at the University of Alaska, Fairbanks, in the Department of Civil & Environmental Engineering.

IAN SNAPE is a Contaminants Geochemist, working for the Australian Antarctic Division in Tasmania.

DAVID BARNES is an Associate Professor at the University of Alaska, Fairbanks, in the Water and Environmental Research Center, part of the Department of Civil & Environmental Engineering.

Bioremediation of Petroleum Hydrocarbons in Cold Regions

Edited by
DENNIS M. FILLER
IAN SNAPE
DAVID L. BARNES

CAMBRIDGE UNIVERSITY PRESS
Cambridge, New York, Melbourne, Madrid, Cape Town,
Singapore, São Paulo, Delhi, Mexico City

Cambridge University Press
The Edinburgh Building, Cambridge CB2 8RU, UK

Published in the United States of America by Cambridge University Press, New York

www.cambridge.org
Information on this title: www.cambridge.org/9781107410503

First published 2008
First paperback edition 2012

A catalogue record for this publication is available from the British Library

ISBN 978-0-521-86970-6 Hardback
ISBN 978-1-107-41050-3 Paperback

Dedicated to Peter J. Williams, whose inspiration and leadership pioneered the study of contaminants in freezing ground

Contents

Contributors

Larry Acomb
Geosphere Inc., 3055 Seawind Drive, Anchorage AK 99516, USA

Jackie Aislabie
Landcare Research New Zealand Ltd., Private Bag 3127, Hamilton, New Zealand

Steve Bainbridge
Contaminated Sites Program, Division of Spill Prevention and Response, Department of Environmental Conservation, 610 University Avenue, Fairbanks AK 99709–3643, USA

David L. Barnes
Dept. of Civil and Environmental Engineering, University of Alaska Fairbanks, PO Box 755900, Fairbanks AK 99775, USA

Kevin Biggar
BGC Engineering, Inc., 207, 5140–82 Avenue, Edmonton, Alberta, Canada T6B OE6

Robert Eno
Department of Sustainable Development, Government of Nunavut, PO Box 1000, Stn 1195, Iqaluit NU X0A 0H0, Canada

Susan Ferguson
Environmental Protection and Change Program, Australian Antarctic Division, Channel Highway, Kingston, Tasmania 7050, Australia

Dennis M. Filler
Dept. of Civil and Environmental Engineering, University of Alaska Fairbanks, PO Box 755900, Fairbanks AK 99775, USA

Julia Foght
Biological Sciences, University of Alberta, Edmonton Alberta, Canada

Walter Fourie
Dept. of Civil and Environmental Engineering, University of Alaska Fairbanks, PO Box 755900, Fairbanks AK 99775, USA

D. Sarah Garland
Dept. of Civil and Environmental Engineering, University of Alaska Fairbanks, PO Box 755900, Fairbanks AK 99775, USA

Ronald A. Johnson
Dept. of Mechanical Engineering, Institute of Northern Engineering Energy Research Center, University of Alaska Fairbanks, PO Box 755910, Fairbanks AK 99775-5910, USA

Kate Mumford
Particulate Fluids Processing Centre (ARC Special Research Centre), Department of Chemical and Biomolecular Engineering, University of Melbourne, Victoria 3010, Australia

Natalie Plato
Department of Sustainable Development, Government of Nunavut, PO Box 1000, Stn 1195, Iqaluit NU X0A 0H0, Canada

John S. Poland
Analytical Services Unit, Queens University, Kingston ON K7L 3N6, Canada

Tania C. Raymond
Environmental Protection and Change Program, Australian Antarctic Division, Channel Highway, Kingston, Tasmania 7050, Australia

John L. Rayner
Environmental Protection and Change Program, Australian Antarctic Division, Channel Highway, Kingston, Tasmania 7050, Australia

C. Mike Reynolds
US Army Engineer Research and Development Center, Cold Regions Research and Engineering Laboratory, 72 Lyme Road, Hanover NH 03755, USA

Martin J. Riddle
Environmental Protection and Change Program, Australian Antarctic Division, Channel Highway, Kingston, Tasmania 7050, Australia

Anne Gunn Rike
Dept. of Environmental Technology, Norwegian Geotechnical Institute, PO Box 3930, Ullevaal Stadion, N-0806 Oslo, Norway

Allison Rutter
Analytical Services Unit, Queens University, Kingston ON K7L 3N6, Canada

Alexis N. Schafer
University of Saskatchewan, 51 Campus Drive, Saskatoon, Canada S7N 5A8

Silke Schiewer
Dept. of Civil and Environmental Engineering, University of Alaska Fairbanks, PO Box 755900, Fairbanks AK 99775, USA

William Schnabel
Golder Associates, 1346 West Arrowhead Road, Duluth MN 55811, USA

Yuri Shur
Dept. of Civil and Environmental Engineering, University of Alaska Fairbanks, PO Box 755900, Fairbanks AK 99775, USA

Steven D. Siciliano
University of Saskatchewan, 51 Campus Drive, Saskatoon SK S7N 5A8, Canada

Ian Snape
Environmental Protection and Change Program, Australian Antarctic Division, Channel Highway, Kingston, Tasmania 7050, Australia

Dale Van Stempvoort
National Water Research Institute, PO Box 5050, Burlington ON, Canada L7R 4A6

James L. Walworth
Dept. of Soil Water and Environmental Science, University of Arizona, 429 Shantz Bldg. #38, Tucson AZ 85721, USA

Daniel M. White
Institute of Northern Engineering, PO Box 755910, University of Alaska Fairbanks, Fairbanks AK 99775-5910, USA

Craig R. Woolard
Anchorage Water and Wastewater Utility, 3000 Arctic Boulevard, Anchorage AK 99503-3898, USA

Preface

Bioremediation of Petroleum Hydrocarbons in Cold Regions is written by a multi-disciplinary group of scientists and engineering professionals working in polar regions. The monograph is designed as a state-of-the-art guidance book to assist industry, environmental practitioners, and regulators with environmental cleanup in cold climates. The book can also be used for environmental science and remediation engineering seniors and graduate students who are preparing for a career in professional environmental practice or applied scientific research. The intent of this book is to articulate conditions unique to our cold regions, and present practical and cost-effective remediation methods for removing petroleum contamination from tundra, taiga, alpine, and polar terrain.

Oil and its refined products represent a significant proportion of the pollution found in the Arctic and Antarctic. This pollution is encountered at former military and industrial sites, scientific research stations, rural communities, and remote airstrips, while recent spills and releases tend to be associated with resource development and transportation mishaps. Bioremediation is recognized as potentially the most cost-effective technology for removing petroleum contaminants from ecosystems in cold regions.

Permafrost, suprapermafrost water, tundra, cold-tolerant microorganisms, short summers and long, dark winters, cold air and ground temperatures, and annual freezing and thawing of the active layer are but a few environmental characteristics of cold regions. Their prevalence limits practical remediation methods and has led to the development of innovative and pragmatic bioremediation schemes for use at contaminated sites in cold climates. Case studies and costs (where known) are integrated in appropriate chapters to aid decision making.

Scope

For the purposes of defining the scope of this volume, we define *bioremediation* as any process that involves biological processes, through the action of biota such as bacteria, fungi, and plants, in the transformation and breakdown of petroleum hydrocarbons. Since abiotic breakdown, such as evaporation, is often an important or even dominant mechanism in some biologically active treatment systems, this process is not excluded from the broad definition of bioremediation used here. However, purely abiotic treatments, such as thermal incineration and *in situ* chemical oxidation, are not considered in detail here.

Petroleum Hydrocarbons are hydrogen and carbon based compounds derived from rock, mineral or natural oils, and any of their derivatives. From a remediation perspective, this book mainly considers crude oil and its refined products such as diesel, kerosene, aviation fuels, and gasoline.

Cold Regions are Arctic and sub-Arctic, Antarctic and sub-Antarctic, and alpine geographic regions that exhibit permafrost or experience seasonally frozen ground. Additionally, maritime subpolar regions that do not freeze but share many other environmental characteristics are also included.

Future initiatives

We recognize that this book is incomplete in many regards. For example, there is little integrated information from the many disseminated mid-latitude and high-altitude fuel spills. Similarly, reliable information from Russia and China is sadly lacking. Furthermore, the focus of this volume is soil treatment; much still needs to be done with respect to treating petroleum-contaminated water in cold regions.

We welcome further information and criticism. Please send comments to Dennis Filler at ffdmf@uaf.edu.

Glossary

Absorption is retention of a solute within the mass of a solid rather than on its surface.

Action value is a contaminant concentration beyond which remediation is considered necessary.

Active layer (or seasonally thawed layer) is the upper realm of soil or permafrost that experiences annual freezing and thawing as a function of temperature.

Adsorption is retention of a solute by the surface of a solid rather than within its mass.

Air freezing index *(AFI)* is the index representing the number of negative ($T < 0\,^{\circ}C$) degree–days between the highest and lowest points on a curve of cumulative daily average air temperatures versus time (degree–days) for a given location. AFI is calculated as the area below the freezing isotherm drawn through such a curve.

Air sparging is an *in situ* technology where air is injected into contaminated water-saturated soils, or aquifers, at high flow rates (10–20 m^3/h) to promote contaminant volatilization.

Air thawing index is the index representing the annual total thawing degree–days (i.e. area above the freezing isotherm) of the average daily temperature curve for a given location.

Allochthonous microorganisms are nonindigenous bacteria.

Amino acids are organic acids that contain a carboxyl group (COOH) and amino group (NH_2), linked together into polypeptide chains to form proteins that are essential to all life.

Ammonia volatilization is loss of ammonia gas (NH_3) from the soil into the atmosphere.

Ammonification is the production of ammonia resulting from the bacterial conversion of organic nitrogenous compounds.

Anthropogenic refers to man-made in origin.

Autochthonous refers to native soil bacteria whose populations do not change rapidly in response to the addition of specific nutrients.

Autotrophs are organisms that produce complex organic compounds from simple inorganic molecules, e.g. soil bacteria that mineralize ammonium into nitrate (i.e., autotrophic nitrification).

Bioaugmentation is soil amendment with non-indigenous microorganisms (inoculation), bioproducts, or engineered microbes to enhance the biodegradation of contaminants.

Bioavailability is the fraction of the total of a chemical in the surrounding environment (i.e. water, sediment, soil, suspended particles) that can be taken up by organisms.

Biodegradation is the breaking of intramolecular bonds (or *breakdown*) of organic substances by microorganisms to derive energy.

Biogenic interference is the fraction of natural organic matter that cannot be distinguished from petroleum in a standard petroleum analysis.

Biopile is a remediation technique involving heaping contaminated soils into piles and stimulating aerobic microbial activity within the soils through aeration and/or addition of minerals, nutrients, and moisture.

Bioreactor is a closed-vessel system in which biological, chemical, and physical processes occur, and that permits control of process parameters (e.g. temperature, moisture, oxygen, pH, and nutrients).

Bioremediation is any biological process, through the action of bacteria, fungi, plants, or managed biodegradation, that transforms or breaks down environmental contaminants.

Biosparging is an *in situ* technology where air is injected into contaminated water-saturated soils, or aquifers, at low flow rates (<5 m^3/h or 3 cfm per point) to promote aerobic biodegradation. Volatilization is typically less than that of the standard air sparging system.

Biostimulation is the modification of the environment (e.g., by the addition of nutrients, oxygen, etc.) to stimulate the rate of biological degradation of contaminants by indigenous microorganisims.

Bioventing is the process of injecting air (i.e., aeration) into subsurface soils to stimulate *in situ* bioremediation using soil vapor extraction systems.

Capillary forces are mechanical forces exerted on the soil–water–gas system at the pore scale. The combined effect of *surface tension* and effective *contact angle* occurs at the interface between water, air and soil in the interconnected soil pores. Surface tension is the tangential force acting at the interface between the water and air vapor, caused by the difference in attraction between liquid and gaseous molecules. The effective contact angle is the equilibrium angle of contact of water on the soil particle surface, measured within the water at the contact line where the three phases (liquid, solid, and gas) meet.

Capillary fringe is the zone of soil immediately above the water table that *wicks* water up from the underlying water table (i.e., soil suction) and retains this water by capillary forces. As most soil pores are completely filled with water at the base of the capillary fringe, but only the smallest soil pores are water filled at the top, moisture content of the capillary fringe decreases with increasing distance above the water table. Furthermore, because the size of the soil pores define the capillary rise of the water column, and since soil pore size is highly heterogenous, the height of capillary rise has a ragged upper surface.

Chromatography is an analytical technique for separating substances by adsorption on media for which they have different affinities. A *chromatogram* is the graphical output of the chromatograph, in which different peaks or patterns correspond to different components of the separated mixture.

Cold regions are Arctic and sub-Arctic, Antarctic and sub-Antarctic, and alpine lands that exhibit permafrost or experience seasonally frozen ground.

Cometabolism is the simultaneous metabolism of two compounds, in which the degradation of the second compound requires the presence of the first compound.

Denitrification is the process of reducing nitrates and nitrites, most commonly by soil bacteria (denitrifying bacteria), into nitrogen-containing gases (e.g., N_2, N_2O, NO, NO_2). Denitrification is primarily an anerobic process.

End Point is a contaminant concentration that once reached, remediation is considered to be complete.

Engineered remediation is the design, analysis, and/or construction of works for the removal of environmental contaminants from media such as soil, groundwater, sediment, or surface water.

Engineered bioremediation is the design, analysis, and/or construction of works for the removal of environmental contaminants from media such as soil, groundwater, sediment, or surface water utilizing microorganisms to enhance bioactivity.

Enzyme is a compound, usually a protein, that acts as a catalyst to a specific chemical reaction.

Evolutionary operation is the process of optimizing a system(s) to maximize its performance. In bioremediation, this process is used to fine-tune systems in order to maximize treatment effectiveness and minimize operation and maintenance costs.

Extraction *(solvent)* is a method to separate compounds based on their solubility in various solvents. Petroleum compounds are extracted out of an aqueous phase into an organic phase for chromatography and quantitation.

Field trial refers to a treatability study or pilot study that is implemented at a contaminated site generally to test the performance of a particular remediation technique under the conditions in which it will be used.

Freeze exclusion is the displacement of petroleum from soil pores as soil water freezes.

Funnel and gate is an *in situ* system of cut-off walls (e.g., sheet pile, slurry, or composite walls) and a central gate (e.g., permeable reactive barrier) equipped with permeable reactive or non-reactive media, used to control and treat contaminated water.

Genotypic adaptations refer to changes in an organism's genome over evolutionary time scales.

Heterotrophs are organisms that are only able to utilize organic sources for biosynthesis.

Homologues are identical compounds except for their frequency of repetition (homologous).

Humic substances comprise the refractory, mainly hydrophilic, high molecular weight, chemically complex and polyelectrolyte-like fraction of natural organic matter (NOM).

Hydrolysis is chemical reaction involving cleavage of a molecular bond by reaction with water.

Hydrophilic means water attracting.

Hydrophobic means water repelling.

Hydrophobic bonds *(or interactions)* occur where compounds possess non-polar properties and are brought into close association in aqueous environments due to their repulsion to water.

Hyperthermophiles *(or extreme thermophiles)* are microorganisms with a thermotolerance range above 70 °C, up to about 115 °C.

Immiscible fluids are liquids or gases that cannot mix to form a homogenous mixture.

Indigenous microorganisms are microorganisms that naturally exist in a particular environment, and that usually have a selective, competitive advantage in that environment.

Insolation is incoming solar radiation, or solar heating.

Intrinsic bioremediation *(or natural attenuation)* is bioremediation by natural processes (e.g., biodegradation, abiotic transformation, mechanical dispersion, dilution, and sorption) that reduces contaminants in the environment.

Landfarming is land treatment of soil for degradation or transformation of contaminants by a combination of volatilization and biodegradation by resident microorganisms. A common practice is to place the soil as a shallow layer within a bermed and lined treatment cell (or biocell), occasionally amend the soil with nutrients and water to stimulate biodegradation, and regularly till (aerate) the soil to mix and aid contaminant volatilization.

Magic number refers to an arbitrary cleanup value arrived at when one considers a trigger value as the trigger, action and remediation target.

Matric potential, one of two major components of soil water potential, is the result of the attraction between soil particles (primarily clay size – less than 0.002 mm in diameter) and water.

Mesocosms are small-scale bioassays designed to simulate natural soil conditions for the purpose of evaluating conditions supportive of bioremediation. A mesocosm treatability study can be conducted in the field, and results can be used to assess biodegradation rates and for estimating treatment duration and remediation costs.

Mesophiles are soil microorganisms with a thermotolerance range of 15 °C to 45 °C, that grow optimally between 20 °C to 40 °C, and that can survive at low temperatures (less than 4 °C) without growing.

Methanogenisis is conversion of short-chain organic compounds by anerobic microorganisms to methane, carbon dioxide, and inorganic byproducts.

Microbial consortia are mixed populations of interacting microorganisms.

Microcosms are small-scale experimental bioassays used to evaluate conditions supportive of bioremediation. This type of treatability study does not attempt to simulate a soil ecosystem, but results can be used to infer bioremediation potential and treatment effectiveness.

Microfauna are microscopic and/or very small animals (e.g., protozoans and nematodes) that live in a particular environment.

Mineralization is the breakdown of organic compounds (e.g., organic residues in soil) into inorganic materials (e.g., mineral nutrients that can be utilized by plants).

Natural attenuation is the unassisted biodegradation, evaporation, adsorption, metabolism, or transformation of contaminants in soil and water by microorganisms.

Nitrification is the process of oxidizing ammonia and organic nitrogen to nitrite (mainly by bacteria of the genus *Nitrosomonas*) followed by the oxidation of these nitrites into nitrates (mainly by bacteria of the genus *Nitrobacter*). Nitrification is an aerobic process.

Nitrogen demand refers to the amount of nitrogen required for degradation of a specific amount of contaminant.

Non-aqueous phase liquid *(NAPL)* is an organic liquid that is relatively insoluble in water.

Oligotrophs are organisms able to survive on very low nutrient concentrations.

Osmotic potential, a major component of soil water potential together with matric potential, that results from the interaction of water molecules and dissolved salts.

Permafrost is soil that remains frozen for two or more consecutive years.

Permeable reactive barrier is an open *in situ* reactor system designed to achieve contaminant mass transfer reactions, as with bioreactors, but with somewhat less control.

Permeability is a measure of the ability of a material to transmit fluids.

Permeases are membrane transport proteins that facilitate the diffusion of a specific molecule in or out of the cell.

Petroleum Hydrocarbons are hydrogen and carbon based compounds derived from rock, mineral or natural oils, and any of their derivatives.

Phenotypic adaptations refer to changes in the structural and physiological properties of organisms in response to a genetic mutation or a change in the environment.

Photooxidation is ultraviolet light-induced oxidation for destruction of organic contaminants.

Phytoremediation is the use of plants to remediate contaminated soils or water.

Pilot study generally refers to a scaled down test of a remediation system or component. The study may take place at the contaminated site to simulate operation under field conditions, or in the laboratory as a *bench-scale study*.

Polar regions are geographic regions above the Arctic Circle (N66°33′39″) and south of the Antarctic Circle (N66°33′38″), characterized by long, dark and cold winters followed by a few light summer months. Land in these regions includes the high-Arctic, coastal plains and archipelagos of North America, northern portions of Greenland, Scandinavia and Russia, and Antarctica.

Pore ice is ice that occurs in the pores of soils and rocks.

Proteins are long polypeptide chains often bonded with nucleic acids, lipids, etc., essential to the diet of microorganisms in the composition of cellular structures and in enzymatic processes.

Psychrophiles are soil microorganisms with a thermotolerance range of −5 °C to 25 °C, that grow optimally around 10 °C, and usually do not grow above 20 °C.

Psychrotolerant microbes *(psychrotrophes)* have the ability to grow at low temperatures (less than 4 °C), but optimally between 20 °C to 30 °C.

Psychrotrophic is the ability to grow best at temperatures below 20 °C.

Recalcitrant means resistant to microbial degradation.

Rhizosphere is the soil region in the immediate vicinity of growing plant roots including the region of the soil modified as a result of the presence of a root.

Rhizodegradation is the transformation of contaminants in soil proximal to plant roots by organisms associated with vegetative species.

Segregated ice is ice that occurs as discrete layers or ice lenses in soil.

Sequestering of a contaminant compound occurs when the compound becomes less available or is wholly unavailable to microorganisms for biodegradation.

Soil warming is passive (solar or water conducted) or active (mechanized) heating of soil to enhance biodegradation of contaminants.

Soil water potential is a measure of physical and chemical potential (or energy) of the soil water. Its two major components are matric and osmotic potentials.

Sorption refers to both absorption and adsorption, or the retention of solutes in solution in the solid mass and on surface of the solid.

Sulfate reduction is the conversion of reducible forms of sulfur (e.g., sulfate, thiosulfate) to reduced forms (e.g., sulfides, elemental sulfur) by bacterial anaerobic respiration.

Suprapermafrost water is water that occurs in the active layer above permafrost.

Surface freezing index is the cumulative number of degree-days below 0 °C for the surface temperature (of the soil, ground, etc.) during a given time period.

Surrogates are organic compounds that are similar to analytes of interest in extraction and chromatography, but which are not normally found in environmental samples. These compounds are spiked into method blanks, calibration and check standard, samples (including duplicates and quality control reference samples) and spiked samples prior to analysis. Analysis typically includes calculations of percent recoveries for each surrogate used.

Taliks are layers or bodies of unfrozen soil in permafrost. An open talik is completely surrounded by permafrost, a closed talik is a thawed zone bordered by permafrost and the ground surface and a through talik is bordered by unfrozen layers beneath the permafrost and the ground surface.

Thermokarst is soil deformation (subsidence) that results from the thawing of ground ice and consolidation of the thawed soil.

Thermophiles are organisms that live under conditions of high heat and acid. In the soil these organisms usually have a thermotolerance range of 40 °C to 70 °C, grow optimally between 50 °C to 60 °C, and usually do not grow above 65 °C.

Treatability studies are *in vitro* microcosms with individual bacterial species or soil consortia incubated in liquid or slurry media, mesocosm studies with soils and natural microfauna, or field trials, used to evaluate conditions supportive of bioremediation.

Trigger value is a conservative soil contaminant value (or concentration) below which environmental impacts are considered unlikely. Soil contamination in excess of a trigger value should instigate further assessment, although such values are often used as action values and remediation endpoints.

Vadose zone is unsaturated soil above saturated soil or the water table, and includes the zone of capillary rise.

Volatilization is the transformation of a solid or liquid contaminant into vapor. In bioremediation, volatilization occurs naturally by wind and as consequence of the relative partial pressures of contaminant and air, or is artificially induced by mechanized systems (e.g., tilling, bioventing).

Water holding capacity is the field capacity of soil to store water after saturation, and once gravitational water is drained.

Weathering of a contaminant refers to the degradation of more biodegradable compounds via biotic or abiotic reactions and the formation of an aged residue.

Zeolites are predominantly alumino-silicate hydrated minerals (e.g., chabaite, clinoptilolite, heulandite, natrolite, stilbite) that have a micro-porous structure that can readily accommodate a wide variety of cations (e.g. Na^+, K^+, Ca^{2+}, Mg^{2+}). There are 48 naturally occurring and more than 1500 synthesized zeolite types known.

ACRONYMS

ABS	acrylonitrile butadiene styrene
AFI	air freezing index
ATI	air thawing index
BTEX	benzene, toluene, ethylbenzene, xylenes
CAF	coarse adjustment factor
CALM	Circumpolar Active Layer Monitoring program
CAM	cold-adapted microorganism
C:N	carbon to nitrogen ratio
C:P	carbon to phosphorus ratio
cPCR	competitive polymerase chain reaction
CRREL	Cold Regions Research & Engineering Laboratory
CRN	controlled release nutrients

CSP	cold-shock proteins
CSR	cold-shock response
DEA	diethylamine extraction solvent
DEG	diesel-electric generator
DNA	deoxyribonucleic acid
DRBO	diesel range biogenic organics
DRO	diesel-range organics
FID	flame ionization detection
GC	gas chromatography
GRO	gasoline-range organics
GRPH	gasoline-range petroleum hydrocarbons
HDPE	high-density polyethylene
LNAPL	light non-aqueous phase liquid
LOI	loss of ignition
m.a.a.t.	mean annual air temperature
MAE	microwave assisted extraction
MS	mass spectrometry
MPN	most probable number
NAPL	non-aqueous phase liquid
NOM	natural organic matter
N:P:K	ratio of nitrogen to phosphorus to potassium
PAH	polycyclic aromatic hydrocarbon
PCB	polychlorinated biphenyls
PCR	polymerase chain reaction
PHC	petroleum hydrocarbons
PRB	permeable reactive barrier
RRO	residual range organics
RRBO	residual range biogenic organics
SAB	Special Antarctic Blend
SFI	surface freezing index
TEB	thermally enhanced bioremediation
TPH	total petroleum hydrocarbons
USEPA	United States Environmental Protection Agency

1

Contamination, regulation, and remediation: an introduction to bioremediation of petroleum hydrocarbons in cold regions

IAN SNAPE, LARRY ACOMB, DAVID L. BARNES, STEVE BAINBRIDGE, ROBERT ENO, DENNIS M. FILLER, NATALIE PLATO, JOHN S. POLAND, TANIA C. RAYMOND, JOHN L. RAYNER, MARTIN J. RIDDLE, ANNE G. RIKE, ALLISON RUTTER, ALEXIS N. SCHAFER, STEVEN D. SICILIANO, AND JAMES L. WALWORTH

1.1 Introduction

Oil and fuel spills are among the most extensive and environmentally damaging pollution problems in cold regions and are recognized as potential threats to human and ecosystem health. It is generally thought that spills are more damaging in cold regions, and that ecosystem recovery is slower than in warmer climates (AMAP 1998; Det Norske Veritas 2003). Slow natural attenuation rates mean that petroleum concentrations remain high for many years, and site managers are therefore often forced to select among a range of more active remediation options, each of which involves a trade-off between cost and treatment time (Figure 1.1). The acceptable treatment timeline is usually dictated by financial circumstance, perceived risks, regulatory pressure, or transfer of land ownership.

In situations where remediation and site closure are not urgent, natural attenuation is often considered an option. However, for many cold region sites, contaminants rapidly migrate off-site (Gore *et al.* 1999; Snape *et al.* 2006a). In seasonally frozen ground, especially in wetlands, a pulse of contamination is often

Bioremediation of Petroleum Hydrocarbons in Cold Regions, ed. Dennis M. Filler, Ian Snape, and David L. Barnes. Published by Cambridge University Press.

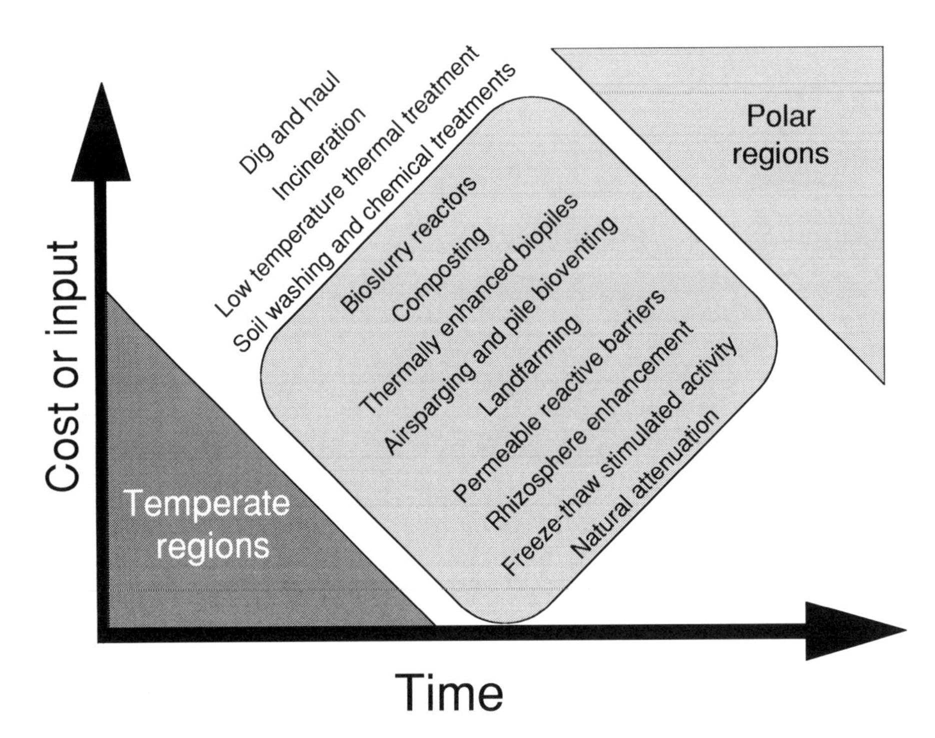

Figure 1.1. Cost-time relationship for remediation options that are suitable for petroleum contaminants. Note that cost and time for a given treatment type are invariably greater in cold regions. Boxed treatments are those considered within the scope of this book (modified after Reynolds *et al.* 1998).

released with each summer thaw (AMAP 1998; Snape *et al.* 2002). In these circumstances natural attenuation is likely not a satisfactory option. Simply excavating contaminants and removing them for off-site treatment may not be viable either, because the costs are often prohibitive and the environmental consequences of bulk extraction can equal or exceed the damage caused by the initial spill (Filler *et al.* 2006; Riser-Roberts 1998). Similarly, in-ground incineration does not effectively treat spills, but rather causes downward migration of contaminants and permafrost degradation through heating (AMAP 1998; Filler and Barnes 2003; UNEP-WCMC 1994).

In situ or on-site bioremediation techniques (Figure 1.1) offer a relatively low-cost approach for managing petroleum-contaminated soils in cold regions, with the potential to achieve reasonable environmental outcomes in a timely manner. The challenge for scientists, engineers, and environmental managers is to derive or refine a range of remedial strategies that are well suited or optimized for cold region conditions. The overall aim of this monograph is to document techniques and scientific principles that underpin good remediation practices so that managers can remediate petroleum hydrocarbons to an appropriate level as quickly and cheaply as possible. This chapter provides an overview of petroleum contamination, regulation, and remediation in cold regions. It describes some of the regulatory frameworks that govern how spills are dealt with in a variety of regions, and provides some recent case studies of how guidelines are developing or evolving, and how petroleum remediation in cold regions is currently viewed in the current regulation context. The chapter concludes by making the case for further quantitative information for development of petroleum remediation guidelines that are more appropriate to cold regions.

1.2 Review – contamination, regulation, and remediation

1.2.1 Contamination: reason for concern[1]

Many petroleum products are used, stored, transported, and subsequently spilled in the cold regions. Colin Campbell of the *Association for the Study of Peak Oil and Gas* (Campbell, Pers. Comm. 2006) noted that there are few reliable estimates of polar reserves, but "guessed" that ~5% of the world's oil is contained in the Arctic fields. Most production and transport from northern oil fields occurs in Arctic Russia and Alaska, although the Alberta oil fields of Canada could eventually prove to be substantial in global terms. Antarctic estimates vary even more wildly, from "poorly prospective, lacking effective source rocks" (Campbell, Pers. Comm. 2006), to reported figures of ~50 G barrels in the Ross and Weddell Seas (similar to Arctic reserves) or more (EIA 2000; Elliot 1988; Shapley 1974). Regardless, oil in Antarctica is unproven, and further exploration is prevented by moratorium.

Crude oil spills from ruptured pipelines in the Arctic are by far the largest sources of terrestrial petroleum pollution, followed by shoreline spills from tankers or resupply vessels. Diesel fuels are the next most common spills. Incidents are typically caused by infrastructure failure, human error during fuel transfer, "third party actions" (e.g. sabotage), or natural hazards. Risks are also known to be higher in some permafrost regions relative to non-permafrost regions within the same country (e.g. Russia (Det Norske Veritas 2003)).

1.2.1.1 Scale of the problem

There are currently insufficient data to define precisely the areal extent or volume of petroleum-contaminated soil in the cold regions. If we consider

[1] AMAP 1998 began their analysis of *Petroleum Hydrocarbons in the Arctic* by outlining the "reason for concern" (p. 661).

the majority of cold regions soils as belonging to one of three broad geographic regions: Arctic/sub-Arctic, Antarctic/sub-Antarctic, and alpine, it is possible to compare selected case studies and get a sense of the scale of the problem. Assessing how much risk petroleum hydrocarbons pose to the various cold regions is more difficult.

To compare geographic regions or countries, a simple evaluation matrix has been compiled to assess the overall progress in the assessment and remediation of petroleum hydrocarbons in some reasonably well-documented sites (Table 1.1). The matrix is based closely on tables presented by the European Environment Agency – Indicator Management Service for Western Europe (EEA-IMS 2005). Reporting is known to be inconsistent between countries and discrepancies result because there is no international or even European consensus regarding legal standards for soil contamination. Only some of the European Union countries have legal standards, and national standards vary by country. The areal extent or volume of contaminated soil is also not often reported.

Whereas some data are available for the Arctic/sub-Arctic and Antarctic/sub-Antarctic, there is currently no synthesis on the extent of contamination in alpine regions, in particular the mid-latitude high-altitude regions, which are widely distributed. The data presented in Table 1.1, and discussed below, are therefore not specifically relevant to the Alpine regions.

1.2.1.2 Arctic

The Arctic Circle comprises the Arctic Ocean, Greenland, Baffin Island, other smaller northern islands, and the far northern parts of Europe, Russia, Alaska, and Canada. There is no consensus in this region on the response to petroleum hydrocarbon contamination and a direct comparison on the level of contamination is difficult. The definitive survey of pollution in the Arctic is the Arctic Monitoring and Assessment Programme (AMAP) Assessment Report of 1998. Much of the information in the report is now outdated, and AMAP are currently in the process of updating the petroleum hydrocarbon chapter (AMAP 2006). However, the 1998 report still offers the best overview of petroleum hydrocarbon contamination in the Arctic region, and it is used extensively in this summary. Below is a brief overview of the extent of contamination in most countries within the Arctic Circle as presented in Table 1.1.

Northern Europe: Finland, Norway, and Sweden

Although there is extensive reporting and coverage for Europe, petroleum hydrocarbon contamination in cold region soils are not specifically categorized. Overall, almost half of all European soil contaminants are petroleum hydrocarbons (20% oil, 16% PAHs, 13% BTEX, excluding chlorinated

Table 1.1 *Petroleum contaminated sites in cold regions or countries with a significant cold regions/permafrost presence. Some countries are not included due to insufficient data. The table structure closely follows the European Environment Agency – Indicator Management Service (EEA-IMS 2005) table on the progress in control and remediation of soil contamination. The numbers presented here are best estimates only*

Country	Preliminary study or investigation			Detailed site investigation			Implementation of remediation measures or development of a plan			Measures completed	
	Estimated total	Completed	%	Estimated total in need	Completed	%	Estimated total	Completed	%	Total	%
Finland[a]								292		1898	
Norway[a]		1521			277			43		246	
Sweden[a]	11 100	2960	12	4400	925	21	2960	185	6	444	15
Iceland[b]	>230	96	42								
Canada[c]	~2400										
Alaska	6400									3400	
Antarctica[d]	~200			~100	~8	30	10+	4		1	

[a] Estimated from EEA-IMS (2005) by multiplying the proportion of contaminated sites where the main contaminant was either mineral oil, BTEX, or PAH by the total number of sites registered.

[b] Environment and Food Agency Iceland (2002); Meyles and Schmidt (2005).

[c] The total estimated number of contaminated sites in Northern Canada (Yukon, Northwest Territories, and Nunavut) is estimated at approximately 2400. The primary source of contaminants at northern sites is petroleum hydrocarbons.

[d] Aislabie *et al.* (2004); Delille *et al.* (2006); Gore *et al.* (1999); Kennicutt (2003); Rayner *et al.* (2007); Revill *et al.* (2007); Roura (2004); Snape *et al.* (2006a); Stark *et al.* (2003).

hydrocarbons) (EEA-IMS 2005), and countries with substantial permafrost or cold soils have a similarly high proportion of petroleum contamination (Norway, 44%, Sweden, 37%, and Finland, 73%). Where known, the estimated number of petroleum-contaminated sites is shown in Table 1.1. When surveyed in 2004, most sites had not been assessed, and only a small proportion of those sites needing remediation had been completed. However, all three Scandinavian countries have ambitious targets for assessment and remediation (these targets are documented in the indicator table, EEA-IMS 2005).

Alaska

Most spills in Alaska are well documented. Many spills have been remediated, and many more are currently under investigation. Approximately 2000 spills per year are responded to, with most of these being cleaned up by the responsible party. A small percentage of these are of such magnitude that they become subject to long-term management and cleanup. A sizeable portion of the long-term remediation work deals with "legacy spills," typically associated with past handling practices that have led to the release of contaminants to the environment. As of June 30, 2006, approximately 6400 contaminated sites had been recorded in the Alaska contaminated sites database, of which approximately 3400 have been "closed."

Iceland

Cataloging of soil environmental problems has centered on soil erosion, and systematic data regarding soil contamination have been lacking. The European Environment Agency – Indicator Management Service (EEA-IMS 2005) estimated that in 2002 there were 100 contaminated sites in Iceland, but it is unclear how many of these sites are significantly contaminated with petroleum hydrocarbons. Preliminary investigations had been conducted on just five of these sites, and detailed site investigation and remediation had been completed on only three sites. No area or volumetric statistics were provided. The Icelandic Environment and Food Agency (UST) website estimates that over 200 fuel filling stations and about 30 fuel storage tanks are potential contaminated sites, accounting for approximately 25% of the contaminated soil in Iceland (Environment and Food Agency Iceland 2002). Meyles and Schmidt (2005) attempted to catalog contaminated soil sites throughout Iceland by surveying regional environmental agencies. They identified over 200 sites nationwide, 96 of which were filling stations and fuel storage tanks.

Canada

One major source of hydrocarbon contamination in the Canadian Arctic is associated with the early warning radar stations constructed across the Arctic during the Cold War. In 1985, an agreement between Canada and the United States was reached to replace the Distant Early Warning (DEW) line with a new satellite-based system and to clean up the old sites. Environmental assessments were conducted at all 42 DEW line stations between 1989 and 1993. The predominant contaminants identified were PCBs and metals (Poland *et al.* 2001; Stow *et al.* 2005). Hydrocarbons were not part of the original cleanup plans but were added at a later date. The contamination is well documented and both the Department of National Defence (DND) and Indian and Northern Affairs (INAC) maintain inventories of the sites for which they are responsible. Many sites have now been cleaned up. DND has completed ten of twenty-one DEW lines sites and INAC four of twenty-one sites.

Chronic spillage within human settlements is another major source of hydrocarbon contamination in the Canadian Arctic. For example, between 1971 and 2006, fuel spills reported in Iqaluit (Nunavut, Canada) totalled 627 000 L of diesel fuel while in Rankin Inlet, diesel fuel spills from tank farms accounted for 289 000 l. Most of these spills occurred during the 1970s and 1980s. During this period, environmental regulations were not as stringent as they are today, thus it is possible that a considerable quantity of fuel was never recovered or cleaned up to today's accepted standards.

Although there are few large mines in the Arctic, some abandoned older sites may have significant petroleum contamination present (e.g. Biggar *et al.* 2006). For example, at Nanisivik (North Baffin Island), it is known that 2000 drums of used oil were disposed of within a landfill. They have not been excavated and the mobility of this fuel is not currently known. For relatively newer operations, such as Polaris (near Resolute Bay), current regulations and water license conditions are such that any contamination would have been remediated before the site was "closed." Contamination associated with current mining operations is far more controlled and less likely to be a source of hydrocarbon contamination than activities associated with the communities.

In addition to these examples, Indian and Northern Affairs Canada (INAC) have identified ~800 sites that require cleanup. Hydrocarbon contamination is the main contaminant in the majority of these sites.

Russian Federation

Little accurate information is available for the Arctic, sub-Arctic, or alpine regions of the Former Soviet Union in the mainstream literature, but several case studies illustrate the scale of the petroleum pollution. The Komi oil spill of 1994 is perhaps the most infamous Russian oil spill and is regarded by some as one of the most serious environmental disasters of the last century (Bazilescu and Lyhus 1996). Following the rupture of an old degraded pipeline, oil (estimates[2] range from 14 000 tonnes to 240 000 tonnes, with a range of 37 000–44 000 tonnes most often quoted) poured across the Siberian tundra (areal estimates vary from 0.3 to 186 km^2) and into the Kolva River, a tributary of the Pechora River that flows into the Barents Sea (Bazilescu and Lyhus 1996; Wartena and Evenset 1997). Including chronic leakage until the mid 1990s, AMAP (1998) estimated the total discharge to be around 103 000–126 000 tonnes. The scale of this spill in comparison to all other world spills is staggering, but the surprising fact is not that it happened at all, but that the Komi spill is but one of many very large Russian oil spills. Det Norske Veritas (2003 p. 3) reported 113 major crude oil spills in the Former Soviet Union between 1986 and 1996 (with 17% in permafrost regions), but these "are widely thought to represent only a fraction of the total number of spills that occurred. Independent but unconfirmed sources indicate that some pipelines have experienced several hundred smaller spills." It is estimated that 20% of all oil that is pumped from the ground in Russia is lost through either chronic leakage or theft (Bazilescu and Lyhus 1996).

The accuracy of spill reporting in the Russian Federation is poor (Det Norske Veritas 2003). The Russian Minister of Natural Resources in the Komi Republic, Aleskandr Borovinskih, is quoted in the Bellona report of Feb 17, 2006 (Bellona 2006) as saying that oil companies often conceal spills. Alternatively, the extent of the spill is massively under-reported. In another case study described on the Bellona web site, an oil gusher occurred that reached 35–40 m into the air. The Natural Resource Ministry web site put the spill volume at 3000 tonnes (many times bigger than the much publicized "largest" Alaska Pipeline spill of ~700 tonnes in March 2006, discussed below). In contrast, Transneft, the company responsible for the oil gusher, estimated the spill to have been only 10 tonnes. The Regional Environment Oversight Agency agreed with the Natural Resource Ministry that the spill was many times greater than the estimate provided by Transneft.

Given the poor level of reporting, it is not possible to estimate reliably the volume of petroleum hydrocarbon contaminated soil in Russia. Many incident reports describe hectare-scale contamination (see Bellona.com), or polluted rivers or plumes that extend hundreds of kilometers. AMAP (1998) noted that almost all water samples taken in a regional survey of northwest Siberian Rivers exceeded

[2] Several of the most widely available sources of Russian information are from non-government organization or US websites. The authors were unable to verify the accuracy of such non-peer reviewed sources.

the maximum permissible concentration of 0.05 mg total petroleum hydrocarbons l^{-1}. A few incidents are reported where contamination is in the region of kilometers square in size, and one or two in the range of 10s–100s km^2. Based on available information, it is not possible to reliably estimate the extent of contamination. Extrapolating from what information is available, we speculate that the total amount of petroleum-contaminated permafrost in Russia could be somewhere in the region of 1–10 billion m^3. If the area of Arctic and sub-Arctic Russia is taken as ≈6 million km^2 (AMAP 1998), that would equate to $\approx 10^{-3}$–10^{-4}% of this type of terrestrial habitat contaminated with petroleum products. However, this figure seems very low, given the reported extent of water contamination noted above.

1.2.1.3 Antarctic

The Madrid Protocol prohibits all Antarctic Treaty member nations from mineral resource activities in Antarctica, including oil exploration and exploitation (Rothwell and Davis 1997). As a result the extent of petroleum contamination is very low in absolute terms, and the Antarctic and sub-Antarctic are undoubtedly the least polluted cold regions in terms of total area polluted, volume spilled or volume/mass of soil/water contaminated. Spills are mostly of kerosene, Antarctic blend or similar light diesel.

In common with the Arctic region there are many countries responsible for petroleum hydrocarbon spills within the Antarctic/sub-Antarctic region. Forty-five countries are signatories to the Antarctic Treaty and 28 of these are Consultative Parties (i.e. may participate in any decision-making process). Reporting requirements for fuel spills have only been required of these signatories since 1998 and documentation for prior spills was country dependent.

Spills near Australian Antarctic stations or sub-Antarctic Islands are generally well documented. Six fuel plumes known from the Australian Casey Station and Macquarie Island areas have been fully evaluated and are scheduled for remediation as part of a national program of works (Rayner *et al.* 2007; Snape *et al.* 2005; Snape *et al.* 2006a). An additional area of chronic spillage (i.e. many small spills) at the abandoned Wilkes Station is partly evaluated (Snape *et al.* 1998). World Park Base operated by Greenpeace has been fully remediated and well documented (Roura 2004). Spills near the New Zealand, French and Argentine stations are typically partly documented to well documented (e.g. Aislabie *et al.* 2004; Delille *et al.* 2006; Waterhouse and Roper-Gee 2002). Many spills under the jurisdiction of the US program are partly documented (Kennicutt 2003; Klein *et al.* 2006), although some have been more fully evaluated, have been remediated, and are well documented (Christensen and Shenk 2006; Kennicutt 2003). The 2003 report on human disturbance at McMurdo Station (by far the largest station in Antarctica) noted 385 spills between 1991 and 2000, mostly of JP8, totalling ~80 000 l. Several larger spills are known from the 1970s and 1980s, including a spill of 260 000 l in 1989 (Kennicutt 2003). Little widely published information is available for fuel spills near Japanese, Chinese or Russian stations. Many of these spills are either not systematically or reliably documented, or only some aspects of spill evaluation have been published and are partly documented (e.g. Goldsworthy *et al.* 2003).

Taken at the largest geographic scale, the Antarctic/sub-Antarctic region is at best partly documented. Most spills are not accurately delineated, and estimating the extent of petroleum contamination from such a patchy dataset is difficult. Nevertheless, an order-of-magnitude has been estimated. Where a spill has been reported or there is a picture of a spill, it has been assumed that the spill reached a depth of 1 m on average (Revill *et al.* 2007; Snape *et al.* 2006a). Where a station has recorded no spill history, but there is anecdotal evidence of contamination, the authors assumed that the extent of contamination was similar to other well-documented stations of similar size. There are approximately 65 stations distributed around the continent (COMNAP 2006), and there is perhaps a similar number of large field camps. Assuming most stations and field camps have some fuel contamination, it can be best estimated that there are 100 000 to 1 million m^3 of soil contaminated[3] at a concentration >100 mg fuel kg^{-1} soil.[4] If the vast glaciated regions are excluded and contamination is considered as a proportion of terrestrial habitat with soil and significant ecosystem development, that equates to $\approx 10^{-3}$% of the ~6000 km^2 of ice-free coastal habitat (Poland *et al.* 2003).

1.2.1.4 *Summary of the extent of contamination in cold regions*

The degree of contamination in cold regions is difficult to assess as the level of documentation is country specific. Even in the Antarctic, where waste is managed under a common international environmental protocol (Madrid Protocol), variation between countries on their reporting of historic spills (prior to 1998) varies greatly. Generally, spills of petroleum hydrocarbons tend to be well documented in developed countries and response to spills more clearly defined when compared to less-developed countries. For perspective, Russian oil spills are categorized on a three-level scale: Localized spills (up to 500 tonnes), regional level spills (500–5000 tonnes) and federal level spills (>5000 tonnes)

[3] Based on observation of fuel spills at Casey Station, Antarctica, every 1 kg of fuel spilled creates between 100 to 1000 times that amount of contaminated soil by mass. This observation is the same order-of-magnitude estimate that we can infer from the AMAP oil spill migration model (AMAP 1998, p. 676).

[4] Note: all soil concentrations throughout this chapter are presented as dry mass.

(Kireeva 2006), but we were unable to ascertain what response is required for spills within each category. Other regions have more clearly defined reporting requirements. In Alaska, spills to land >40 l need to be reported to the Alaska Department of Environment and Conservation (ADEC) within 48 hours of release/discovery. In the Antarctic, all spills >200 l are now reported to the Council of Managers of National Antarctic Programs (COMNAP) within 30 days. The Australian national program reports any spill >20 l.

Presenting the area of petroleum hydrocarbon contaminated land as a proportion of habitat offers an interesting perspective. Russia is probably the most polluted cold region, Antarctica is certainly the least; but in a comparison on the basis of habitat disturbance, it is possible that the proportion of affected area is broadly similar. This highlights the relative rarity of terrestrial Antarctica, and the disproportionate impact that a spill can have on Antarctic terrestrial soil habitats.

1.2.2 *Regulation*

1.2.2.1 *Overview*

Currently there are no universally accepted guidelines specifically developed for petroleum hydrocarbons in cold regions. Those nations that regulate Arctic or sub-Arctic regions (mainly Canada, Finland, Greenland, Norway, Russia, Sweden, and the United States) do so through a number of domestic guidelines and acts of legislation (Table 1.2). Of those nations that also operate stations in the Antarctic (Norway, Russian, Sweden, and the United States), domestic legislation is generally not applied with the same rigor as in the north. Often this is because contaminated sites are regulated at state/regional level rather than national level. There are no contaminated site guidelines that are specifically set for use in the Antarctic other than the requirement to report all spills >200 l (or <200 l if considered significant) to COMNAP.

Several Arctic countries have developed or are in the process of establishing soil quality guidelines based on risk assessment science. Country-wide contaminant cleanup levels are generally developed on the basis of toxicity and exposure to humans, land classification (e.g. agricultural, residential, commercial, and industrial, proximity to natural water bodies), and other environmental considerations (e.g. ecotoxicity). However, there are few initiatives specifically designed to derive guideline values that accommodate the characteristics of cold regions. As one example, in Alaska soil cleanup levels specific to manmade pads and roads are in place (18 Alaska Administrative Code 75, DEC 2005). However, Alaska Department of Environmental Conservation (ADEC) has the authority to determine cleanup levels for undisturbed tundra on a site-specific basis if a cleanup

Table 1.2 *Summary of soil quality guidelines for petroleum contamination by country. The values were compiled between 2002 and 2006 from the regulations listed below and information supplied by regulators. Some country soil-quality guidelines are solely based on human exposures; others base cleanup criteria on more sensitive biota (e.g. soil invertebrates and flora) (concentrations are mg* kg^{-1} *on a dry mass basis). Information derived from the following organisations:* ***Canada*** *Environment Canada and Canadian Council of Ministries of the Environment,* ***Finland*** *Finnish Ministry of the Environment,* ***Greenland*** *Greenland Home Rule, Ministry of Health and the Environment,* ***Iceland*** *Ministry for the Environment,* ***Norway*** *Norway Pollution Control Authority,* ***Sweden*** *Swedish Environmental Protection Agency,* ***United States*** *Alaska Department of Environmental Conservation. Note:* ***Iceland*** *and* ***Greenland*** *have not established soil-quality guidelines for land-based contaminated sites and therefore are not included in this table. BTEX–benzene, ethyl benzene, toluene, and xylenes; PAH–polynuclear aromatic hydrocarbons; TPH-total petroleum hydrocarbons.*

Product	*Finland*	*Norway*	*Sweden*	*Canada*	*Alaska[a] (Arctic)*
Gasoline	100	7	5–100	1000	100
Diesel	200	100	100	2000	200 (500)[b]
Crude oil		100	100	2500	200
Residual	600	100	100	2500	2000
BTEX Analytes					
Benzene	2.5	0.005	0.6	0.0068–0.03[e]	13[c]
Ethyl benzene	50	0.5	12	0.018–0.082[e]	89[c]
Toluene	120	0.5	10	0.08–0.37[e]	180[c]
Xylenes (total)	25	0.5	15	2.4–11[e]	81[c]
PAH Analytes					
Acenaphthene					8200[c]
Anthracene	50				41 000[c]
Benzo(a)anthracene	40		0.4		15[c]
Benzo(b)fluoranthene			0.7		15[c]
Benzo(k)fluoranthene	40		0.4		150[c]
Benzo(a)pyrene	40	0.1	0.4	0.1–0.7[e]	1.5[c]
Chrysene	40		0.5		1500[c]
Dibenzo(a,h)anthracene	20				1.5[c]
Fluorene	20	0.6			5500[c]
Indeno(1,2,3-c,d)pyrene	40		0.4		15[c]
Naphthalene	100	0.8		0.1–22[e]	5500[c]
Pyrene	40	0.1	0.6		4100[c]
$\sum$16 PAH[d]		2	20.3		

(*cont.*)

Table 1.2 *(cont.)*

Product	*Finland*	*Norway*	*Sweden*	*Canada*	*Alaska*[a] *(Arctic)*
Petroleum Hydrocarbon Range					
C_6–C_{10}-volatile	500	7	100	30–230[e]	1400[c]
C_{10}–C_{25}-light extractable[f]	1000	100	100	150–2200[e]	12 500[c]
C_{25}–C_{36}-heavy extractable[g]	2000	100	100	400–5000[e]	13 700[c]

[a] Based on human exposures; applicable to contaminated manmade pads and roads in the Arctic (no migration to groundwater pathway).

[b] Applicable to a diesel spill with total BTEX concentration less than 15 mg kg^{-1}, benzene concentration less than 0.5 mg kg^{-1}, and other site conditions are favorably protective of human health, safety, welfare, and of the environment.

[c] Human ingestion critical exposure pathway.

[d] The 12 PAH analytes listed plus four more, mostly carcinogens, are the most toxic.

[e] Range reflects criteria for four land-use categories (agricultural, parkland/residential, commercial to industrial).

[f] For Canada the petroleum hydrocarbon range is $>C_{10}$.

[g] For Canada, the petroleum hydrocarbon range is $>C_{16}$.

action is deemed to be more damaging than the release itself. Canada, Finland, Norway, and Sweden do not have soil quality guidelines specific to their Arctic regions, although they are all attempting to accommodate Arctic-specific factors in their guidelines. Greenland and Iceland also do not have soil quality regulations for land-based petroleum-contaminated sites.

1.2.2.2 *Arctic regulation*

The guideline values for countries that regulate petroleum spills in the Arctic or sub-Arctic are summarized in Table 1.2. These values are generally environmental investigation guidelines (also known as trigger values), but are often used as remediation targets (frequently referred to as "magic" numbers) (cf. NEPC 2005). Values vary considerably by country and land use and are discussed in more detail below.

Canada

The Canadian Council of Ministers for the Environment (CCME 2001) published Canada-wide guidelines relating to levels of contaminants for water and soil depending on land use. Cleanup guidelines specific to the Arctic were first developed in Canada in the 1990s (ESG 1993) as part of the DEW Line remediation program, although these cold-regions specific criteria (e.g. freezing soil, ecosystem function) did not include hydrocarbons.

In 1998, an agreement was reached between the Department of National Defence and Nunavut Tunngavik Incorporated (DND-NTI 1998) regarding the cleanup of DEW Line sites that were the responsibility of the Department of National Defence. As part of this agreement, hydrocarbons were to be remediated in a number of possible ways (including landfilling, landfarming, soil washing, and bioremediation) if the concentration was greater than 2500 mg kg^{-1}. However, the working group that oversaw implementation of the agreement often raised the limit to take into account site-specific considerations. The Government of the Northwest Territories (GNWT 1998) also has a guideline of 2500 mg kg^{-1} to evaluate hydrocarbon contamination.

In 2002, the Department of Indian Affairs and Northern Development developed a Contaminated Sites Program Management Framework (DIAND 2002), which followed recommendations in the Treasury Board Federal Contaminated Sites Management Policy. In this framework it was recommended that CCME Guidelines should normally be followed unless site conditions, land use, receptors, or exposure pathways differ from the protocols used to develop the guidelines.

The Canada-Wide Standards for Petroleum Hydrocarbons in Soil (CCME 2001 p. 7) is "a tiered framework offering the proponent the option to comply with a set of reasonably conservative risk-based standards corresponding to a number of defined land uses, exposure scenarios and site characteristics (Tier 1) or to use additional site-specific information to assess and manage the risks through a more precise knowledge of actual or potential exposures (Tiers 2 and 3)." Tier 1 numerical levels are generically applied to contaminated sites; Tier 2 cleanup levels are adjustments to Tier 1 levels, based on site-specific information; Tier 3 levels require a comprehensive risk assessment study based on site-specific information in developing site cleanup levels and related management options. The Tier 3 process is guided by CCME (1996). Because of the cost involved, the Tier 3 option typically applies to large sites undergoing long-term remediation.

The CCME 2001 guidelines are described in more detail in Section 1.3.2 entitled Recent Advances with a case study from Nunavut.

Greenland and Iceland

The environmental focus in Greenland[5] and Iceland has been protection of the oceans from pollution. Greenland's Ministry of Health and the Environment is working on developing petroleum cleanup standards for marine and land environments (A. Engraf 2002, pers. comm).

[5] Greenland is a special cultural community of Denmark; it established home rule in 1979.

Iceland has yet to establish soil quality guidelines for cleanup of contaminated land sites; however, draft regulation on soil contamination developed by Iceland's Environmental Agency (UST), from 2002, described risk-assessment methodologies and remediation limits (Meyles and Schmidt 2005; UST 2005). Iceland's National Waste Management Plan, released in April 2004, also requires "compulsory management of . . . contaminated soil" (UST 2006).

Furthermore, Iceland's participation in the European Economic Area requires it to implement environmental legislation. The European Economic Community Act on Waste Management 55/2003 requires member states to "take the necessary measures to ensure the prohibition of . . . any deposit and/or discharge of waste oils harmful to the soil . . ." (Council of the European Communities 2004).

Norway and Sweden

Norway and Sweden are developing or revising environmental regulations for application to the Arctic in conjunction with the European Committee for Standardization and International Organization for Standardization protocol. Norway ratified the Svalbard Environmental Protection Act in June 2001. However, no specific soil quality standards were instituted for Svalbard (the Norwegian Pollution Control Authority is responsible for regulating contaminated sites in Norway and Svalbard). The Norwegian Pollution Control Authority first published technical guidelines for environmental site investigations in 1991, and in 1995 they published preliminary guidelines for management of contaminated lands as SFT Report 95:09, *Management of Contaminated Land – Preliminary Guidelines for Executive Procedures*. Norway established *Guidelines for the Risk Assessment of Contaminated Sites* (SFT 1999) as an improvement to SFT Report 95:09. The revised Norwegian soil quality guideline is a compilation of sub-reports from Aquateam AS, the Norwegian Geotechnical Institute, GRUF, the contaminated soil program arm of the Norwegian Research Council, and the final report from the European Union collaboration project CARACAS (Concerted Action on Risk Assessment for Contaminated Sites in the European Union). Work by the Total Petroleum Hydrocarbon Criteria Working Group (Edwards *et al.* 1997; Gustafson *et al.* 1997; Potter and Simmons 1998; TPHCWG 1998b; Vorhees *et al.* 1999) and USEPA (1996) are incorporated in the regulation. The guideline is a three-tiered decision model: Tier 1 – source exposure, with generic soil quality criteria established for the most vulnerable land use, Tier 2 – site-specific assessment criteria considers multiple points of exposure, and Tier 3 – modeling consideration for contaminant fate and transport. The new generic soil quality criteria were developed based on Dutch and Swedish cleanup values for the most vulnerable land use. The Norwegian model required pragmatic adjustments to some target values that were set below background levels in some Norwegian soils.

Sweden's *Environmental Quality Standards* (Swedish EPA 2002), a work in progress, will supplement their existing environmental regulation with Arctic consideration. Sweden implemented the *Environmental Code* in January 1999, based on risk assessment without concern for cost and technology. Guideline values for 36 contaminants or contaminant groups in soil, including petroleum constituents, were based on land-use categories: (1) land with sensitive use (e.g. residential, kindergartens, agricultural, together with groundwater abstraction); (2) land with less sensitive use (e.g. offices, industries, roads, car-parks, together with groundwater abstractions); and (3) even less sensitive lands with no groundwater abstraction. The generic cleanup values were derived using a Swedish exposure model similar to those of other European countries and suggestions from Naturvårdsverket Research, the Swedish Petroleum Institute, and other international organizations.[6] Soil quality criteria for contaminated sites are based on health and environmental risk posed by hazard and level of pollutant(s), migration potential, risk to human exposure, proximity to valuable natural features, and consider Swedish geology and related policy.

Finland

Finland's national contaminated land approaches are largely developed from CARACAS research, in particular Ferguson and Kasamas (1999). Finland does not have separate legislation for protection and remediation of contaminated soil. Contaminated soil is defined as waste and is therefore regulated under the Waste Act of 1994 (lands contaminated prior to 1994) and Waste Management Act, legislation predominately focused on liability issues for contaminated sites. Through the Finnish Environmental Institute (SYKE), risk assessment at Finnish contaminated sites routinely compares observed concentrations with guideline values for site cleanup. The Finnish Ministry of Environment published preliminary guideline values for some 170 compounds as part of a country-wide contaminated site inventory and cleanup project. The guidance values are based mainly on Dutch criteria. Updated cleanup values based on ecotoxicity and human health considerations have been proposed by SYKE, structured as a tiered multi-criteria system similar to that of other Nordic countries.

[6] Research groups include COLDREM, a Swedish case study of soil remediation at two cold climate industrial sites and ISO Technical Committee 190, a world-wide federation of national standardization bodies charged with developing international environmental standards; CLARINET, a Concerted Action within the Environment & Climate Program of the European Commission DG Research, with the purpose of developing technical recommendations for practical rehabilitation of contaminated sites in Europe.

Alaska

Since 1980, the United States government has enacted comprehensive environmental regulations. Each state has an environmental department charged with environmental enforcement of the federal mandate. In addition, the state agencies have authority to establish, subject to federal approval, alternative cleanup levels based on state-specific criteria. ADEC has established alternative cleanup levels for petroleum-contaminated sites based on unique geography and climate. For example, they distinguish between natural and manmade contaminated sites (e.g. gravel pads, roads, and runways) in establishing applicable cleanup levels. In addition, ADEC distinguishes between soil type (i.e., gravel and sand vs. organic soil). For sub-Arctic regions cleanup levels are based on precipitation, exposure, and land use criteria. In some instances, site-specific environmental risk assessment is applied in establishing alternative cleanup levels. The mandate for environmental regulation in Alaska is the *State of Alaska Department of Environmental Conservation 18 AAC 75, Oil and Other Hazardous Substances Pollution Control.*

The current trend in Alaska is to consider "risk-based corrective action." This involves greater examination of exposure pathways, and site-specific soil data are collected to represent accurately the soils present at a particular site. Then a conceptual site model is evaluated to determine if remediation of the contaminated soil is warranted, with the option to leave contaminated soil in place, given a certain set of assumptions regarding land use and placement of engineered controls.

Russia

We were unable to obtain Russian fuel spill assessment or remediation guidelines. Similarly AMAP (2006) has found it very difficult to update its Assessment and Monitoring report for Russia. In their report on Russian oil spills, Det Norske Veritas (2003 p. 10) noted that the regulatory and monitoring policies of the Former Soviet Union, and the agencies responsible for implementing policies concerning oil spill contingencies, did not compare favorably with regimes in Western Europe, North America, or Canada. Their "regulatory and monitoring regimes are more fragmented, less accountable and less able to quickly delegate responsibility . . . when spills occur. The result is that oil spills in the Former Soviet Union create more environmental damage than in similar environments . . . ."

1.2.2.3 *Antarctic regulation*

Petroleum spill regulation in the Antarctic/sub-Antarctic is currently a complex mix of broad international obligations, variously applied domestic policies, and legal instruments of those nations that maintain stations on the continent and offshore sub-Antarctic islands. In the Antarctic Treaty area, the main international obligations are provided by the Protocol on Environmental Protection to the Antarctic Treaty (1991) (known as the Madrid Protocol) that was ratified in 1998. Ratification required that all parties establish domestic legislation to implement the Protocol (Rothwell and Davis 1997). For sub-Antarctic islands outside of the Antarctic Treaty area the legal instruments are those of the country with sovereignty for that island. For Macquarie Island, the legislative requirements for contaminated sites in the State of Tasmania (Australia) are discussed below in further detail as part of a case study.

The Madrid Protocol

The environmental principles outlined in Article 3 of the Madrid Protocol essentially impose two considerations on activities in Antarctica.[7] The first objective is to ensure the protection of the Antarctic environment and associated and dependent ecosystems. The second objective is to ensure the protection of the intrinsic nature of Antarctica, including its wilderness and aesthetic values and its value as an area for the conduct of scientific research. To meet these two objectives, Article 3.2 states that activities in the Antarctic Treaty area must be planned and conducted to limit adverse impacts on the Antarctic environment and dependent and associated ecosystems and to avoid six specific consequences:

1. adverse effects on climate or weather patterns
2. significant adverse effects on air or water quality
3. significant changes in the atmospheric, terrestrial (including aquatic), glacial or marine environments
4. detrimental changes in the distribution, abundance or productivity of species or populations of species of fauna and flora
5. further jeopardy to endangered or threatened species or populations of such species
6. degradation of, or substantial risk to areas of biological, scientific, historic, aesthetic, or wilderness significance.

Petroleum hydrocarbon spills could potentially affect most of these criteria except climate and weather. Unfortunately, terms such as *significant* and *substantial* are not defined and as such are open to a range of interpretations or opinions.

[7] An important aspect of the Madrid Protocol is a prohibition on mineral exploration/exploitation, principally as a means of preventing pollution of the type seen in the Arctic.

These basic principles are given more operational substance by the other provisions of the Protocol. In particular, Article 15 imposes an obligation to respond to environmental emergencies. It requires responsible parties to provide for prompt and effective response action to environmental emergencies, such as petroleum spills, which might arise in the course of government and other activities in the Antarctic Treaty area.

Annex III imposes obligations with respect to waste disposal and removal, and it defines fuel as a potential waste. Article 1.5 of Annex III states that past and present waste disposal sites on land and abandoned work sites must be cleaned up by the generators of such wastes and the users of such sites. This obligation has a number of exceptions. Most importantly, it does not require the removal of waste material when such action would involve greater adverse environmental impact than leaving it in its existing location. For soil rehabilitation, this is an important consideration.

Although Annex III does not impose any deadlines for remediation, it does require that all waste to be removed from the Antarctic Treaty area be stored in such a way as to prevent its dispersal into the environment (Article 6 of Annex III).

Compliance and liability in Antarctica

The Liability Annex was a matter of heated debate for more than a decade, and when finally ratified in 2005 it was the most significant addition to the Antarctic Treaty regime since the Madrid Protocol was adopted in 1991. The Annex applies to environmental emergencies arising from national Antarctic programs, tourism, and any other activity in the Antarctic Treaty area. The Annex establishes who has responsibility for cleaning up after an environmental emergency. It also creates a mechanism for other operators to recover the costs of cleanup if the party creating the environmental emergency does not do so.

1.2.2.4 *Free-product dispersal and sheen*

For most countries, the observation of free product floating on surface water is a trigger for remediation. For the application of Alaskan guidelines, for example, there is a requirement that contaminated groundwater must not cause a surface water quality violation (water quality standards set in ADEC 18 AAC 70). Groundwater entering surface water must not cause a visible sheen upon the surface of the water, and surface waters and adjoining shorelines must be virtually free from floating oil, film, sheen, or discoloration *(18 AAC 75.340 f: Groundwater that is closely connected hydrologically to nearby surface water may not cause a violation of water quality standards in 18 AAC 70 for surface water or sediment).*

This latter clause from Alaskan regulation is similar to the approach taken by Canada and Tasmania (the Australian State responsible for regulation of sub-Antarctic Macquarie Island). In Canada, this is addressed in the Tier 2 adjustment criterion for the *Protection of groundwater for aquatic life* (see Table 1.4, discussed further in Section 1.3.2) and also by the Fisheries Act which states that no deleterious substance shall be discharged to waters that contain fish. Under Tasmanian legislation (EMPCA 1994), sheen fits within the broad definition of environmental harm (Tasmanian Environment Division, pers. comm., 2006).

1.2.3 Remediation

The various remediation treatment regimes that are commonly used in temperate regions were presented in Figure 1.1. The central theme of this monograph is that bioremediation offers a good compromise between cost and timeliness. However, bioremediation has been successfully applied to soil on relatively few occasions in cold regions.

1.2.3.1 *Arctic remediation*

Russian Federation

There is very little mainstream information about remediation of the many fuel spills in cold regions managed by the Russian Federation. However, it is likely that Table 4.2 of the Det Norske Veritas 2003 report[8] (page 47) provides a typical breakdown of the types of remediation response. Removal of surface product through either mechanical or manual approach (51 events) is the favored approach, followed by containment and berming (17 events). *In-situ* burning, used in eight out of 65 incidents, is more prevalent than in Western Europe, the United States, or Canada. There is only one recorded incident where bioremediation was used as a treatment method. The other methods are known to cause significant damage to soil and the underlying permafrost (Barnes and Filler 2003; UNEP-WCMC 1994), although recovery methods might be appropriate where the aim is to prevent further spill migration. Excavation and disposal by landfarming, landfilling, or incineration are noted as methods used in several spills; however, cleanup criteria are not readily available (Det Norske Veritas 2003). Similarly, Det Norske Veritas (2003 p. 47) noted that "reports rarely include any information on the disposal method used but merely note that contaminated soil was removed."

[8] The sites reported on are not all from soils underlain by permafrost, but at least six incidents involved managing snow.

Canada

Much of the remediation work in the Canadian Arctic relates to the cleanup of the DEW Line. Landfarming of contaminated soils has frequently been used, though highly contaminated soils, particularly those contaminated with heavy oils, have been landfilled or shipped off site. Large landfarms have been set up at several of these sites across the Arctic. In addition to landfarming, research has been performed by Queens University and the Environmental Sciences Group at the Royal Military College on biopiles, permeable reactive barriers, and geocomposite liners for spill containment (Bathurst *et al.* 2006; Mohn *et al.* 2001a; Mohn *et al.* 2001b; Paudyn *et al.* 2006; Reimer *et al.* 2003; Reimer *et al.* 2005) (see also Chapter 9).

Alaska

In Arctic Alaska, industry most often employs dig-and-haul and off-site remediation. It is only relatively recently that *in situ* bioremediation has been used at remote gravel pad and tundra sites. Notable industry examples include comparative biostimulation (with use of commercially available fertilizer) and bioaugmentation (with use of allochthonous microorganisms or bioproducts) projects at Service City Pad, Haliburton Pad, and Deadhorse Lot 6 sites (Braddock *et al.* 2000). Successful landfarming research trials have also been completed in Barrow and at the Toolik Field Station (Braddock *et al.* 1999; Woolard *et al.* 2000) (see also Chapter 9).

1.2.3.2 *Antarctic remediation*

Although much fundamental bioremediation research has been undertaken in Antarctica, there have not yet been any full-scale remedial operations other than dig-and-haul activities by the United States and Greenpeace. The United States, Britain, New Zealand, France, and Australia in particular have studied the microbial response to fuel spills and the opportunities for accelerating degradation through various bioremediation treatments (see Chapters 8 and 9 and references therein).

To some extent the lack of any full-scale remedial operations is due to scientists and engineers taking a cautious approach to applying technologies developed for quite different environments. For example, there are concerns about the effects that uncontrolled release of nutrients could have on nutrient poor soils or lakes. Also, bioaugmentation with allochthonous microorganisms is occasionally used in the Arctic, but introductions are prohibited under the Madrid Protocol (Article 4 of Annex II). However, it is likely that the lack of an independent enforcement mechanism, relatively weak and at best broadly defined cleanup criteria, coupled with extreme financial costs and unproven low-cost bioremediation technologies are the main reasons why petroleum-contaminated soils have not yet been remediated in Antarctica.

The first challenge for scientists, engineers, and policy makers is to derive a set of guidelines or protocols suitable for Antarctica that can be used to protect the environment. In parallel and in conjunction with this research, acceptable low-cost bioremediation technologies need to be assessed, developed and optimized for Antarctica so that the financial burden becomes more palatable to the managers of national programs.

1.3 Recent advances

1.3.1 Assessing and managing petroleum contamination in cold regions

Many of the recent advances in assessing petroleum hydrocarbon contamination in cold regions are based on approaches developed elsewhere. Significant advances include the work of the TPH Working Group on standard methodologies for measuring total petroleum hydrocarbons (Edwards *et al.* 1997; Gustafson *et al.* 1997; Potter and Simmons 1998; TPHCWG 1998a; TPHCWG 1998b). Other significant advances concern improvements in information management and the adoption of web-based environmental management systems. The quality of spatial information regarding fuel spills at Antarctica's McMurdo Station stands out as an example of recent improvement in reporting and evaluation (Kennicutt 2003; Klein *et al.* 2006) using GIS as a tool to accurately record the extent of fuel spills. Similarly the ADEC contaminated sites website offers a clear, relatively easy-to-access interface for the user (www.dec.state.ak.us/SPAR/csp/index.htm). This website is particularly useful because it offers intuitive navigation and a high level of transparency regarding contaminated site management.

One of the goals of this monograph is to establish guidelines or recommendations, in the broadest sense, for bioremediation practice in cold regions. Bioremediation technology, in its many forms, offers a remediation approach that has a high probability of success, and where cost and time can be balanced with risk and regulatory pressure. However, to evaluate this balance for a particular site requires appropriate knowledge of risks and values. Some of these will be site specific, but many sites are likely to have issues that are common to cold regions in general. The following two case studies illustrate some of the recent advances in the development of holistic approaches to assessing remediation triggers and targets that incorporate physical and chemical properties of soil, water, and contaminant type, and relate these to toxicological and ecological response. The case studies also illustrate some of the current uncertainties in the application

Table 1.3 *Canadian Council of Ministers for the Environment (CCME) Tier 1 default guidelines (mg kg^{-1}). The fractions F1, F2, F3, and F4 relate to the properties of the hydrocarbon contamination in a particular carbon fraction range. F1 would be the predominant fraction for gasoline, F2 the predominant fraction for diesel, and F3 and F4 would apply to lubricating oil and grease respectively. ECN refers to Effective Carbon Number as defined by the TPH Working Group (Gustafson et al. 1997)*

Soil type	Land use	F1 ECN C_{6-10}	F2 ECN $>C_{10-16}$	F3 ECN $>C_{16-34}$	F4 ECN $>C_{34-50+}$
Fine-grained surface	Agricultural	180	250	800	5600
	Residential/Parkland	180	250	800	5600
	Commercial/Industrial	180	250	2500	6600
Coarse-grained surface	Agricultural	130	150	400	2800
	Residential/Parkland	30	150	400	2800
	Commercial/Industrial	230	150	1700	3300
Fine-grained subsurface	Agricultural	180	250	3500	10 000
	Residential/Parkland	750	2200	3500	10 000
	Commercial/Industrial	180	2500	5000	10 000
Coarse-gained subsurface	Agricultural	200	150	2500	10 000
	Residential/Parkland	40	150	2500	10 000
	Commercial/Industrial	230	150	3500	10 000

of broadly defined regulations, and a profound lack of environment-specific data for the cold regions.

1.3.2 A Canadian case study from Resolute Island, Nunavut

For the Canadian Arctic, most areas of contamination outside of communities or industrial activity sites are defined as parkland (Table 1.3). The majority of Arctic soils are coarse-grained in that the median grain size is greater than 75 μm. Soils are generally thin, particularly if permafrost is considered to be the lower boundary layer. In the context of the guidelines, the subsurface soil category applies if contamination is <1.5 m deep. Therefore, for most remote Canadian Arctic sites applications contamination is of the surface soil. The most common terrestrial spills in Canada are diesel range fuels. Therefore, when using Table 1.3 a level of 150 mg kg^{-1} would apply to most situations as the lowest default guideline or trigger value. If the measured level exceeds the trigger value, there is then an option to use Tier 2 (Table 1.4) or Tier 3 risk assessment derived criteria.

Table 1.4 *Detailed examination of CCME Tier 2 guidelines for F2 (C_{10}-C_{16}) (~diesel range fuel). Discussed further in the Nunavut case study (Section 1.3.2). Key issues discussed further in the text are shown in bold. TBD = to be decided*

Exposure pathway	Concentration mg kg^{-1}	Comments
Soil ingestion (by children and animals)	8000	Ingestion by children only likely near communities. Ingestion by animals is limited by seasonal snow cover – but is of concern to indigenous groups.
Dermal contact	Free product	Fencing the site could remove this pathway. Unlikely to be of concern in remote sites.
Vapor inhalation (indoor)	240	Pathway can be removed at remote sites by preventing entry into buildings This is problematic where there are heritage issues or prevention of entry cannot be enforced.
Vapor inhalation (indoor, slab on ground)	150	Only relevant where buildings present.
Protection of potable groundwater	1200	Of concern where the contamination is in the watershed of a drinking water source.
Protection of groundwater for aquatic life	150	Highly variable pathway in the Arctic where groundwater is ephemeral and of limited extent. Preventing sheen in surface water is important and distance to water body is the key determinant.
Nutrient cycling	TBD	No data available to guide or assess impacts.
Ecological soil contact	TBD	Vascular plants are scarce in most Arctic sites. Faunal distribution is highly variable, but burrowing animals or insects could be of concern.
Produce	TBD	Not important for Arctic Canada.

Canadian researchers from Queens University considered the application of Tier 2 guidelines for a remote DEW Line site on Resolution Island, Nunavut (Paudyn *et al.* 2006; Poland *et al.* 2004). Contamination was measured with mean concentrations of 2140 mg kg^{-1} and a maximum of 19 000 mg kg^{-1}, with both mean and hotspots considerably above the 150 mg kg^{-1} trigger value. The site is hydraulically active each summer thaw, and a small stream passes through an area of petroleum-contaminated soil adjacent to an imploded fuel tank, although fuel in water was <1.0 mg l^{-1} when sampled each year since 2001. Water flows in the stream 300–400 m before flowing over a cliff top and into the sea. One active water course flows through a pond in the centre of the area. Fauna at this site are very limited; there are no earthworms, and even mosquito larvae are not numerous in comparison to other Arctic locales. It is not known whether fuel spills have changed the ecology of this area. The various risk factors and exposure pathways in the Tier 2 assessment are summarized in Table 1.4, although only *soil ingestion, protection of groundwater for aquatic life, ecological soil contact*, and *nutrient cycling* are applicable for remote sites with no inhabited buildings.

For *soil ingestion*, the site is remote and ingestion by humans or animals is likely to be minimal. The animals in contact with the soil are polar bears, arctic fox, and birds such as guillemots, ptarmigan, raptors, and migratory geese. Much of the contamination is <10 cm deep and the site is covered in snow for eight months each year. Thus animals would be in contact with the contaminated areas for only a short period of time. In any case, very little of the soil contamination is as high as 8000 mg kg^{-1} (see Table 1.4).

With respect to the *protection of groundwater for aquatic life*, groundwater in the conventional sense is essentially absent in this part of the Canadian Arctic. Seasonally unfrozen subsurface water to depths of ~1.5 m is important if there is a link between the contaminant source and surface water because of the necessity to prevent free product or sheen on surface waters. This value is adjustable depending on the distance to a water body and factors such as: soil bulk density, soil moisture content, organic carbon fraction, depth to groundwater, saturated hydraulic conductivity of the underlying aquifer, hydraulic gradient, and distance to surface water. For instance, the remediation criterion of 150 mg kg^{-1} can be changed to 360 mg kg^{-1} if the organic carbon fraction is 0.7% rather than the default value of 0.5%. The distance to surface water is the most significant parameter. The default is 10 m, but at 25 m, the calculated criterion is 890 mg kg^{-1} if all other default values are kept constant. For distances >55 m the criterion exceeds the soil ingestion value of 8000 mg kg^{-1}.

The *ecological soil contact* pathway is non-adjustable (450 mg kg^{-1}). For ecological soil contact to be important there must be flora and fauna present. At many sites in the Canadian Arctic vascular plants are scarce and mosses and lichens are the only species present. Fauna at many sites may be limited to bacteria and small insects, whereas at other locations larger animals (e.g. lemmings, ground squirrels) could inhabit a limited territory where exposure could be high. The relative importance of these criteria, however, is open to interpretation.

The *nutrient cycling* pathway currently has “insufficient data available for Tier 2 modification” and cannot therefore contribute to the Tier 2 adjustment.

Ecological soil contact and *nutrient cycling* may be excluded at the discretion of the jurisdiction, provided that the contaminated area is found to have “no adverse effect on the ecosystem.” In some cases, jurisdictions have allowed for the exclusion of soil organisms and related pathways from consideration where “no productive use of the soil system is anticipated or required” (CCME 2001).

For all sites, the collection of soil samples, the measurement of soil properties, and the determination of various site characteristics can easily be obtained and used to re-calculate remediation criteria quantitatively for a Tier 2 adjustment. However, the issue considered by researchers at Queens University concerned the qualitative attribution of "importance," especially whether the soil ecosystem in the area is important (ecological soil contact pathway), whether nutrient cycling might be significant if it was investigated as part of a Tier 3 investigation, and assigning a degree of significance to localized and transient surface water sheen. In their assessment of the various risk factors, the group also considered much broader factors. Of key importance, trials indicated that low-cost bioremediation by landfarming (with tilling and nutrient addition) readily increased biodegradation rates at this site. A full-scale quantitative risk assessment was undertaken which calculated a cleanup criterion of 26 860 mg kg^{-1}. In the end, however, a precautionary approach mitigated by socio-political considerations of perceptions regarding pollution of indigenous land, coupled with access to low-cost bioremediation techniques established a remediation criterion of 8000 mg kg^{-1} and the construction of a landfarm in 2003. Actual residual levels are much lower than 8000 mg kg^{-1} (Paudyn *et al.* 2006).

1.3.3 An Australian Antarctic/sub-Antarctic case study

Six fuel plumes from the Antarctic Casey Station and sub-Antarctic Macquarie Island Station areas are scheduled for remediation (Rayner *et al.* 2007; Snape *et al.* 2005; Snape *et al.* 2006a). The spills are all diesel range products (kerosene, Arctic blend diesel, special Antarctic blend diesel, and diesel) with typical concentrations of ~5000 ranging up to ~40 000 mg fuel kg^{-1}. Soil types range from coarse-grained cryosols characterized by low organic carbon content and seasonally variable ice-filled pores in Antarctica, to cold water-logged peaty soils that do not freeze on sub-Antarctic Macquarie Island. Although both areas come under Australian jurisdiction, legislatively they are treated quite differently.

The Australian Commonwealth Antarctic Treaty Environment Protection (ATEP) Act 1980 and regulations made under that Act implement the Madrid Protocol (see Section 1.2.2.3) into Australian domestic law. The law aims to limit adverse impacts on the Antarctic environment and dependent and associated ecosystems. Through a number of regulations, there is an obligation on the polluter to clean up spills.

Unlike the Antarctic Territory, sub-Antarctic Macquarie Island is part of the State of Tasmania in Australia and comprises Crown Land owned by the

Tasmanian State Government. The relevant legislation in this case is the Environmental Management and Pollution Control Act 1994 (EMPCA). Under this act there is a general environmental duty for the protection of the environment (23A.1) and a requirement not to allow a pollutant to cause serious or material environmental harm (51A.1&2). In addition, the National Environment Protection Council (NEPC), a national body established by State, Territory and Federal Governments, developed the National Environment Protection (*Assessment of Site Contamination*) Measure in 1999 (NEPM). Each state and territory in Australia is responsible for implementing the NEPM and in Tasmania it has been made state policy under the State Policies and Projects Act 1993.

The development of this NEPM fulfilled, in part, Australia's commitment as a signatory to the Rio Declaration on Environment and Development which is committed to conserving, protecting, and restoring the health and integrity of Australia's ecosystems. More specifically, the aim of the NEPM was to establish a nationally consistent approach to the assessment of site contamination to ensure adequate protection of human health and the environment. The NEPM does have some guideline values for investigation of specific substances, but does not include values for petroleum hydrocarbons. In addition, the NEPM does not deal with the management and remediation of contaminated sites. This is dealt with under the broader requirement of EMPCA which does not allow a pollutant to cause serious or material environmental harm.

Thus, under Australian Federal and State laws there is a clear aim to protect the environment, but there are no explicit guideline values for investigating a site contaminated with petroleum hydrocarbons.[9]

To accommodate the broad requirements of the various Federal and State laws and regulations, and to begin the process of establishing suitable guidelines for the Antarctic, sub-Antarctic and cold regions environments more generally, the Australian Antarctic Division has developed a partnership with Canadian and Alaskan researchers to provide quantitative data for a "weight-of-evidence" approach (Batley *et al.* 2002; Burton *et al.* 2002a; Burton *et al.* 2002b; Chapman *et al.* 2002b; Smith *et al.* 2002; Snape *et al.* 2006b; Snape *et al.* 2006c). Once developed, this approach will then be used for the remediation of the six fuel plumes from the Antarctic Casey Station and sub-Antarctic Macquarie Island Station areas described above.

This plan involves a simple decision tree involving three principal lines of evidence: physical and chemical; ecotoxicological; and community level

[9] In Australia, contaminated sites are regulated by the states and territories. There are no national guidelines for petroleum contamination in soil, though most states and territories use the Dutch guidelines.

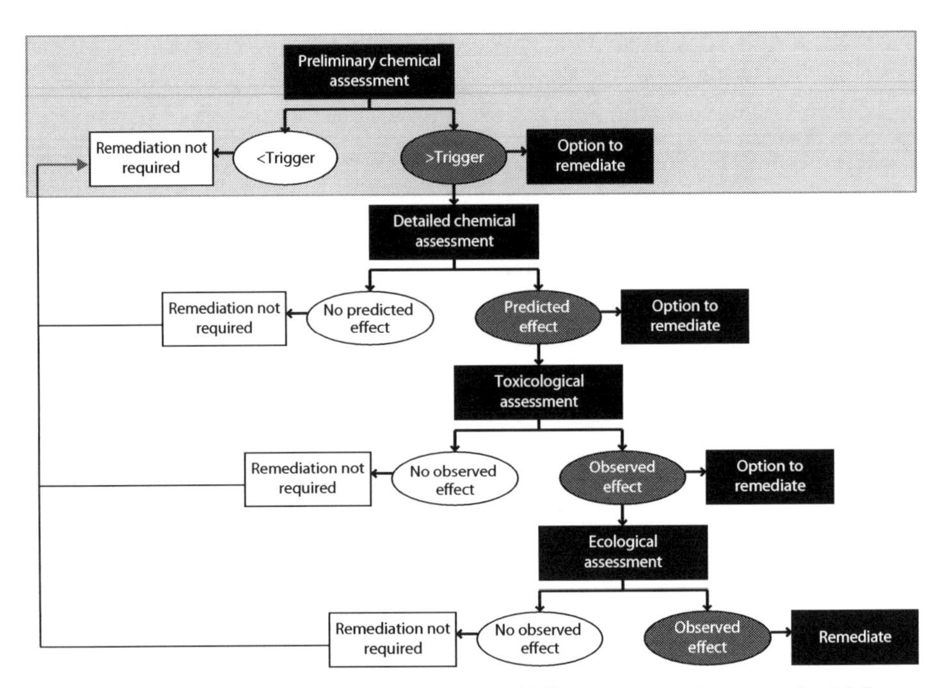

Figure 1.2. Several recent contaminant guidelines use an environmental guideline investigation or trigger value as a screening tool (shaded), followed by the option to either remediate or undertake further studies to determine an appropriate action value. Three principal lines of evidence can be used to determine an appropriate remediation action value; these involve chemical/physical observations of soil properties, ecotoxiciology, and ecological investigations. All three lines can also be used to determine a conservative set of trigger values for the most sensitive soil type in a region.

ecological assessment (Figure 1.2). In many ways the approach is similar to the Canadian tiered risk-based methodology, although the aim differs by developing appropriate methods for Tier 2-type adjustments based on environment-specific data for cold regions soils.

The physical and chemical assessment (Figure 1.2) can be used at its simplest level as a trigger value similar to the Canadian Tier 1 guidelines. These numbers can then be modified on a risk basis again using physical and chemical data but also incorporating ecotoxicological and/or community level ecological data (this approach has many parallels with the Sediment Quality Triad (Chapman 1986; Long and Chapman 1985)). This will ultimately allow inclusion of factors that are recognized as being important, such as low temperatures, freezing water, and the extent of subzero unfrozen water, organic carbon, oxygen, and salt content. Due to the remote location, risks related to human health are generally minimal, so the predominant issue is environmental protection. To evaluate how these properties and processes might impact on the various protection criteria outlined above (see also Section 1.3.2 and Table 1.4), criteria that need addressing were first identified and are discussed in more detail below.

1.3.3.1 *Physical and chemical criteria*

To derive an environmental investigation level or trigger value that would prevent non-aqueous phase liquid (NAPL) or sheen on water, Rayner *et al.* (2007) began by modeling to determine at what concentrations of fuel NAPL might form by using the Mariner *et al.* (1997) model. The model uses a simple partitioning equation that first apportions the mole fraction of compounds or classes of compounds into soil, air, and water phases, then attributes any excess as NAPL and calculates the proportion of pore-filled NAPL. For highly insoluble products like diesel, the dominant control is the amount of organic carbon present. Preliminary results indicate that a precautionary value of ~100 mg fuel kg^{-1} would prevent the occurrence of NAPL formation for most polar/subpolar soils (Figure 1.3). This estimate might eventually be lowered if freezing promotes desorption, but that estimate will likely be revised upwards when there is better knowledge of what proportion of NAPL saturation is needed before movement is possible in frozen or freezing soils (see Chapter 3 and Barnes and Adhikari 2006 for more detail). The Canadian Tier 2 *protection of ground water* criterion also addresses the mobility issue, although it does not specifically consider the effect of ice-filled pores or freeze-thaw effects.

1.3.3.2 *Ecotoxicology*

The second line of evidence concerns ecotoxicology and the disruption of biogeochemical cycles. Quantitative information on the toxic effects of petroleum can be derived from observed effects on individuals, individual or ecosystem function through biogeochemical processes (e.g. a change in rates), microbial community analysis (e.g. using a variety of genomic techniques), or an evaluation of effects on specific animal/plant functions. For cold regions soils, those bacteria, fungi, and invertebrates that have important roles to play in biogeochemical cycling offer the best opportunity for quantitative assessments.

To begin the process of evaluating the sensitivity of polar and subpolar soils to petroleum contamination, Schafer *et al.* (2007) evaluated a number of soil biogeochemical toxicity endpoints for petroleum hydrocarbon contamination in a sub-Antarctic soil. Soil from Macquarie Island was collected and exposed to various doses of SAB diesel fuel for a 21-day period. The sensitivity of nitrification, denitrification, carbohydrate utilization, and total soil respiration to hydrocarbon contamination were then assessed. Potential nitrification activity was the most sensitive reliable indicator of contamination, with an estimated

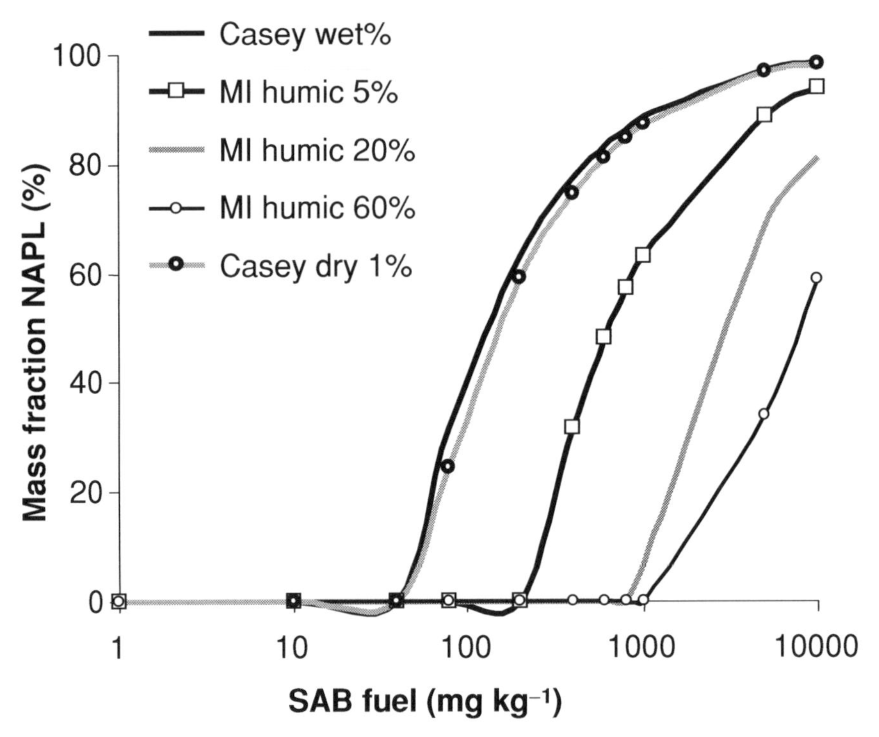

Figure 1.3. Preliminary results from modelling NAPL formation in Antarctic and sub-Antarctic soils from Casey Station and Macquarie Island. Model conditions were 1, 5, 20 or 50% organic carbon, and either wet (35% vol) or dry (7%) conditions (after Rayner *et al.* 2007).

conservative inhibition concentration (IC_{20}) of 190 mg fuel kg^{-1}; other tests ranged up to 950 mg fuel kg^{-1}. These results indicate that some Arctic cleanup guidelines derived from temperate zones might be too high for sub-Antarctic Islands (see Table 1.2).

1.3.3.3 Community level ecological investigations

At this stage, ecological investigations of the effects of petroleum hydrocarbons on soil ecosystems and the receiving environment are limited to broad observational studies and a few recruitment experiments (Banks and Brown 2002; Clarke and Ward 1994; Moles *et al.* 1994; Powell *et al.* 2005; Powell *et al.* 2007; Thompson *et al.* 2006). The intention is to develop these preliminary studies into integrated manipulative experiments that can be related to a critical threshold of contaminant concentration. There are currently no quantitative ecological models or data that can be used to set an environmental investigation level or trigger values for the cold regions.

1.3.3.4 *Summary on guideline development for the Antarctic/sub-Antarctic region*

Although much has been written about the importance of soil contamination, the German Act on Protection against Harmful Changes to Soil and on Rehabilitation of Contaminated Sites (German Law translated, Federal Soil Protection Act, 1998) nicely articulates the importance of maintaining soil health. The act recognizes that the soil and soil functions are:

- the basis for life and a habitat for people, animals, plants, and soil organisms
- a part of natural systems, especially by means of water and nutrient cycles
- a medium for decomposition, balance, and restoration as a result of its filtering, buffering, and substance-converting properties, and especially groundwater protection.

Petroleum is known to be toxic to soil organisms that are interdependent parts of complex communities. Communities contain organisms that differ in size, number, habits, life cycle, food sources, interactions, and interconnectedness. It is these soil organisms that facilitate soil formation and organic matter breakdown, and they are crucial in biogeochemical cycles (e.g. N, C). Hence the Australian Antarctic Division approach to determining appropriate cleanup guidelines is based on the concept of maintaining soil ecosystem health in its broadest context. To do this they are considering three principal lines of evidence that incorporate environment-specific observations of physicochemical processes, ecotoxicology, and ecological data (Figure 1.2). This is similar to Canadian Tier 2 criteria for protection of groundwater, and maintaining nutrient cycling and ecological soil health (Table 1.4).

1.4 Guidelines and recommendations

For highly regulated countries, most notably Alaska and Canada, guidelines and recommendations are clearly established. Both countries also have an active research capability in risk assessment and are currently investigating environmental properties and ecosystem characteristics that could influence the relative risk of petroleum spills in their Arctic regions. It is likely that their future developments will lead to regulation that is based on science that is more relevant to the cold northern region.

For regions such as Antarctica where there is very limited environment-specific data, and no international framework for even deriving guidelines or recommended procedures, some form of interim measures or procedures are urgently needed. Based on limited geochemical modeling and toxicological data from Antarctic and sub-Antarctic soil, and where diesel range organics (DRO) are the primary concern, it appears that the 100 mg fuel kg^{-1} guideline value in use by Norway and Sweden offers a realistic but relatively precautionary Tier 1 guideline investigation level or trigger value. For most soils, there is unlikely to be much NAPL present at concentrations $<\sim$100 mg DRO kg^{-1}. Adopting a relatively low trigger value then offers the potential to consider a more site-specific risk assessment, such as determination of important soil properties that could be used in an adjustment. In this respect the Canadian 3-Tier scheme offers a useful framework, although several important criteria do not currently have sufficient data to be usable, and many of the Tier 2 criteria might be affected by freezing water.

The Alaska Department of Conservation is evaluating a set of spreadsheet tools that may be used to characterize risk and the benefit of different remediation approaches given site-specific input data. One tool is a "hydrocarbon risk calculator" (Geosphere & CH2MHILL 2006) that characterizes spilled hydrocarbons using 16 aromatic and aliphatic equivalent carbon fractions. It solves the phase partitioning equations to assess the concentration of each fraction in the dissolved, vapor, adsorbed, and NAPL phases (this is similar to the NAPLANAL model used by Rayner *et al.* (2007) for the Antarctic and sub-Antarctic). The calculator uses the phase partitioning data to characterize human health risk via the soil ingestion, groundwater ingestion, migration to outdoor air, migration to indoor air, and migration to groundwater routes (ecological risks are not calculated). Other spreadsheet calculation tools (Acomb, unpublished resources) help assess the source longevity and change in concentration of the 16 hydrocarbon fractions within and downgradient of NAPL source areas during remediation by vadose zone and saturated zone intrinsic remediation processes and active air based remedial approaches.

For the Australian Antarctic remediation program, the development of quantitative risk-based guidelines is occurring in parallel with remediation research and development with the hope that the guideline targets for protection of soil health will be appropriate for the polar environment and be achievable at reasonable cost.

1.5 Future research

1.5.1 The need for scientifically informed policy development

Petroleum contamination in cold regions is managed by many different countries using a variety of regulatory frameworks. So far there has been relatively little attempt to accommodate environment-specific factors into the assessment–regulation–remediation process. However, there is growing recognition that the environmental risks associated with petroleum spills in such areas are influenced by a number of common characteristics. For this reason, development of quantitative guidelines would benefit from international collaboration between government, industry, and the academic research sectors. This is perhaps the single most important issue for improving all aspects of petroleum pollution management. The importance of this issue was also noted by the World Bank *Russian Pipeline Oil Spill Study* (Det Norske Veritas 2003 p. 1): "although pipeline oil spills cannot be completely eliminated, their frequency and severity can be reduced by . . . setting adequate standards and regulations and enforcing them . . . ." The caveat that we would add is that standards and regulations need to be based on scientific knowledge that is relevant to the cold regions. To do this, it will be necessary to collaborate extensively.

For the Arctic region, the work of AMAP stands out as a model of what can be achieved through international cooperation. For the Antarctic, it is less clear who is best placed to coordinate and lead new environmental initiatives, although it is likely that the Committee on Environmental Protection (CEP), which reports to the Antarctic Treaty Consultative Meetings, could be the most influential. Coordinated support for an international initiative in the area of contaminant regulation, possibly facilitated through AMAP and CEP, would greatly accelerate improved environmental protection. Such a coordinated approach would also drive improvements in the remediation industry, and facilitate the flow-through from scientific understanding to practical environmental management measures.

1.5.2 Research needs for the development of guidelines and regulations

In the Antarctic case study (see Section 1.3.3), three principal lines of evidence were presented as components that could be used for establishing trigger values, remediation guidelines, or site-specific risk assessment: physical and chemical; ecotoxicological; and community level ecological data (Figure 1.2). These lines of evidence are broadly compatible with the regulation frameworks used in Australia, Alaska, and Canada, and offer a simple conceptual model that allows for a mix of relevant quantitative data and professional judgement (Batley *et al.* 2002; Chapman *et al.* 2002a). However, each of the lines of evidence requires further research both in the development of suitable methods and in the application of quantitative comparisons from a wide range of soils and environments.

1.5.2.1 *Modeling oil and NAPL dispersal in freezing soils*

Several experimental empirical studies have documented the behavior of spilled oil and fuel on frozen soil. Research by Chuvilin and coworkers (Chuvilin *et al.* 2001; Chuvilin and Miklyaeva 2003) has examined the factors that affect spreadability and transport of crude oil in frozen soil. They concluded that the amount of pore filled ice/water, oil composition, soil temperature, salinity, soil mineralogy and cryogenic structure all affect oil dispersion. Preliminary work has examined concentrations which indicate the presence of NAPL (DRO) in a range of Antarctic and sub-Antarctic soils through application of a partitioning model developed by Mariner *et al.* (1997). The model predicts that the hydrocarbon mass held in the soil in the dissolved, vapor, and adsorbed phases is predominantly controlled by soil organic carbon, and that NAPL is present at concentrations in the range 50–1000 mg DRO kg^{-1} (Rayner *et al.* 2007). However, these models have yet to be extended to estimate at which point NAPL movement is possible. Charbeneau (1999) developed spreadsheet tools that assess the mobility of NAPL near the saturated capillary fringe. Qualitative observations indicate that soil type and the amount of pore filled ice/water have a significant influence on the spreadability (Barnes *et al.* 2004 and Chapter 3; Barnes and Adhikari 2006). However, there is virtually no quantitative information that can currently be used to predict what proportion of pore space needs to be filled by NAPL before movement is possible in soils that undergo freezing.

The most pressing future research need is to derive further experimental information to allow the development of a holistic dispersal model that can accommodate a range of freezing soil properties to enable calculation of dispersal distances and rates for a range of oil types. This is needed for schemes such as the Canadian Tier 2 adjustment for the *protection of groundwater for aquatic life* so that dispersal in freezing ground can be accurately estimated. It is central to the "risk-based corrective action" proposed for Alaska. It is also needed where monitored natural attenuation is proposed, or where engineered controls such as permeable reactive barriers are to be implemented, such as in Antarctica.

1.5.2.2 *Soil ecotoxicology*

The dynamic relationships that exist between the biotic and abiotic components of the soil ecosystem are often changed by petroleum hydrocarbons, and soil ecosystem function is strongly linked to properties such as soil composition and structure. The current goal in polar ecotoxicology is to test and evaluate how sensitive organisms are to contaminants, and to relate toxic responses to ecosystem resilience (discussed below). The ideal way to undertake such sensitivity evaluations is to use organisms or criteria that can be quantified across a wide range of environment types (tropics to poles). In high polar environments, the biotic components of the soil ecosystem are often relatively simple and of limited diversity (Wall and Virginia 1999), and the types and abundance of potentially suitable test organisms are low. For this reason soil microorganisms and invertebrates are likely to be the best indicators of petroleum hydrocarbon toxicity in polar soils.

The terrestrial ecological toxicology of hydrocarbons in polar regions needs to take into account unique polar attributes that influence exposure. For example, the effect of temperature on toxicity of petroleum hydrocarbons is not known. It is commonly assumed that decreased temperature will result in decreased exposure due to lower biological activity (i.e. decreased contaminant uptake). However, some soil invertebrates (e.g. oribatid mites) remain active at temperatures well below that of other invertebrates (Hayward *et al.* 2003), and the effect of soil water freezing on the concentration of contaminants in the residual unfrozen water, or NAPL, and the second-order effects that this might have on toxicity is not known. In addition, hydrocarbons have a narcotic mode of toxic action (they disrupt cellular membranes). Regulation of cellular membrane composition is critical to the survival of some polar organisms, and it is not clear if this disruption could yield lower toxicity thresholds. There are currently no data on terrestrial toxicological temperature dependence and further understanding of soil variability and its influence on toxicity is required to develop experimental design parameters that will result in protection of polar environments. The priority is to identify native species that are suitable test organisms, and expose these organisms to petroleum hydrocarbon contamination for a range of soil types and environmental parameters, which will allow quantitative derivation of toxicity thresholds.

1.5.2.3 *Ecosystem resilience to contamination*

The toxicological effect of petroleum contaminants on one or more particular species can have follow-on effects at community or ecosystem levels. In the cold regions, microorganisms and invertebrates drive ecosystem processes. They influence and are influenced by landform development, with features such as slope angle, aspect, cryoturbation, and snow/ice accumulation being critical to soil formation and the development of fluvial systems. The polar regions share a number of landscape similarities, however differences between the Arctic and Antarctic regions do exist, with one of the most important differences being that the Arctic has a much larger proportion of ice-free terrestrial habitat. Terrestrial Antarctica comprises 99.5% monotonous ice. Much of the remaining 0.5% comprises barely weathered rock in the Transantarctic Mountains, with perhaps only 0.05–0.1% of the continent being the small and rare patches of ice-free land where soils form (Poland *et al.* 2003). Such ice-free oases are important habitats for the vast majority of terrestrial Antarctic life.

Patch size and proximity to remnant populations are important factors influencing the abilities of disturbed populations to recover from disturbance (contamination) events. The physical environment (e.g. substratum type, aspect, snow accumulation, drainage) provides the background mosaic of habitat variability on which is superimposed the variability caused by biological processes (recruitment, competition, predation and grazing, biogeochemical processing, and atmosphere and hydrosphere interactions).

At small spatial scales, non-sorted landscapes are often linked in terms of chemical weathering products and the flow of biological "resources" that are moved down-slope by water migration. This contrasts with sorted landscapes where cryoturbation often leads to frost boils and associated sorting patterns. In such landforms there is tentative evidence that measures of biodiversity might be highly disconnected and heterogeneous at small scales of ~1–10 m (Kade *et al.* 2005; Virginia and Wall 1999). In the Antarctic, these two different landform types typically occur in the landscape separated by continental ice, periodically frozen fjords, or ephemeral snow, and are potentially disconnected at scales of 100s of meters to 10s of kilometers. At the very largest scale, the individual ice-free areas of Antarctica are separated from each other by enormous distances (more than 1000 km in some places). Thus it is quite possible that there are a series of self-contained populations isolated from each other around the coast of Antarctica.

Temperate ecosystems have been shown to be robust against random perturbations (habitat or species loss) but can be extremely fragile when critical elements in the system are selectively removed (Dunne *et al.* 2002b; Rhodes *et al.* 2006; Solé and Montoya 2001). It is not known whether high-latitude ecosystems show similar general characteristics. Examples of this are extinctions of species that cause secondary extinctions of other species (Dunne *et al.* 2002a), indirect effects that can be propagated through links of several inter-species interactions or via non-trophic interactions such as competition (Jordán 2001), and species that directly interact only weakly with other species but in fact have a strong role in maintaining overall ecosystem stability (Berlow 1999). All these system-based interactions are likely to be influenced by petroleum contamination, and relating these changes back at a range of spatial scales that correspond to the landscape could be important. Integrating information on the predominant scales of patchiness, variability, and connectedness of polar terrestrial soil ecosystems will help determine whether these properties of ecosystem structure could potentially influence the susceptibility of Antarctic soil, lake, and nearshore marine ecosystems to the effects of petroleum contamination.

1.5.3 Concluding remarks

The tendency has been to extend temperate guideline values to cold regions. We have highlighted three common assumptions made by regulators: (1) similar soil development, (2) similar toxicological potency, (3) similar ecosystem or landscape structure. Reliance on such assumptions reflects a dearth of scientific data on contaminant dispersal processes in cold regions soils, and a complete absence of relevant polar toxicology.

2

Freezing and frozen soils

WALTER FOURIE AND YURI SHUR

2.1 Introduction

Frozen soil is defined as a soil where the soil moisture has turned totally or partially into ice. On the other hand, permafrost is defined solely on the basis of soil temperature. If the soil temperature remains below 0 °C for at least two years, the soil is considered permafrost. The upper layer of the permafrost undergoes a cyclic temperature change during the year from frozen in the winter to thawed in the summer. This layer is called the *active layer* or seasonally thawed layer. The active layer in a permafrost region can extend from as little as 20 cm to about 2 m (Shur *et al.* 2005) depending on climate, soil texture, and organic content above mineral soil. In areas without permafrost the layer of soil which is frozen in the winter is called the seasonally frozen layer. Most permafrost on earth is thousands of years old, but some can be quite new. In permafrost regions, contaminant impacts generally initiate at or near the soil surface and affect the active layer, suprapermafrost water, and uppermost permafrost (Chapter 3). It is this realm that most concerns environmental scientists and engineers tasked with environmental cleanup. A thorough understanding of properties of the active layer and the upper permafrost is necessary for planning and implementing effective remediation of cold media.

Bioremediation of Petroleum Hydrocarbons in Cold Regions, ed. Dennis M. Filler, Ian Snape, and David L. Barnes. Published by Cambridge University Press.

2.2 Review and recent advances

2.2.1 Thermal and physical properties of frozen ground

2.2.1.1 Thermal conductivity of soils

The thermal conductivity of soil is the measure of its ability to conduct heat. Soil thermal conductivity is a function of the thermal state of the ground (frozen or unfrozen), water content, dry density, gradation, and mineralogy. In 1949, Kersten developed empirical equations that relate the thermal conductivity of soil to moisture content and dry density (Kersten 1949). The equations do not take into account variations in quartz content or whether the moisture in the soil is completely frozen or not, but have an accuracy of ±25%. Farouki (1981) converted Kersten's equations to SI units (W/m-K), and stated that their use is adequate for most practical applications. The thermal conductivity for frozen and unfrozen coarse-grained soils (silt-clay content <20%) can be expressed as:

$$k_u = 0.1442 \cdot (0.7 \log w + 0.4) \cdot 10^{0.6243\rho_d} \tag{2.1}$$

and

$$k_f = 0.01096 \cdot 10^{0.8116\rho_d} + 0.00461 \cdot 10^{0.9115\rho_d} \cdot w \tag{2.2}$$

where k_u and k_f are unfrozen and frozen soil thermal conductivity, w is soil moisture content (%), and ρ_d is the dry density of the soil (g cm^{-3}). For fine-grained soils (>50% silt-clay) the thermal conductivities are:

$$k_u = 0.1442 \cdot (0.9 \log w - 0.2) \cdot 10^{0.6243\rho_d} \tag{2.3}$$

and

$$k_f = 0.001442 \cdot 10^{1.373\rho_d} + 0.01226 \cdot 10^{0.4994\rho_d} \cdot w \tag{2.4}$$

Johansen (1975) developed more complex expressions for thermal conductivity for frozen and unfrozen mineral soils which also accounts for unfrozen water content in the soil. Gavril'ev (2004) developed equations characterizing thermal properties of mineral and organic soils.

2.2.1.2 Soil heat capacity, unfrozen moisture content, and the latent heat of fusion

The heat capacity (or specific heat) of a soil is the amount of heat required to elevate the temperature of one unit of mass (or volume) by one degree. Heat capacity can be calculated as the sum of the constituent heat capacities of mineral soil, water, ice, and air (negligible) that comprise the subsurface matrix. Johansen (1975) expressed the volumetric heat capacities for unfrozen and frozen mineral soils as:

$$c_{vu} = \left(\frac{\rho_d}{\rho_w}\right)\left(0.17 + 1.0\frac{w}{100}\right)c_{vw} \tag{2.5}$$

and

$$c_{vf} = \left(\frac{\rho_d}{\rho_w}\right)\left[\left(0.17 + 1.0\frac{w_u}{100}\right) + 0.5\left(\frac{w - w_u}{100}\right)\right]c_{vw} \tag{2.6}$$

where c_v is the volumetric heat capacity of water (4.187 MJ m^{-3}) and ρ_w the density of water (g cm^{-3}). To calculate the heat capacity for organic soils, replace 0.17 with 0.4 in Equations (2.5) and (2.6). For coarse-grained soils, the unfrozen moisture content (w_u) is typically negligible, and for very fine-grained soils Equation (2.6) is more precisely evaluated as soil moisture freezes over a range of temperatures around the freezing isotherm. In other words, one must consider latent heat and w_u as a function of temperature as soil moisture undergoes phase change in association with the freezing isotherm. The latent heat of fusion of a soil is that amount of energy absorbed by a unit mass (or volume) to create a phase change (melting) at the phase change temperature. The volumetric latent heat of soil is given by

$$L_v = \rho_d \cdot L' \cdot \frac{w - w_u}{100} \tag{2.7}$$

where L′ is the mass latent heat of water (333.7 kJ kg^{-1}). The phase change temperature is generally considered that of freezing (0 °C), although we recognize that soil moisture continues to freeze down to the *eutectic* temperature (about −70 °C).

All soil water does not necessarily freeze, even if the temperature is well below 0 °C. This phenomenon may be due to the presence of solutes like NaCl, or capillary forces associated with very small soil particles. In a fine-grained soil like silt or clay, capillary pressures are extremely high and inhibit the freezing process. Although freeze inhibition of water is not yet clearly understood, it is thought that for a fine-grained soil, ice forms from the center of the pore space and a thin layer of liquid water is left at the soil particle boundaries. It is this unfrozen film of water that accounts for water *wicking* towards the freezing front in fine-grained soils. Tice, Anderson and Banin (1976) compiled experimental data for unfrozen moisture content in various soils, and fit the data to a simple power curve:

$$w_u = \alpha \cdot \theta^{\beta} \tag{2.8}$$

where α and β are characteristic soil parameters and θ is absolute temperature below freezing in degrees Celsius. Values for α and β for a range of soils can be found in Andersland and Ladanyi (2004).

2.2.1.3 *Freezing saline soils*

Salt in water decreases the freezing point of a soil and increases the amount of unfrozen water. During the freezing process, salt is excluded from the ice phase and thus the solute is redistributed through the soil (Hallet 1978).

Mahar *et al.* (1983) reported that the rate of freeze to a certain depth increases with an increase in salinity. They attributed this phenomenon to the gradual release of latent heat over a range of temperature. Yen *et al.* (1991) provided an approximation for the latent heat as a function of ice salinity, which shows that the latent heat released is less than that of pure water. Visualization studies by Arenson and Sego (2004) showed that the frozen fringe becomes thicker with an increase in salt concentration, and they hypothesized that needle-like ice formations in a saturated coarse-grained soil could adversely affect soil shear strength. Chamberlain (1983) gave evidence of reduced soil hydraulic conductivity under freezing conditions.

Experiments done on saline sand columns by Baker and Osterkamp (1988) showed that significant salt rejection occurred when the columns were frozen from the top down, but that this does not occur when the columns froze from the bottom up. They attributed this contrast to gravity drainage of the brine. Cryogenic structure of saline soils is generally characterized by the same types of cryogenic structure which are typical for soils which do not contain salts. But, as was noticed by Khimenkov and Brushkov (2003), the greater the salinity of soil the more prominent become vertical ice lenses in frozen soil. Phase equilibrium models of saline fine-grained soils have been developed (Grechishchev *et al.* 1998; Aksenov 1998). Studies indicate that the soil-water-salt system is dynamic, and that hydraulic conductivity in saline cold soils is a function of temperature and salt exclusion.

2.2.1.4 *Permeability*

The permeability of a soil is its ability to accommodate liquid flow. In the past three decades it has been shown that layers of ice-rich soil (and permafrost) are not impervious to the flow of liquids, whether it be water or non-aqueous phase liquids (NAPL) (discussed in Chapter 3). Susceptibility to liquid flow is a function of the soil type, temperature, and moisture/ice content. Measuring hydraulic conductivity and permeability of frozen soils is difficult and only a few experimental methods have been developed. Burt and Williams (1976) and Andersland *et al.* (1996) studied lactose and decane as fluid permeants in soil. Other methods, such as referencing soil infiltration by using the air permeability (Seyfried and Murdock 1997; Olovin 1993), evaluating soil hydraulic conductivity from measured thermal conductivity (van Loon *et al.* 1988), and using super-cooled liquid with a nucleating agent (Aguirre-Puente and Gruson 1983) have also been used. It has also been shown that water molecules can be transported through ice by regelation, which can be a significant moisture transport mechanism in saturated soils (Horiguchi and Miller 1980; Wood and Williams 1985). The infiltration of NAPL into frozen soils has been studied by

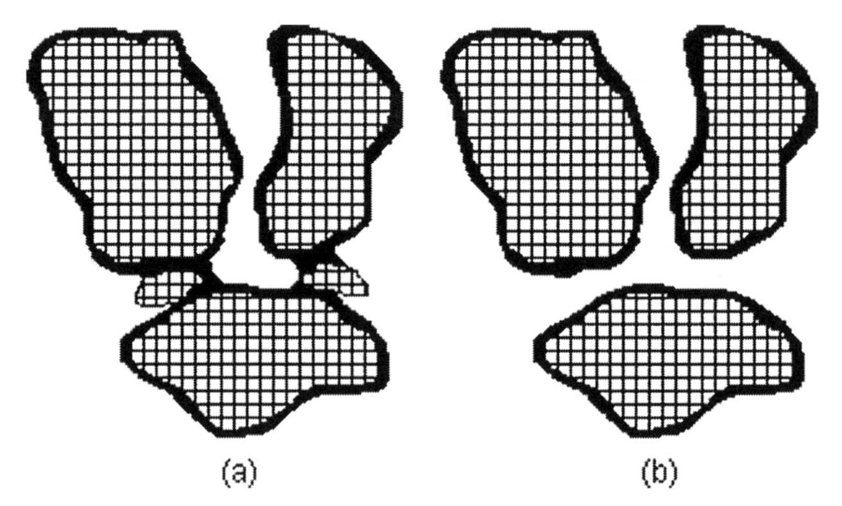

Figure 2.1. Comparison of pore ice formation in coarse-grained soils with (a) and without (b) the presence of smaller particles. Cross hatched areas represent soil grains and black areas represent water held by capillary forces. The scenario shown in (a) represents the creation of a dead end pore with minimal pore ice content in comparison to the scenario shown in (b) where pore channels remain open to flow. Further additions of water to the pore space shown in (a) will result in the pore becoming either filled with ice or entrapped air (Fourie *et al.* 2007).

Wiggert *et al.* (1997) and McCauley *et al.* (2002), amongst others. Both conclude that the infiltration of fuel into a frozen soil decreases with increasing ice saturation.

In Olovin's study (1993), the results from over 3000 tests generally showed that permeability decreased by approximately two orders of magnitude with an increase in saturation of up to 0.5. Overall the results from his studies showed that the permeability of a frozen soil is an uncertain parameter that depends on initial water content of the soil prior to freezing, soil temperature, and structure.

The gradation of a soil has a strong influence on soil permeability. In a coarse-grained soil, the average pore space diameter is large, and water can flow unheeded through the soil matrix. Upon freeze-up, water freezes along soil grain boundaries, thereby decreasing the average pore diameter and altering the flow of water. In a system that includes fine particles, the average pore diameter is drastically reduced and dead end pores can easily be created (Fourie *et al.* 2007). This process is schematically shown in Figure 2.1.

2.2.1.5 Vapor flow

Vapor flow in frozen soils is induced by a temperature gradient. Smith and Burn (1987), using moisture traps, demonstrated that moisture movement through frozen soils overestimates soil diffusivity by orders of magnitude. They hypothesized that vapor harvested in the traps was due to liquid water movement and vapor diffusion. This result concurs with work done by Dirksen and Miller (1966). However, a study by Jackson (1965) concluded that in coarse-grained soils vapor movement is the predominant transport mechanism, while liquid movement was negligible. Research done by Nakano *et al.* (1984) supports the belief that vapor transport plays a significant role in dry granular soils, but that vapor flow becomes insignificant as pore-ice volume increases. Goering and Kumar (1994) showed by simulation that highly porous embankments induced a large enough air flow to significantly alter the thermal regime of the underlying ground.

2.2.2 The active layer

The active layer is that part of the soil that undergoes annual freezing and thawing as a function of temperature. In a tundra environment underlain with continuous permafrost, subsoil conditions can be characterized based on time of year and precipitation (Figure 2.2).

In the northern hemisphere, from January to March (Figure 2.2(a)), winter prevails and snow accumulates with the maximum thickness occuring in depressions. Soil may not be completely frozen in the depressions as snow is a good insulator. If the soil is not completely frozen, soil water may redistribute under pressure from the advancing freeze-front. Between April and May (Figure 2.2(b)), the increase in solar radiation causes some initial melting and surface runoff may occur. Precipitation as rain or snow occurs during this period. Late May and June (Figure 2.2(c)) marks early summer, when precipitation is generally in liquid form and evapotranspiration from the ground increases markedly. Water collects in the depressions and the resulting higher thermal conductivity increases the thaw rate. During July to September (Figure 2.2(d)), precipitation is predominately liquid and evapotranspiration decreases. Extreme temperature variations occur in surficial soils and this realm may dry out completely. From late September through October (Figure 2.2(e)), winter sets in and precipitation transitions to snow. During the early part of this time period the maximum depth of thaw exists and evapotranspiration becomes negligible. The winter period of November and December (Figure 2.2(f)) is marked by snowfall, deeply frozen soils, and little, if any, unfrozen soil moisture.

Suprapermafrost water occurs in the active layer above permafrost. It exists during a part of the year when the active layer thaws and is recharged by precipitation and melt water. Some authors (Reinuk 1959; Tolstikhin and Tolstikhin 1973) consider condensation inside the active layer as a sufficient source of

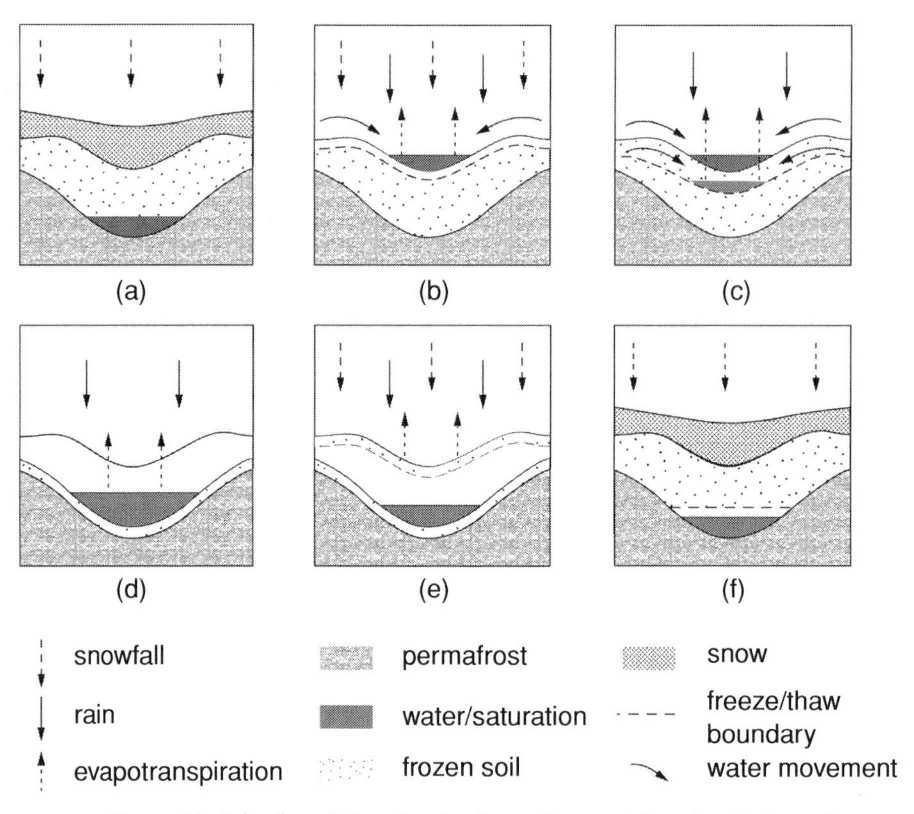

Figure 2.2. Subsoil conditions in a tundra environment (based on Ryden and Kostor 1977)

recharge of suprapermafrost water if the active layer is comprised of coarse soil. Freezing of the active layer causes elevation of the pressure in suprapermafrost water, which migrates with advance of the freezing front. Freezing of suprapermafrost water of the active layer is accompanied by frost heave and sometimes by the creation of frost mounds. In natural arctic settings, suprapermafrost water typically has low mineral and high organic contents. The converse is true for gravel pads and roads where a layer of fine sediment develops at the base of these manmade features, in direct proximity with suprapermafrost water. Here, suprapermafrost water may have a high mineral content.

Suprapermafrost water is a very limited source of water supply and is mainly used for technical needs. It is particularly susceptible to contaminants in general, and liquid and solid contaminants at human settlements and industrial sites. At industrial sites, this water is usually confined within or limited to the fringes of earthen pads and roads, and only later exposed after infrastructure decommission.

2.2.2.1 *Heat transfer in the active layer*

The thermal regime of the active layer describes the heat flux in and out of the ground system. The overall heat flux on the ground surface comprises the radiation transfer on and from the ground surface, convective heat transfer between the soil surface and air above, heat conducted through the ground surface, and latent heat effects such as evaporation. Each of these fluxes is dependent on the physical and thermal properties of the soil, water, and air, and environmental conditions.

On a clear day, short wave solar radiation received by the ground is a function of the relative orientations of the sun and the receiving surface, as well as surface reflectance (or albedo). Several models have been developed over the years to estimate incoming solar radiation (Duffie and Beckman 1991), though most models have limited applicability in arctic regions. In the northern hemisphere, south facing surfaces receive, on average, more radiation than north facing surfaces, with east and west facing surfaces receiving a dose somewhere in between. Thus the ground will typically be colder on the northern sides of hills and mountains. The albedo of the ground also greatly impacts the ground temperature; the darker the ground, the greater the amount of heat that is absorbed.

Convection at the ground surface is a function of soil and air temperatures and a heat transfer coefficient. The air thawing index is a summation of the air temperature through all the days in a year where the temperature is above freezing, thus having units of Celsius degree-hours (or degree-days). The driving force for convective heat exchange to take place at the ground surface is the difference between the air and surface thawing indices (or freezing indices), depending on location and surface composition. The relationship between these indices and surface composition have been studied and reported as *n-factors* (Lunardini 1978). The *n-factors* are a function of the surface and environmental conditions, and account for effects such as air film resistances, surface albedo, and vegetation.

Conductive heat transferred through frozen, freezing, or thawing soils is a function of a temperature gradient and soil thermal diffusivity (soil thermal conductivity divided by specific heat capacity). The thermal conductivity of a soil determines the resistance of a soil to heat transfer as a function of soil type and density, moisture content, and frozen/thawed state.

Heat can be transferred out of the ground by evaporation. In summer, evapotranspiration in a low-Arctic watershed can be appreciable and remove up to 66% of the precipitation (Kane *et al.* 1990). In the higher Arctic, this figure reduces drastically to about 20% of the annual precipitation. In winter, snow covers the ground and evaporated soil moisture condenses in snow.

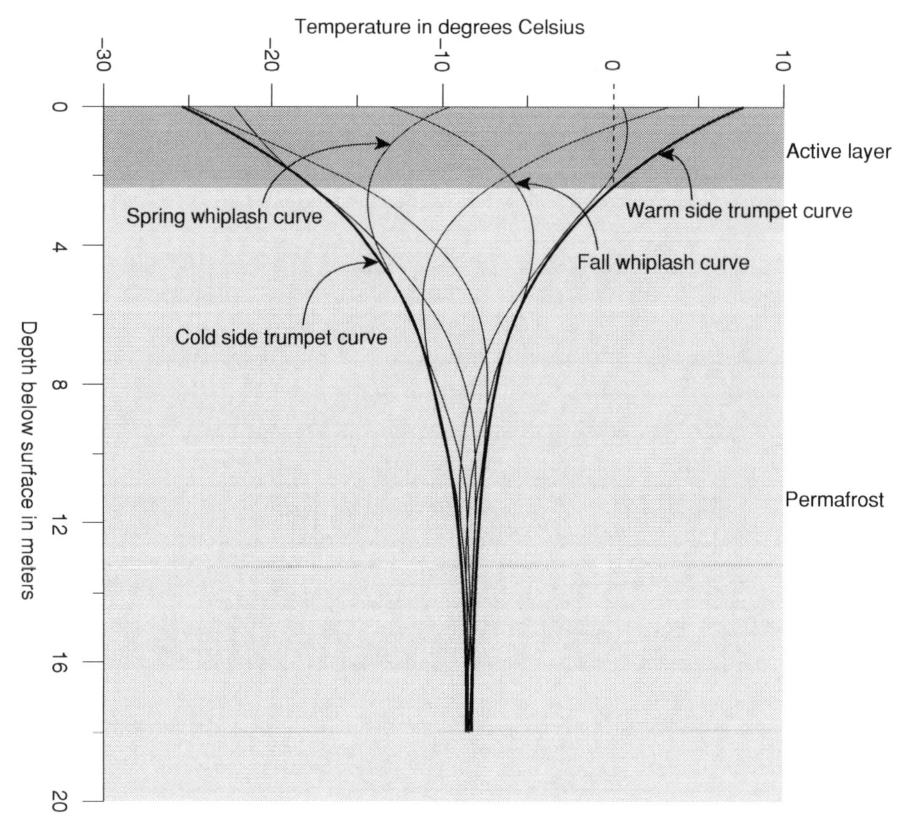

Figure 2.3. Whiplash and trumpet curves.

2.2.3 Depth of freeze and thaw

Figure 2.3 models ground temperatures at an Arctic setting for one freeze-thaw cycle in a permafrost soil.

The figure depicts oscillating soil temperature as a function of depth. For a depth up to about 2.1 m, soil temperature changes throughout the year predominately as a function of air temperature variations. This realm is characteristic of the active layer. Below 2.1 m the soil stays continually frozen as permafrost. The oscillating soil temperature curves are called whiplash curves and the cumulative maximum and minimum limits are referred to as trumpet curves.

To understand how the active layer is formed and how deep it reaches, consider the annual air temperature cycle of Nome, Alaska (Figure 2.4).

From the figure it is apparent that for a certain period of the year, the air temperature is above freezing, and for the rest of the year it is below. By integrating the area between the actual temperature (dotted line) and the freezing temperature (zero-degree isotherm), we calculate the total amount of degree-days

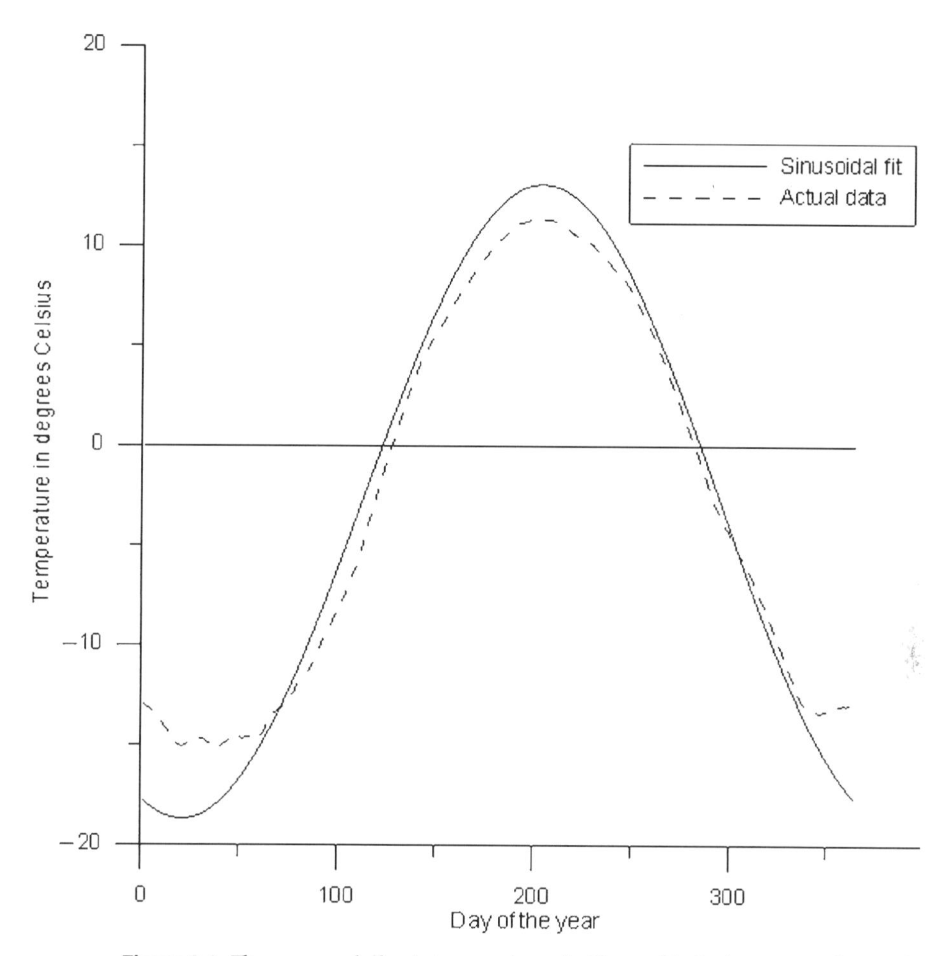

Figure 2.4. The average daily air temperatures in Nome, Alaska (representative year).

that influences soil thawing. Similarly, we can calculate freezing degree-days (the area below the freezing line). These values are called the air thawing index (ATI) and air freezing index (AFI). These indices represent the driving force by which soil freezes and thaws, thus for example, the bigger the AFI the deeper soil will freeze. Weather services in most Arctic countries provide data on the internet for numerous locations. From these data an analysis of air freezing and thawing indices can be performed. To use the air indices to calculate soil freeze and thaw depths, we then relate *air* indices to soil *surface* indices as a ratio, or "*n-factor*" (Lunardini 1978). The *n-factors*, which account for effects such as air film resistances, albedo affects, and vegetation, are expressed as

$$n_f = (\textit{soil freezing index})/(\textit{air freezing index}) \quad (2.9)$$

$$n_t = (\textit{soil thawing index})/(\textit{air thawing index}) \quad (2.10)$$

Table 2.1 summarizes *n-factors* for common Arctic surfaces.

Table 2.1 *Sample of n-factors from published literature (Lunardini 1978)*

Condition	Freezing	Thawing
Asphalt pavement	0.74–1.0[a]	1.8–2.3
Asphalt pavement painted white		1.0–1.2
Gravel	0.63–1.0[b]	1.4–2.0
Spruce trees and brush over moss	0.29	0.37
Moss and peat	0.25	0.73
Asphalt with insulation below	1.0	1.7–2.0
Turf	0.5	1.0
Snow	1.0	
Sand and gravel	0.9	2.0[b]

[a]For cold regions use 1.0.
[b]Use 1.25 to 1.4 for cold regions.

The Stefan equation predicts the depths of freeze and thaw (x) in soils as a function of surface freezing index (SFI) and soil thermal properties:

$$x = \sqrt{\frac{2 \cdot k_f \cdot \mathrm{SFI}}{L_v}} \tag{2.11}$$

The Stefan equation does not take into account soil sensible heat and assumes that soil temperature beneath the freezing or thawing front is equal to 0 °C. It does not take into account the insulation effect provided by vegetation.

The modified Berggren Equation was developed by Aldrich and Paynter (1966) by simplifying the Neumann equation. Similar to the Stefan equation, the Berggren equation includes a factor, lambda (λ), which accounts for soil sensible heat. In this equation the average thermal conductivity, k_{ave}, is used. The λ-factor can be found from Figure 2.5, as a function of a thermal ratio and fusion parameter (also called the Stefan number). The thermal ratio (α) for a freezing soil is computed as:

$$\alpha = \frac{|T_f - T_i|}{\frac{n_f \mathrm{AFI}}{d_f}}$$

with T_i = the initial temperature of the soil
T_f = the freezing temperature
n_f = *n-factor* for the frozen surface
AFI = air freezing index
d_f = duration of freeze period in days

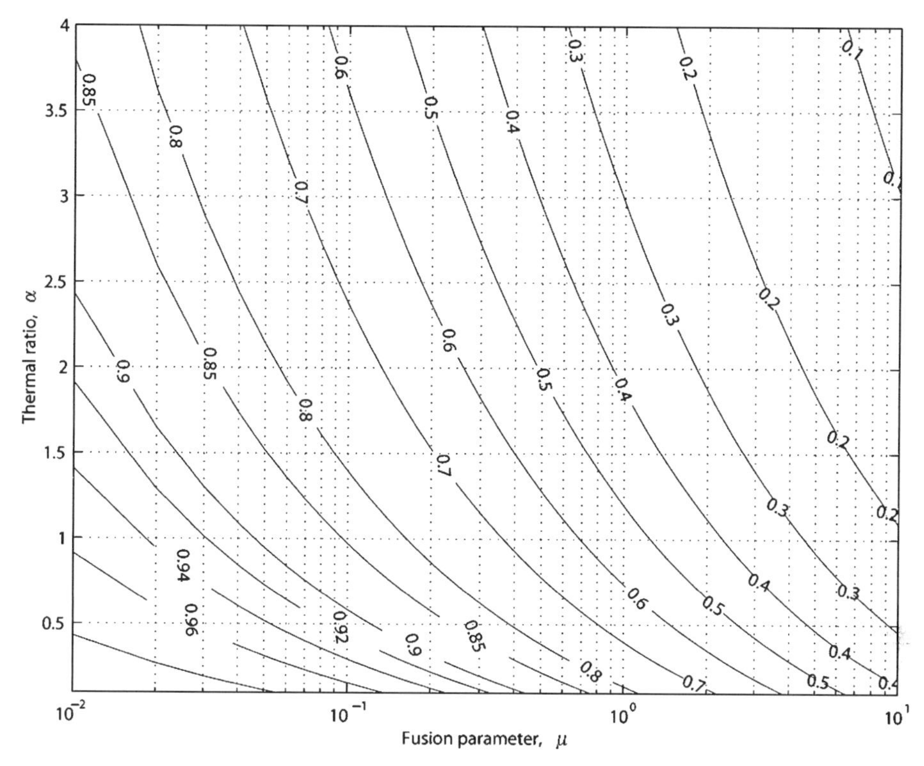

Figure 2.5. λ-coefficient (based on Aldrich and Paynter 1966).

and the fusion parameter:

$$\mu = \frac{n_f \cdot \mathrm{AFI} \cdot C_{ave}}{d_f \cdot L_v}$$

where C_{ave} is average soil volmetric heat capacity.

The depth of freeze, x, can then be calculated as:

$$x = \lambda \sqrt{\frac{2 \cdot k_{ave} \cdot \mathrm{SFI}}{L_v}} \tag{2.12}$$

2.2.4 *Phenomena of freezing ground*

The movement of water in freezing soil is caused by disequilibrium in the soil-water-ice system that results from a complex combination of differences in temperature, pressure, concentration, and other internal factors such as humidity, and electrical and magnetic potentials. Considerable effort has been made to understand freezing ground phenomena such that we can predict the parameters that influence ice segregation, frost heave, and thaw settlement.

2.2.4.1 *Frost heave and thaw settlement*

In a freezing fine-grained soil, such as a sandy-silt, silt, and clay, water is drawn to the freezing front from the surrounding soils. This water movement (wicking) occurs because of the temperature and moisture gradients in the soil. A soil can wick moisture if water is able to flow under negative pressures in the thin capillaries between the soil particles. A sandy-gravel for instance, is a non-frost susceptible soil, as it cannot wick moisture through its relatively large pore spaces. Therefore, the three necessary conditions for frost heave are sub-freezing temperatures, the availability of moisture, and the ability of water to wick through the soil. Wicked water freezes in a segregated fashion, forming ice lenses. Ice lenses push soil upward differentially, causing damage to the overlying engineered structure. Similarly, during the thawing period, heaved soils now settle differentially, also causing damage. Telephone poles, foundation posts, and fences are jacked out of the ground by a similar mechanism.

2.2.4.2 *Thermokarst*

The thawing of ground ice and consolidation of thawing soil results in the deformation of soil and the formation of specific forms of relief called thermokarst. As thawing ice-rich soils consolidate by squeezing out the available moisture, the resulting depression in the ground will either drain away water or water will accumulate in the depression and create a thermokarst lake.

2.2.4.3 *Taliks*

It is usual that permafrost limits the possibility of sub-permafrost water recharge. According to van Everdingen (1974), in permafrost regions, rates of groundwater recharge are orders of magnitude lower than in non-permafrost regions. Taliks are layers or bodies of unfrozen soil in permafrost. In relation to permafrost integrity, taliks are divided into two groups: open and closed. An open talik penetrates the entire thickness of permafrost, while a closed talik is a thaw bulb bordered by permafrost and the soil surface. Taliks occur both in continuous and discontinuous permafrost regions, and many areas of unfrozen soil in discontinuous permafrost were developed as taliks. In areas of continuous permafrost, taliks beneath lakes are usually closed and prevent interaction between suprapermafrost and sub-permafrost waters. Rare open taliks beneath big rivers and deep lakes can provide connection between surface water and sub-permafrost water. Open taliks also occur below mountainous rivers where the channel deposits are presented by coarse material. High hydraulic conductivity provides extensive water flow in channel deposits and thermal conditions favorable for talik existence.

Numerous shallow lakes are abundant in low-lying continuous Arctic permafrost regions, but not every lake has a talik. Taliks occur only beneath lakes whose depths are greater than a critical depth (Kudriavtsev 1978). In the Arctic Coastal Plain of Alaska, the critical depth of lakes is about 1.5–1.8 m. Only beneath lakes with a greater depth is soil perennially unfrozen to some depth. As previously mentioned, open taliks are very rare. For example, to develop an open talik in continuous permafrost, where mean annual temperature of water in lakes is about 1 °C and the permafrost temperature is about −8 °C, the lateral extent of a lake has to be an order of magnitude greater than the permafrost thickness (Grechishchev *et al.* 1980). And in the northern parts of Alaska, the permafrost thickness is greater than 500 m!

In a discontinuous permafrost region, size of a lake in the lateral direction is less critical. Here, the age of a lake is an important factor in formation of the talik. For example, if a section of ice-saturated permafrost is about 50 m thick, a young thermokarst lake (with an age to hundreds of years) could have a closed talik. Lakes which are thousands of years old could be underlain with open taliks. Soil stratigraphy is also a very important factor as a thinner layer of ice-saturated soil will result in the faster formation of an open talik. Also the permeability of lake sediment is usually very low. Some authors (Kane and Slaughter 1974; Tishin 1983) acknowledged the possibility of interaction of lake water and sub-permafrost groundwater.

2.3 Guidelines and recommendations

An evaluation of a contaminated site in cold regions depends not only on the physical properties of the site but also on the time of the year. An oil spill on tundra in winter will exhibit different impact behavior than a spill that occurs in summer. Two test sites near Fairbanks were contaminated with crude oil, one during the winter of 1976 and the other during the summer of 1976 (Collins *et al.* 1994). The purpose of these trials was to determine the fate of the environment after a crude oil spill on forest underlain by permafrost. Fifteen years later, the sites were re-examined to determine the long-term effects of the spill. It was found that the winter spill had affected a larger area as the oil traveled over the frozen ground. There were also indications that some of the volatile compounds still existed in the subsurface and that little biodegradation took place during the 15 years (see Chapter 3, Section 3.1 for more discussion). The population of hydrocarbon-degrading bacteria may increase at such a spill, while overall microbial diversity may decline (Aislabie *et al.* 2004).

The upper permafrost is almost impermeable (Shur 1988a; Olovin 1993). But seasonal freezing of soils and their thermal contraction form cracks of different sizes in the upper permafrost. Fractures and cracks in frozen soil allow contamination to penetrate into the upper permafrost and impact the root systems of vegetation (Filler and Barnes 2003). Removal of trees and compaction and other disturbances to the surface organic layer can result in an increase in the active layer depth and melting of the highly impermeable intermediate layer of the upper permafrost. In discontinuous permafrost regions, contaminants may follow a lowering permafrost table as it degrades.

Lessons learned through mitigation of numerous petroleum spills and contaminated sites has resulted in a *Tundra Treatment Manual* (Athey *et al.* 2001) in Alaska. This guidance document considers season and physicochemical properties of petroleum contaminants relative to environmental media in the development of early response and recovery tactics. Management of permafrost, or permafrost control, will be integral to remediation planning at Arctic and Antarctic contaminated sites. Where engineered bioremediation is used with heating schemes, particular attention must be given to heat transfer to the subsurface to prevent settlement and further intrusion of contamination with depth.

2.4 Future research

At present we have little understanding of the effects of repeated freezing and thawing of soils on contamination in the ground. A key area of future research concerns the impact that petroleum spills have on unfrozen water content, which in turn can have an effect on properties such as frost heave. Furthermore, since climate change is anticipated to have significant impact on permafrost and the active layer, it stands to reason that it will also influence contamination within these soil regimes.

2.4.1 Permafrost and climate change

Climate change can induce a shift in the thermal equilibrium of soil and change the depth of the active layer. Such a change could have pronounced effects on infrastructure, the release of greenhouse gases, and other soil phenomena of frozen and freezing ground. Climate change is also inextricably linked with the chemical, biological, and physical conditions in the Arctic (Hinzman *et al.* 2005). To better understand the connection between climate and permafrost, Shur and Ping (1994) divided all climates into three categories: climate sufficient for permafrost formation (permafrost must exist), climate neutral for permafrost formation (permafrost can exist), and climate unfavorable for permafrost formation (permafrost cannot exist except under special conditions; in caves, for example). In a sufficient climate, permafrost forms everywhere under land and under shallow water. Here, climate modifiers (Shur 1988a) like vegetation and relief influence the depth of the active layer and permafrost temperature but not its existence except in some extreme situations (for example, deep water). In a neutral climate, permafrost forms or can survive only in landscapes that have special combinations of snow cover, relief, vegetation, and soil. As a result, modifiers determine permafrost existence and their change can lead to permafrost degradation. In some areas with a propitious climate, the modifiers protect ice-rich relict permafrost with massive ice that was formed more than 10 000 years ago (Shur 1988a).

Vegetation and snow are the most efficient modifiers of a soil thermal regime. Vegetative succession and accumulation of organics (peat) leads to a decrease in the active layer depth with formation of an organic layer, lowering of permafrost temperature, and accumulation of aggradation ice in the intermediate layer (Shur 1988b). This kind of evolved system is efficient in protecting permafrost. Permafrost becomes somewhat independent of atmospheric climate, but at the same time becomes highly dependent on such modifiers as vegetation, soil organic horizon, and snow cover. Changes in even one of these modifiers can alter the thermal equilibrium between permafrost and air climate and lead to permafrost degradation.

Permafrost change is a function of soil condition, irrespective of climatic change or alteration of modifiers. Change in climatic conditions can lead to changes in modifiers, which can even reverse the direction of impact of air on a soil regime. For example, Shur (1988a) reported that layers of short-term permafrost (pereletok) were formed in the Nadim River Valley (West Siberia, Russia) in extremely warm and dry summers but not in cold summers. In the warm and dry summers, thermal conductivity of dry moss and peat is very low, which has greater influence in preserving frozen soil than potential thaw influence from increased air temperature.

Vegetation and peat are extremely powerful modifiers of a soil regime and the most vulnerable among modifiers. They can be destroyed by fire, animal activities, or can be stripped or altered by human activities such as oil spills and contamination. It is of interest to compare the effect of vegetation removal with a change in the thawing index which can produce an equivalent impact on the active layer depth. It has been shown that if stripping the vegetation changes the depth of the active layer from 0.8 m to 1.5 m, the thawing index would have to be increased about 2.5 times to provide the same change in depth of the active layer under vegetation (Shur 1988a).

In the discontinuous permafrost zone, vegetation removal usually leads to an increase in the active layer depth and to permafrost degradation. According to long-term studies by CRREL in Fairbanks (Linell 1973), the permafrost table at a site with stripped vegetation has been lowered to almost 7 m during 26 years. At the field of the University of Alaska experimental farm, which was cleared in 1940, the permafrost table is at a depth of 10 m (Osterkamp and Romanovsky 1999).

The Circumpolar Active Layer Monitoring program (CALM) is currently collecting data on active layer depths at more than 100 sites across the globe including Arctic, Antarctic and sub-Arctic mountainous regions (Nelson *et al.* 2004). CALM will provide valuable information that can be used to better understand the transitional state of permafrost and the effect of natural and anthropogenic effects on the active layer.

Future research must include consideration for the effects of climate change on the state of petroleum contamination that presently exists in cold regions. We pose the following questions to consider:

- How does petroleum contamination influence the active layer and permafrost?
- Will climate change affect the hydraulic connectivity between surface- and groundwater (e.g., suprapermafrost water and tundra marsh or lake waters)?
- If links are established or altered between surface- and groundwater, what will their role become as potential pathways for contaminant transport?

These and other questions bear answering with consideration for past indiscretions (e.g. oil pipeline spills in the Komi Region, Russia, and Prudhoe Bay, Alaska), and future decommissioning of depleted oilfields and development of new ones in cold regions.

3

Movement of petroleum through freezing and frozen soils

DAVID L. BARNES AND KEVIN BIGGAR

3.1 Introduction

Movement of petroleum through non-freezing soils has been studied extensively over the last several decades. Little work has been done on understanding how petroleum moves through seasonal freezing soils (active layer) and frozen soil (permafrost). Petroleum migration through active layer and permafrost soils is influenced by the formation and presence of ice at all scales. At the millimeter scale, ice in pore spaces will either interrupt downward migration causing petroleum to spread laterally, or impede petroleum movement altogether due to the lack of open pore space. Segregated ice at centimeter-to-meter scales will most likely cause the contamination to spread laterally in frozen soils. Segregated ice formation in the active layer can also generate fissures that will enhance petroleum movement when the soil is thawed. At larger scales, discontinuous and continuous permafrost will slow, redirect, or impede contaminant migration.

Understanding the impact freezing and frozen soil conditions have on petroleum movement through soils is necessary to regulation, assessment, and cleanup of contaminated soil and groundwater. A good example of this impact is provided when considering natural attenuation. Seasonal ice and post-cryogenic structure present in active layer soil will influence the movement of petroleum and dissolved compounds, thereby impacting the design of monitoring systems to track natural attenuation. Moreover, cold soil temperatures will slow the physical weathering of compounds in the subsurface. Cleanup levels established for cold regions contaminated soil (Chapter 1) and any remediation plan developed

Bioremediation of Petroleum Hydrocarbons in Cold Regions, ed. Dennis M. Filler, Ian Snape, and David L. Barnes. Published by Cambridge University Press.

for these sites must account for these impacts. The purpose of this chapter is to review what is currently known about the movement of petroleum through cold region soils and to make recommendations on bioremediation strategies.

3.2 Review

Several authors have presented thorough descriptions of immiscible fluid, commonly known as non-aqueous phase liquids (NAPL), movement through unsaturated soils (Mercer and Cohen 1990; Wilson *et al.* 1990; Poulsen and Kueper 1992). Their findings are relevant to the movement of petroleum through unfrozen active layers, and form the basis of understanding and predicting how transport processes might be different in freezing and frozen soils. Once released at or near the ground surface petroleum will move downward through unsaturated soil toward the water table. Due to the immiscibility of petroleum, the fluid migrates as a distinct liquid separate from the air and water present in the unsaturated soil. Water and petroleum (characterized as a NAPL with a density less than water; or a light non-aqueous phase liquid, LNAPL) are held in the pore space of partially saturated soils by capillary forces. Soil water will preferentially wet soil grains with respect to petroleum. Correspondingly, petroleum is then the non-wetting fluid with respect to water, but the wetting fluid with respect to soil gas. As petroleum migrates downward, air and possibly some water are displaced from the pore space.

Petroleum spreads laterally during downward migration due to spatial variability in soil properties and capillary forces. If the petroleum release is sufficiently large, some of the petroleum will reach the nearly saturated zone, commonly called the capillary fringe, above the groundwater table. Relatively high water saturations in this zone cause the petroleum to spread laterally along the top of the capillary fringe. For spill volumes that generate sufficient head to displace capillary fringe water, petroleum that migrates further downward to the water table may displace water from saturated pores and cause depression of the water table. As the water table rises and falls seasonally some petroleum is immobilized or entrapped in the capillary fringe and possibly below the water table during high water level conditions. This immobilized petroleum consists of small pockets (or blobs) of liquid disconnected from the main body of organic liquid (Wilson *et al.* 1990) and is often referred to as residual saturation. A dissolved phase plume results in the saturated zone below the water table from petroleum contained above and below the water surface.

Farr *et al.* (1990) discussed the distribution of LNAPL, water, and air in porous medium. Considering each fluid to be in static equilibrium and continuous throughout a homogeneous porous medium that is devoid of a fluctuating water

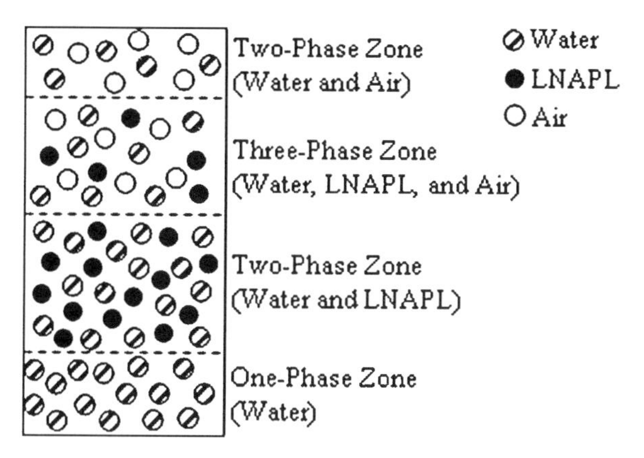

Figure 3.1. Conceptual model of the distribution of water, LNAPL, and air in an unfrozen porous soil (after Farr *et al.* 1990).

table, the fluids will distribute as shown in Figure 3.1. By assuming static equilibrium and a constant water table depth, the existence of entrapped LNAPL is ignored in this conceptual model. As stated by Farr *et al.* (1990), what is evident in this figure is lack of a distinct LNAPL layer "floating" on top of the water table. Instead, a zone of water and LNAPL exists above a zone saturated with water.

Once in contact with water contained in soil pore space, individual petroleum compounds will dissolve into soil water according to the specific solubility of each compound and its mole fraction. Solubility of these compounds is characteristically low since most petroleum hydrocarbons are non-polar. The non-polar nature of these compounds results in partitioning (or sorption) onto natural organic matter in the soil. The high volatility of relatively low molecular weight petroleum hydrocarbons dissolved in soil water results in partitioning of a fraction of these compounds into the gas phase. The mixture of gaseous petroleum hydrocarbons and air comprise soil gas in the pore space.

As released petroleum migrates downward through unsaturated soil the path followed by the liquid is dictated by the properties of the soil encountered – specifically permeability and pore structure. Results from field studies performed by Poulsen and Kueper (1992) illustrate how small variations in permeability result in extreme heterogeneous distribution of NAPL. Their study also showed that the downward progression of a NAPL is influenced by the orientation of soil strata. NAPL generally migrates parallel to bedding, often above impermeable units or lithologies, which is evidence that the migration of the liquid is predominantly controlled by capillary forces.

Wilson *et al.* (1990) visually showed that in unsaturated soils small fingers of NAPL tend to migrate ahead of the advancing larger body of NAPL. This advancement takes place as thin films of NAPL move between the water phase and the air phase. Considering porous media to be made up of relatively large diameter pore spaces (pore bodies) joined together by relatively smaller diameter pore spaces (pore throats), these researchers also illustrated that soil water held by capillary forces in the narrower pore throats is often bypassed by the NAPL as it migrates downward through larger pore spaces in the unsaturated soil (Wilson *et al.* 1990).

As the main body of NAPL moves downward through porous medium, capillary forces immobilize some of the liquid in the pore space. The Wilson *et al.* (1990) visualization study showed that immobilized NAPL was mostly contained in pore throats and in thin films between soil water and soil gas. Soil water was also contained in pore throats that were bypassed by infiltrating NAPL, and soil gas filled the larger pore bodies.

Many researchers describe immobilized NAPL existing as disconnected "blobs" in pore space throughout the unsaturated soils as the residual NAPL saturation (Mercer and Cohen 1990; Poulsen and Kueper 1992; Pankow and Cherry 1996). Others consider residual saturation to be the quantity of NAPL that is immobile but still continuous throughout the porous medium, with pockets of NAPL being connected by thin films in pore spaces occupied by all three fluids (soil water, NAPL, and air), and disconnected blobs in pore spaces only occupied by soil water and NAPL (Wilson *et al.* 1990). However residual saturation is defined, the petroleum held by capillary forces in the unsaturated soils acts as a continuing source of dissolved phase petroleum hydrocarbons as water infiltrates through the soils, and as a source of gaseous phase pollutant as the relatively lighter molecular weight compounds partition from the liquid phase to the gas phase.

3.3 Recent advances

3.3.1 Field observations of petroleum migration in unfrozen active layers

Distribution and fate of released petroleum in polar and alpine regions is a function of the season in which the petroleum is released. The thawed active layer in the summer months is typically shallow in areas of undisturbed permafrost. In areas of poor drainage, a seasonal aquifer (suprapermafrost water) will form above the permafrost table and any unsaturated soils present will most likely have relatively high soil-water saturations (Chapter 2). Thus, the downward flow of petroleum will be impeded due to low relative permeability to petroleum, resulting in increased flow along the ground surface and through the near surface layer, which is typically a partially decayed vegetation layer that is typically present in many Arctic ecosystems. This flow pattern leads to relatively large aerial distributions of petroleum tempered by entrapment of the petroleum onto organic matter present in the uppermost layer of soil. This is not to imply that petroleum does not move downward through underlying mineral soil. In areas of large accumulations of petroleum, soil water will be displaced and petroleum will progress into lower mineral soils. Furthermore, over time, it may migrate deeper into the soil horizon as the active layer freezes and thaws.

To study the movement of petroleum through Arctic soils resulting from a release in the summer months, several controlled releases of crude oil to unfrozen active layers have been conducted (Mackay *et al.* 1974a; Mackay *et al.* 1974b; Mackay *et al.* 1975). In a series of controlled summer releases in taiga ecosystems with varied terrain, Mackay *et al.* (1974a and b) described a reoccurring trend of petroleum rapidly penetrating the upper organic layer and subsequently flowing through this layer. Downward movement of petroleum into underlying unfrozen mineral soils was observed in areas containing relatively deep water tables.

Mackay *et al.* (1975) conducted a controlled release of crude oil on unfrozen tundra. Once again the released petroleum rapidly penetrated the upper vegetation layer. In contrast to the controlled releases in taiga, the released petroleum penetrated to the top of the frost line (depth at which the soil temperature is 0 °C), or the water table (where present), and flowed down gradient through a relatively thin horizon of very permeable soils directly above the frost line. The cause of this zone of high permeability was not known, however the researchers speculated that soil modification through ice segregation might have produced this layer. In a laboratory study White and Williams (1994; 1996) measured increases in permeability due to cyclic formation and thawing of pore ice, which may explain the presence of the layer noted in Mackay *et al.* (1975). To the authors' knowledge there was no long term monitoring at these sites.

In a similar study to the one performed by Mackay *et al.* (1974a; 1974b), Johnson *et al.* (1980) released 7600 l of warm (57 °C) Prudhoe Bay crude oil on an unfrozen soil slope underlain by permafrost. The ecosystem was characterized as an open black spruce forest with moss, lichen, and cotton grass tussock ground cover (Johnson *et al.* 1980). A cross section of the area showed vegetative ground cover overlying a layer of peat followed by a layer of organic silt loam, overlying a layer of gray silt loam resting on permafrost. As in the Mackay study, released petroleum rapidly penetrated the ground cover flowing downslope through the peat layer above the mineral soils. Downslope movement continued until winter freeze-up. Samples retrieved from the impacted area approximately two years after the release showed that petroleum resided predominantly in the peat layer, however several samples obtained close to the release point showed that petroleum had migrated into underlying mineral soils.

Long-term monitoring of the release by Johnson *et al.* (1980) showed further downslope migration of the petroleum fifteen years after the release (Collins *et al.* 1994). High water saturations in the areas sampled by Collins *et al.* (1994) limited the downward migration of petroleum to a maximum depth of 30 cm. Results from further long-term monitoring of this study site showed increased thaw depths since the controlled release occurred (White *et al.* 2004). The increase in thaw depths was most significant at the winter release site, and was attributed to lowered albedo of the oiled ground surface, a decrease in vegetation and surface insulation, resulting in a greater influx of solar radiation, and increased thermal diffusivity in the surface organic layer.

In summary, migration through an unfrozen active layer is typically restricted by relatively high soil-water saturations, relatively low permeable soils near the permafrost table, and possibly enhanced by post cryogenic structure resulting in complex distributions of petroleum. While a majority of the petroleum may exist in shallow plumes within the active layer, deeper penetration may occur over time due to permafrost degradation.

3.3.2 Field observations of petroleum migration in frozen active layers and permafrost

In contrast to petroleum released to unfrozen soils, the flow of petroleum released to frozen active layer soil will be influenced by the presence of ice at different scales. At the millimeter scale, ice present as pore ice will act as a solid, changing the pore geometry and thus, the capillarity and permeability of the soil. In the extreme, the ground surface will be nearly impermeable and downward migration will be minimal for the most part. Under these conditions surface flow will dominate upon release, though the higher viscosity at cold temperatures will inhibit lateral movement. In contrast to a release of petroleum to an unfrozen active layer, the increased exposure of the petroleum to the surface elements leads to greater losses of petroleum hydrocarbons by physical weathering (evaporation and photochemical oxidation (Snape *et al.* 2006a).

Winter releases of petroleum were studied in two separate field tests (Mackay *et al.* 1975; Johnson *et al.* 1980). Mackay *et al.* (1975) applied known volumes of crude oil to frozen ground at two separate field test sites. The ecosystem of each site was characterized as mature black spruce forests. Crude oil released in each of these controlled tests was at different temperatures (heated and ambient). Areas sampled soon after the release showed minimal penetration below the top moss layer. Penetration below the moss layer did not occur until spring thaw (Mackay *et al.* 1975). Downslope migration of oil was limited in both of these controlled releases owing to increases in viscosity as the crude oil cooled shortly after the releases. Downslope migration recommenced during spring thaw. Similar results were found in a controlled release of 7600 l of Prudhoe Bay crude oil to frozen ground reported in Johnson *et al.* (1980). As expected, the summer release resulted in greater downslope migration in comparison to the winter release (Johnson *et al.* 1980).

Due to high pore-ice contents commonly found in the upper few meters of permafrost, the migration of petroleum into permafrost should be minimal in most cases. Nevertheless, petroleum hydrocarbons and liquid petroleum have been measured at depths of meters in permafrost (Biggar *et al.* 1998; McCarthy *et al.* 2004). In both these instances, movement was attributed to free phase petroleum movement through interconnected air voids in the frozen soil. These air voids may result from unsaturated compacted soil, fissures resulting from thermal contraction, or naturally occurring air voids in granular material (such as beach deposits) due to natural processes.

Biggar *et al.* (1998) measured significant petroleum hydrocarbon concentrations (1200–17 000 mg TPH kg^{-1} in soils at depths ranging from 0.5 to 1.5 m below the permafrost surface at old spill sites at Canadian Forces Station Alert and Isachsen High Arctic weather station in Canada's high Arctic. They attributed the contaminant migration in the permafrost to be free phase NAPL movement through air voids in compacted fill or fissures in the native permafrost, depending on the site. The air voids in the fill at Alert would have been a consequence of fill placement. The air voids in the native silty clay (Isachsen) and weather rock (Alert) would likely have been the result of contraction induced fissures during coldest soil temperatures; a similar mechanism to that attributed to ice wedge growth.

Migration of uncontrolled releases of refined petroleum occurring over several years near Barrow, Alaska was reported by Braddock and McCarthy (1996) and discussed further in McCarthy *et al.* (2004). The site was a sandy gravel beach deposit adjacent to the ocean and a nearby lake. These researchers found migration to be strongly influenced by the non-uniform nature of thaw depths brought about by the heterogeneous nature of snow depths and ground cover. They also found isolated "reservoirs" of petroleum at depths greater than 3.0 m, where the active layer was 0.5 to 2.0 m thick and highly variable. Air voids and hydrocarbon seeps were observed in the frozen soil with a down-hole camera. The existence of petroleum hydrocarbons below the permafrost table was attributed to both the presence of air voids and unfrozen brine at concentrations approximately three times that of seawater.

Frozen fine soils can contain unfrozen water at the soil surface boundary (Chapter 2). Lacking sufficient displacement pressures for petroleum to flow into ice rich permafrost, a possible transport mechanism is diffusion of petroleum hydrocarbons through the unfrozen water content. Aqueous phase diffusion is a relatively slow transport process in comparison to advection. The contribution of diffusion to the overall movement of contaminants into permafrost soils is minimal.

In summary, results from studies conducted by Biggar *et al.* (1998) and McCarthy *et al.* (2004) illustrated that migration of petroleum into permafrost is possible most likely through air voids in the frozen soil. However, from a health risk perspective, a potential risk might be associated with the contaminant actually altering the active layer and/or permafrost, and leading to greater lateral migration.

While the case histories are very informative providing a relative "big picture" understanding of petroleum releases to frozen ground, they do not provide detailed insight into the mechanisms controlling the subsurface movement of petroleum in freezing and frozen soils. It is instructive therefore to examine theoretical concepts and available laboratory results that better define these mechanisms.

3.3.3 Theoretical concepts of petroleum movement in freezing and frozen soils

The fact that unfrozen water will exist in fine-grained soil at temperatures below 0 °C is well known (Schofield 1935; Williams 1968; Anderson and Tice 1972). Thus, in partially saturated frozen fine-grained soil, pore ice will most likely separate soil water from petroleum with soil gas remaining in the larger part of the pore space. Unfrozen water is not expected to exist in frozen coarse-grained soils unless there are dissolved solutes in the water at concentrations approaching 1 g l^{-1}. Under these conditions, ice will form in the pore space closest to the soil grains with petroleum and soil gas extending into the pore space.

It is well understood that dissolved salts depress the freezing point of water, resulting in unfrozen water in soils below 0 °C. This unfrozen water may act as a conduit for diffusive contaminant flux in frozen soil (discussed above). What has not been well addressed to date is whether dissolved petroleum hydrocarbons generate significant unfrozen water in frozen soils.

In the literature on strength of frozen soils, little attention has been paid to the effect of salinities below 1g L^{-1} (1000 mg L^{-1}, 0.1%; 1 part per thousand or ppt). Benzene, toluene, ethylene, and xylenes are the most soluble components in petroleum spills. Approximate solubility of individual BTEX compounds at room temperature are: benzene – 1 800 mg l^{-1}; toluene – 500 mg l^{-1}; ethylbenzene – 150 mg l^{-1}, and xylenes – 130 to 200 mg l^{-1}. These values decrease with decreasing temperature. Moreover, these high individual concentrations are never seen in field groundwater samples at petroleum spills. At most field sites, typical total BTEX concentrations seldom exceed a few hundred mg l^{-1}, and are more commonly less that 20–30 mg l^{-1}. Concentrations in this range (30–300 mg l^{-1}) are thus 3% to 30% of the lowest salt concentrations considered significant in saline frozen soil studies. In the absence of a measured increase in unfrozen water content due to dissolved petroleum hydrocarbon constituents, the level of BTEX impact is considered of little significance as it relates to migration of hydrocarbons in permafrost.

To gain a better understanding of how petroleum migrates through frozen soils, two-dimensional soil flumes were used to examine the influence pore ice has on the infiltration of refined petroleum through uniformly graded silica sand, and the retention of the petroleum in the pore space (Barnes and Wolfe, in press). The flumes were filled with compacted wet sand at uniform water contents, and placed in a cold room at −5 °C. Two sand gradations, coarse (0.60 mm diameter) and medium (0.212 mm diameter), were studied at two different ice saturations (26% and 54%). Once the soil had reached −5 °C, 100 ml of colored diesel at −5 °C was introduced into the top of the flume. Progression of the resulting immiscible-fluid plume was tracked using time-lapse photography. Image processing was used to compare the plume development between the two different ice saturations for the two soil types. This process involved quantifying the gray level of each pixel in the digital images of the plume development. Photographs shown in Figure 3.2 illustrate the influence of degree of saturation and sand grain size on the flow of petroleum 20 minutes after petroleum was introduced into the flumes.

As shown in Figure 3.2(a), petroleum migration through the coarse sand (0.60 mm diameter) is not greatly influenced by the presence of ice at the lower ice saturation. Under these conditions the ice is most likely concentrated in sparsely distributed small clusters of connected pore spaces resulting in minimal interference with downward petroleum migration. At the higher ice saturation (Figure 3.2(a), right frame), a greater number of clusters of ice-saturated connected pore spaces possibly forces the petroleum to migrate laterally until open pore channels are found, allowing further downward migration. The same effect can be seen in the medium sand flume (Figure 3.2(b)). However, the similarity in the plume widths at both ice contents indicates that the clusters of ice-saturated pores were relatively the same size, but possibly more numerous in the sand with the greater overall ice content. The greater depth reached by the main body of the plume (darker portions of the plumes shown in Figure 3.2(b))

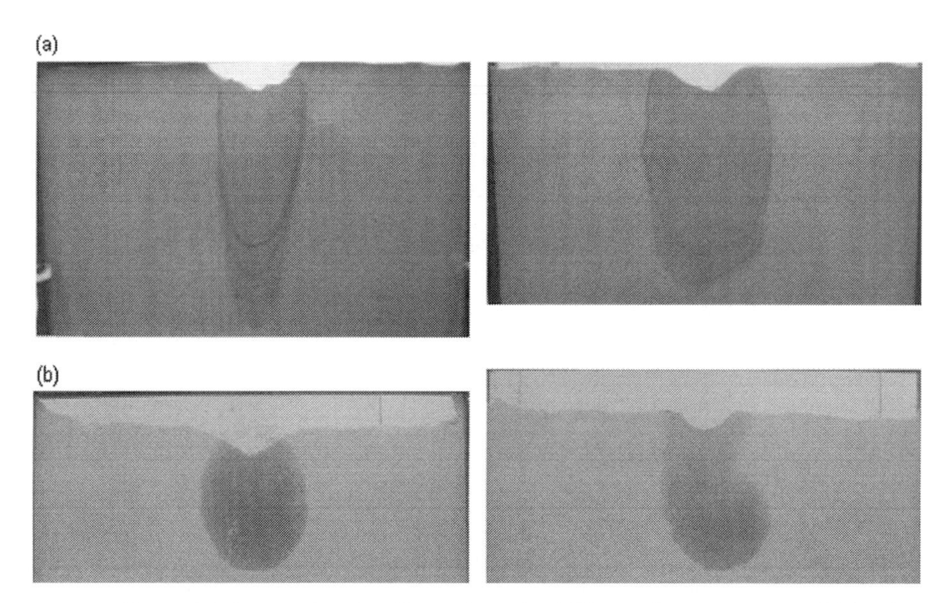

Figure 3.2. Progression of cold petroleum (−5°C) into a frozen coarse sand (a) and a medium sand (b) at two different pore ice contents (26% in left-side flumes and 54% in right-side flumes). Photographs were taken approximately 20 minutes after introduction of petroleum.

in the sand with the greater ice saturation provides the evidence for this last assertion. A greater number of ice-saturated pore clusters may have resulted in the formation of preferential flow paths allowing the petroleum to infiltrate to a greater depth.

A close examination of enlarged and contrast-enhanced images of each plume shown in Figure 3.3 reveals the presence of clusters of ice-saturated pores resulting in preferential pathways in the sand with the greater ice saturation. With an accidental release to frozen ground, formation of clusters of ice-saturated pore space creating dead-end pores and preferential flow paths for petroleum migration is a likely scenario. Fourie *et al.* (2007) proposed a conceptual model for the formation of pore ice in coarse-grained soil. They presented computed tomography evidence of the creation of dead-end pore space resulting from ice-saturated pores.

Overall, from these results the most notable impact pore ice will have on petroleum spills in frozen ground will be the following: reduced overall infiltration rates resulting in increased weathering of petroleum; increased infiltration depths due to the creation of preferential flow paths; and increased lateral distribution of contamination due to increased flow across the ground surface and in shallow soils as blocked pore space is encountered. Field results by Mackay *et al.* (1975), and Johnson *et al.* (1980) presented earlier support this conclusion.

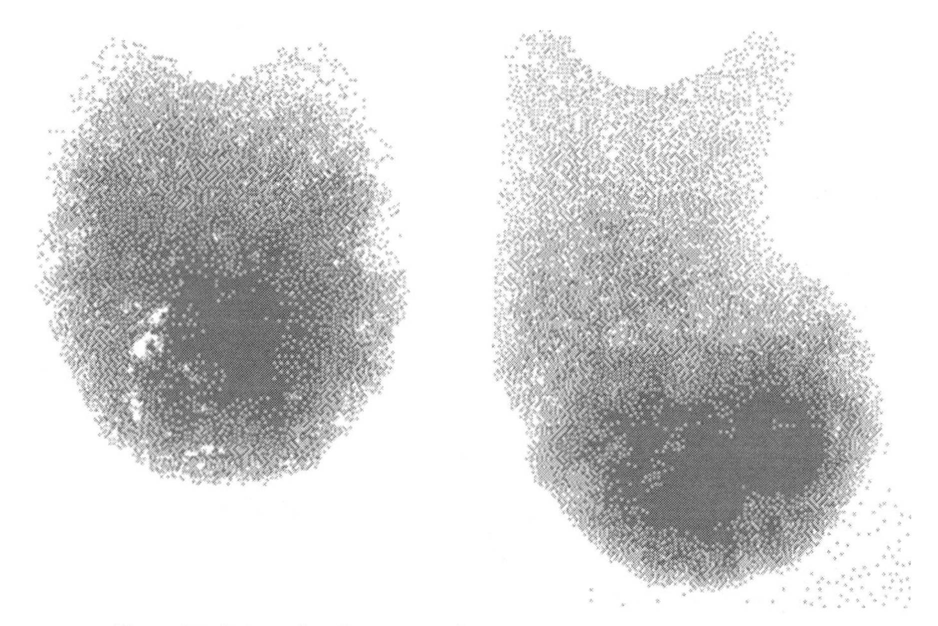

Figure 3.3. Enlarged and contrast-enhanced images of the plumes developed in the medium sand for each ice saturation. Left-side image has an ice saturation of 26% and the right-side image has an ice saturation of 54%. Light-shaded areas in the plume area of both images are ice-saturated clusters of pores. Preferential flow paths can be distinguished in the right-side image as petroleum flowed around clusters of ice-saturated pore space.

As the first thaw of a soil contaminated by petroleum after a winter release occurs, petroleum originally held in pore spaces by capillary forces, or by pore blockage, will once again become mobile as the size characteristics of the pore spaces change as a result of melting pore and segregated ice. The flux of each fluid will be, in part, controlled by the volume of each fluid contained in the pore space, the thawing rate, and, for petroleum, the continuity of the fluid throughout the pore space. If the petroleum is continuous in the pore space, the fluid will eventually drain to a new level of saturation in the thawed soils. Resulting distribution of petroleum after thawing of the active layer is characterized by increased concentrations of petroleum with depth and an increase in lateral spreading (Mackay *et al.* 1975 and Collins *et al.* 1994). Given the typically shallow seasonal water found in a thawed active layer underlain by permafrost, petroleum saturations may be relatively high in the two-phase zone shown in Figure 3.1. The level of saturation in this zone will be dependent upon the volume of infiltrated petroleum per unit area.

There have been a few laboratory tests on prepared soil samples or glass beads that have shown free-phase NAPL exclusion forward of a freezing front

(Konrad and McCammon 1990; Konrad and Seto 1991; Soehnlen 1991; Lehner 1995; Chuvilin *et al.* 2001; and Biggar and Neufeld 1996). Lehner (1995) conducted freezing tests on a mixture of dodecane, water, and glass beads to evaluate the effectiveness of freeze exclusion to remove free-phase hydrocarbons from soils as a remedial option. The media was nearly saturated with water and residual dodecane. Concentration of free-phase dodecane forward of the freezing front was observed, and with varying degrees of success, was attributed to expulsion of the NAPL residuals from pore throats as the water at the freezing front turned to ice, expanded, and forced the NAPL out ahead of the freezing front. However, the freezing front was also observed to jump past and entrap dodecane when the NAPL layer became thick or when the temperature gradient was high.

Chuvilin *et al.* (2001) conducted downward freezing tests on saturated soil samples (sand, silt, clay) that were mixed with distilled water and oil, where the initial oil content varied from approximately 6% to 10%. Reduced oil concentrations in the frozen soil, and increased oil concentrations in the underlying unfrozen soil were observed for the sand and clay samples, indicative of oil expulsion forward of the freezing front. In the silt sample, significant migration of water to the freezing front occurred, resulting in increased moisture concentrations in the frozen soil and variable oil concentrations in frozen and unfrozen soil.

Biggar and Neufeld (1996) conducted cyclic freeze-thaw laboratory tests on a silty sand contaminated with diesel fuel. Increased TPH concentrations were not observed forward of the freezing front, however the freezing rates in the laboratory may have been too rapid. Konrad and McCammon (1990) suggested that exclusion during freezing is controlled by the rate of cooling at the freezing front, expressed as

$$\frac{dT}{dt} = \frac{dT}{dX} \cdot \frac{dX}{dt} \tag{3.1}$$

where T is the temperature, t is time, and X is the depth of the freezing front. Thus dT/dt is the calculated rate of cooling at the freezing front, dT/dX is the temperature gradient at the freezing front, which can be measured with thermistors, and dX/dt is the rate of advance of the freezing front, which can also be measured with thermistors.

Konrad and McCammon (1990) recommended that for exclusion to occur, the cooling rate at the freezing fronts should be less than $3\,^{\circ}\mathrm{C\ day^{-1}}$. These rates are often encountered in the field, but difficult to obtain without very careful temperature control in laboratory experiments. The rates of cooling in the experiments by Biggar and Neufeld (1996) were $10\text{–}40\,^{\circ}\mathrm{C\ day^{-1}}$, and were perhaps too fast to develop freeze exclusion.

Thus, it is conceivable that at the temperature gradients experienced in the field, exclusion of free-phase petroleum may occur forward of an advancing freeze front. However, given the large number of environmental factors that can affect petroleum movement and distribution in soils (e.g. soil structure, pore distribution – as a result of grain size distribution, water table fluctuations, petroleum head, and gradient), it has not been possible to verify whether freeze exclusion forward of a freezing front is a significant process in the field. Even if it is, subsequent redistribution by processes such as water table fluctuation and soil structure modification due to freezing and thawing would likely lead to further redistribution and reduce or negate the effects of concentration forward of the freezing front. This is a question whose practical significance remains to be solved by field investigation, despite evidence in controlled laboratory tests showing that the process can occur under certain limited conditions.

In summary, the movement of petroleum through frozen soils is influenced heavily by the presence of ice. At the extreme, minimal downward penetration of petroleum released to a frozen ground surface will occur until soil thaws. Where petroleum does penetrate frozen ground, the presence of ice will control the lateral and vertical spread of petroleum. Finally, as petroleum-contaminated soils experience freeze and thaw, redistribution of petroleum may result through a combination of displacement from pore space by the formation of pore ice and fluctuations in the seasonal groundwater level.

3.4 Guidelines and recommendations

Obviously, knowing the extent of contamination is vital in any remediation strategy. In freezing and frozen soils the presence of ice in the pore space and post cryogenic structure impact the movement of petroleum. Furthermore, field test results have shown that the time of year in which the petroleum is released influences the extent of subsurface contamination.

Remediation strategies and technologies used in warmer climates require adaptation to polar and alpine regions. For *ex situ* remediation strategies, theoretical, laboratory, and field results all indicate that it is particularly important to complete excavation of impacted soil before the active layer thaws. *In situ* remediation strategies are possible as well (discussed in Chapters 9 and 10). In non-freezing soils natural attenuation has been shown to be a viable treatment strategy for petroleum-contaminated soils. Natural attenuation takes advantage of dilution, biodegradation, and physical degradation through volatilization and photooxidation. In polar and alpine soils, if downward migration of the released petroleum is hampered by pore ice, a significant fraction of the lighter compounds will volatilize from the petroleum in a short period of time. However, volatilization becomes less effective as the petroleum progresses deeper into the soil, and biodegradation is limited by low temperatures and high contaminant concentrations that create toxic conditions. Finally, photooxidation potential is significantly reduced as the petroleum moves into the soil. The combination of these limiting influences will result in longer treatment duration and more frequent monitoring if natural attenuation is relied upon at petroleum-contaminated polar and alpine sites.

3.5 Future research

To properly develop engineered strategies for remediation of petroleum-contaminated soils in cold regions, as well as appropriate regulations, a better understanding of petroleum mobility in soils that experience seasonal freezing is required. This requirement particularly applies to soils that are underlain by permafrost. Specific areas of study include: formation of preferential pathways, the influence of ice formation on the creation of disconnected zones with relatively high saturations of petroleum, and the effects of freezing on soil permeability in seasonally frozen ground.

4

Hydrocarbon-degrading bacteria in contaminated cold soils

JACKIE AISLABIE AND JULIA FOGHT

4.1 Introduction

Bioremediation is a viable option for the cleanup of hydrocarbon-contaminated soils. Although this technology has proven effective for various temperate soils, extrapolation to cold soils is hindered by the lack of information about specific microbes, genes, and enzymes involved in hydrocarbon biodegradation in cold soils. These environments present multiple challenges to bioremediation besides low temperature and concomitantly lower enzymatic reaction rates: for example, cold soils are often poor in nutrients, low in available water, and may exhibit extremes of pH and salinity. Also in such environments the physical nature of the contaminant(s) is affected, with increased viscosity of liquid hydrocarbons and reduced volatilization of toxic, low molecular weight hydrocarbons. Despite these constraints, the biodegradation of many of the components of petroleum hydrocarbons by indigenous cold-adapted microbial populations has been observed at low temperatures in hydrocarbon-contaminated soils (e.g. Braddock *et al.* 1997; Aislabie *et al.* 1998; Margesin and Schinner 1998; Whyte *et al.* 1999a; Mohn and Stewart 2000). However, because hydrocarbons tend to persist in polar soils, there are obvious limitations to the activity of the indigenous microbes *in situ*.

In this chapter we review the literature describing hydrocarbon-degrading bacteria indigenous to cold soils, with a focus on polar soils. We discuss their adaptations to environmental parameters that challenge their activity *in situ*, including cold and fluctuating temperatures, limited nutrient availability, extremes in pH and salinity, and desiccation. We provide recommendations for

Bioremediation of Petroleum Hydrocarbons in Cold Regions, ed. Dennis M. Filler, Ian Snape, and David L. Barnes. Published by Cambridge University Press.

methods to determine whether a soil contains the appropriate microbial community for applying bioremediation.

4.2 Literature review and recent advances

4.2.1 Hydrocarbon-degrading microbes in cold soils

4.2.1.1 Numbers of hydrocarbon-degrading microbes

Spillage of hydrocarbons on cold soils can result in enrichment of hydrocarbon-degrading microbes *in situ* (Atlas 1981; Aislabie *et al.* 2001; Rike *et al.* 2002; Delille and Pelletier 2002) and therefore enhanced potential for hydrocarbon degradation. Numbers of hydrocarbon degraders are often low or below detection limits in pristine polar soils, whereas $>10^5$ hydrocarbon degraders g^{-1} have been detected in contaminated soils in both surface and subsurface layers (Aislabie *et al.* 2001; Rike *et al.* 2002). In general, total and viable counts in both pristine and contaminated Antarctic soils decrease with depth to the ice-cemented layer (Aislabie *et al.* 2001). The distribution of heterotrophic and hydrocarbon-degrading bacteria throughout contaminated soil profiles is a significant factor influencing the potential for bioremediation of these soils.

Enhanced numbers of hydrocarbon degraders in contaminated cold soils provide indirect evidence for their growth and activity under *in situ* conditions, and they have proven to be persistent. High numbers have been detected in contaminated polar soils more than 30 years after the initial hydrocarbon spill (Atlas 1981; Aislabie *et al.* 2001). However, the time required to establish a significant hydrocarbon-degrading community in polar soils after a spill is unknown. For example, 42 days after applying JP-5 jet fuel in summer to contained soil cores *in situ* at Scott Base, Antarctica, no increase in hydrocarbon degraders was detected (Balks *et al.* 2002).

Although the cultivation of hydrocarbon-degrading microbes from cold soils typically results in the isolation of bacteria, fungi may also play a role in degradation under *in situ* conditions. Fungal biomass was enhanced in hydrocarbon-contaminated soils compared to control soils in both the Arctic (Sexstone and Atlas 1977) and Antarctic (Aislabie *et al.* 2001), and contaminated soils may become dominated by fungi reported to degrade hydrocarbons, such as *Phialophora* spp. and *Hormoconis resinae* (Kerry 1990; Aislabie *et al.* 2001).

4.2.1.2 Alkane-degrading bacteria: pure cultures

Alkane-degrading bacteria from cold soils frequently belong to the Gram-positive genus *Rhodococcus* or the Gram-negative genus *Pseudomonas* (Table 4.1). *Rhodococcus* 7/1, 5/1 and 5/14 from Antarctic soil utilized a number of hydrocarbons as sole source of carbon and energy, including alkanes ranging in chain length from hexane (C_6) through at least eicosane (C_{20}) and the isoprenoid compound pristane (2,6,10,14-tetramethyl-pentadecane) (Bej *et al.* 2000), whereas *Pseudomonas* BI7 and BI8 from Arctic soil utilized C_5 to C_{12} *n*-alkanes (Whyte *et al.* 1997). Some alkane degraders have been reported to produce biosurfactants during growth on alkanes (Yakimov *et al.* 1999). Rhodococci, for example, produce cell surface-associated biosurfactant(s) with activity at cold temperatures and which directly adhere to solid alkanes at low temperature (Whyte *et al.* 1999b). The Antarctic alkane degrader *Arthrobacter protophormiae* produces a biosurfactant stable over a wide range of temperatures and pH values (Pruthi and Cameotra 1997). Under *in situ* conditions, production of biosurfactants should increase the bioavailability of hydrocarbons for degradation.

Phylogenetic analysis of 16S rRNA genes has indicated that the alkane-degrading *Rhodococcus* spp. cultivated from Antarctic soils grouped with *R. erythropolis* or *R. fascians*, and were most similar to other alkane degraders from cold climates such as *Rhodococcus* sp. Q15 (Bej *et al.* 2000). In addition to alkanes, many of the biodegradative *Pseudomonas* isolates from polar soils also degrade aromatic compounds (Table 4.1).

4.2.1.3 *Alkane degradation pathways and genes*

Bacteria typically initiate the degradation of *n*-alkanes by the terminal oxidation of the alkane substrate to 1-alkanol (Figure 4.1), catalyzed by alkane monooxygenase. The alkanol is then metabolized to an aldehyde, then a carboxylic acid. Further degradation of the carboxylic acid proceeds by β-oxidation with the subsequent formation of shorter fatty acids and acetyl coenzyme A, with eventual liberation of carbon dioxide. Alkane monooxygenase (alkane hydroxylase), the key enzyme in the bacterial alkane degradation pathway, is usually part of a three component alkane hydroxylase complex consisting of a particulate, integral-membrane alkane monooxygenase, and two soluble proteins, rubredoxin and rubredoxin reductase. Whyte *et al.* (1998) determined that the psychrotolerant hydrocarbon-degrader *Rhodococcus* sp. Q15 oxidizes alkanes at both the terminal and sub-terminal positions; it is reasonable to expect that related psychrotolerant alkane-degraders may also use this combination of pathways.

Whyte *et al.* (1996) screened 135 psychrotolerant bacteria for mineralization (i.e. complete oxidation to CO_2) of alkanes, aromatics, and chlorinated hydrocarbons. Genes amplified from the alkane-degraders by the polymerase chain reaction (PCR), using primers derived from the alkane monooxygenase *alkB1* gene of *Pseudomonas putida* Gpo1 (formerly *Pseudomonas oleovorans*), failed to hybridize strongly to an *alkB1* gene probe derived from *Pseudomonas putida* Gpo1. However,

Table 4.1 *Psychrotolerant hydrocarbon-degrading bacteria isolated from polar, Arctic, and Antarctic soils. All bacteria included in this table have been identified to genus level by 16S rRNA gene sequence analysis*

Bacterial strains	Hydrocarbon growth substrates	Reference
	Arctic	
Rhodococcus Rho10	Jet A-1 fuel, dodecane	Thomassin-Lacroix *et al.* 2001
Pseudomonas BI7 and BI8	C_5-C_{12} *n*-alkanes, toluene, naphthalene	Whyte *et al.* 1997
Pseudomonas Cam-1	Biphenyl	Masters & Mohn 1998
Pseudomonas Sag-50G		
Pseudomonas DhA-91	Jet A-1 jet fuel, octane, dodecane	Yu *et al.* 2000
Pseudomonas IpA-92	Toluene	Yu *et al.* 2000
Pseudomonas IpA-93	Toluene, benzene	Yu *et al.* 2000
Pseudomonas Ps8	Jet A-1 fuel, hexadecane, 2,6,10,14-tetramethyl-pentadecane	Thomassin-Lacroix *et al.* 2001
Pseudomonas PK4	Pyrene, dodecane, hexadecane	Eriksson *et al.* 2002
Pseudomonas K319	Pyrene	Eriksson *et al.* 2002
Sphingomonas DhA-95	Jet A-1 jet fuel, dodecane, pristane	Yu *et al.* 2000
	Antarctic	
Rhodococcus 5/1, 5/14 and 7/1	JP8 jet fuel, C_6-C_{20} *n*-alkanes, pristane	Bej *et al.* 2000
Pseudomonas Ant 5	JP8 jet fuel, naphthalene, 2-methyl-naphthalene	Aislabie *et al.* 2000
Pseudomonas Ant 9	JP8 jet fuel, *p*-xylene, 1,2,4-trimethyl-benzene, naphthalene, 1 and 2-methyl-naphthalene	Aislabie *et al.* 2000
Pseudomonas 7/22	JP8 jet fuel, toluene, *m*- and *p*-xylene, 1,2,4-trimethyl-benzene	Aislabie *et al.* 2000
Pseudomonas 30–3	JP8 jet fuel, C_8-C_{13} *n*-alkanes, toluene, *m*- and *p*-xylene, 1,2,4-trimethyl-benzene	Panicker *et al.* 2002
Pseudomonas stutzeri 5A	JP8 jet fuel, benzene, toluene, *m*-xylene	Eckford *et al.* 2002
Pseudomonas 5B	JP8 jet fuel, hexane	Eckford *et al.* 2002
Sphingomonas Ant 17	JP8 jet fuel, *m*-xylene, naphthalene, 1 and 2-methyl-naphthalene, dimethyl-naphthalene, 2 ethyl-naphthalene, fluorene, and phenanthrene	Aislabie *et al.* 2000 Baraniecki *et al.* 2002
Sphingomonas Ant 20	JP8 jet fuel, 1-methyl-naphthalene, and phenanthrene	Aislabie *et al.* 2000

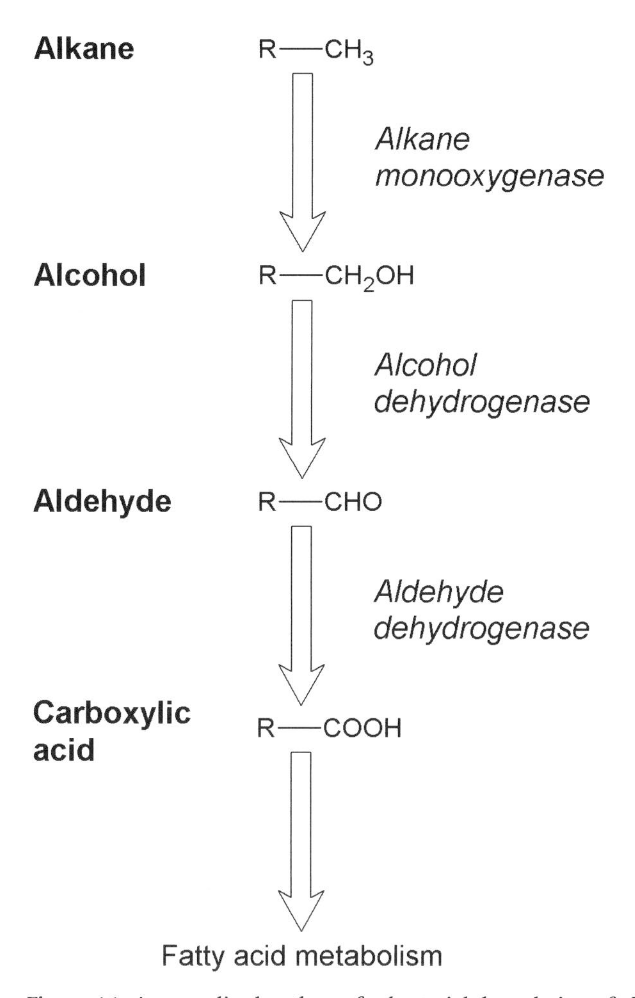

Figure 4.1. A generalized pathway for bacterial degradation of alkanes.

most of the amplified products hybridized to an *alk* gene probe from the psychrotolerant isolate *Rhodococcus* sp. Q15. Recently, that strain was found to harbor at least four alkane monooxygenase homologues (Rh *alkB1* through *alkB4*; Whyte *et al.* 2002a), which may explain its broad alkane substrate range (Whyte *et al.* 1998). The prevalence of four alkane monooxygenase genotypes from *P. putida* (Pp *alkB*), *Rhodococcus* spp. (Rh *alkB1* and Rh *alkB2*) and *Acinetobacter calcoaceticus* (Ac *alkM*) was investigated in hydrocarbon-contaminated and control Arctic and Antarctic soils (Whyte *et al.* 2002b). Rh *alkB1* and Rh *alkB2* homologues were found to be common in contaminated and control soils, whereas Pp *alkB* homologues were common in contaminated soil and Ac *alkM* homologues were rare. Furthermore, Rh *alkB1* was more prevalent in culturable cold-adapted bacteria. Based on these hybridization results, Whyte *et al.* (2002b) proposed that *Rhodococcus* is the predominant alkane-degrading bacterial genus in pristine and contaminated polar soils, while *Pseudomonas* may become enriched in contaminated soil and *Acinetobacter* are rare. In contrast to polar soils, *alkM* was prevalent in contaminated cold alpine soils (Margesin *et al.* 2003).

4.2.1.4 *Aromatic-degrading bacteria: pure cultures*

Aromatic-degrading bacteria isolated from polar soils typically belong to the Gram-negative bacterial genera *Pseudomonas* or *Sphingomonas* (Table 4.1). *Pseudomonas* isolates tend to degrade a narrower range of aromatic substrates than *Sphingomonas* spp. For example, *Pseudomonas* PK4 and K319 isolated from PCB-contaminated Arctic soil grew slowly on pyrene but not naphthalene, fluorene, or phenanthrene (Eriksson *et al.* 2002). *Pseudomonas* BI7 and BI8, isolated from petroleum-contaminated Arctic soil, degraded naphthalene and toluene but not fluorene (Whyte *et al.* 1997). In contrast, the Antarctic isolate *Sphingomonas* Ant 17 degraded numerous compounds in the aromatic fraction of several crude oils, jet fuel, and diesel fuel (Baraniecki *et al.* 2002) and was subsequently shown to utilize many aromatic compounds for growth including *m*-xylene, naphthalene and its methyl derivatives, and fluorene and phenanthrene.

Phylogenetic analysis of 16S rRNA gene sequences indicates that cold-tolerant hydrocarbon degraders, including the Arctic isolates *Pseudomonas* BI7 and BI8 and Antarctic strains Ant 5, Ant 9, 7/22 and 30-3, cluster with *P. syringae*, whereas *Sphingomonas* isolate Ant 17 is related to aromatic degrading *Sphingomonas* spp. isolated from throughout the world (Aislabie *et al.* 2000).

4.2.1.5 *Aromatic degradation pathways and genes*

In contrast to alkane degradation described above, there is no reason to believe that genes or pathways used for aromatic degradation by psychrotolerant and psychrophilic bacteria differ significantly from those described in mesophilic genera. In fact, Whyte *et al.* (1996) found that several aromatic-degrading psychrotolerant strains carried catabolic genes with homology to those described in mesophilic bacteria (although other isolates appeared to have novel genes). Furthermore, a TOL plasmid carrying genes for toluene degradation was successfully transferred from its mesophilic host to a psychrotolerant *P. putida* strain where it was expressed (Kolenc *et al.* 1988).

Bacteria commonly initiate the degradation of aromatic hydrocarbons using a dioxygenase to incorporate both atoms of molecular oxygen into the aromatic ring; this produces a *cis*-dihydrodiol that is then dehydrogenated to give a catechol (Figure 4.2). The aromatic dioxygenases are multi-component enzyme systems consisting of at least three proteins. For example, the naphthalene dioxygenase of *P. putida* consists of a flavoprotein, a ferrodoxin, and a terminal oxidase.

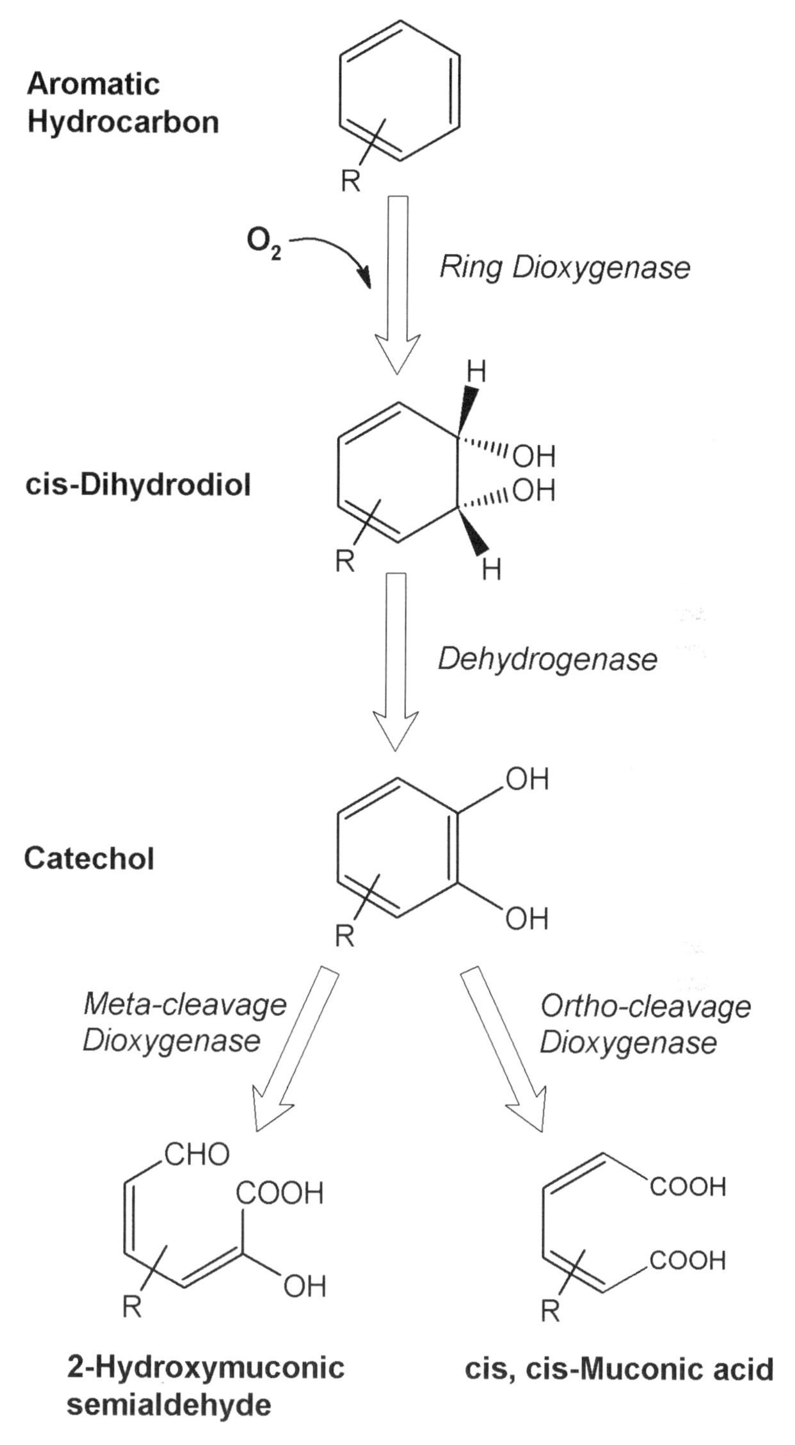

Figure 4.2. A generalized pathway for bacterial degradation of aromatic hydrocarbons.

Cleavage of the dihydroxylated aromatic compound can occur either between the two hydroxyl groups (*ortho*-fission) or adjacent to one of the hydroxyl groups (*meta*-fission). The aliphatic intermediates feed into central metabolic pathways where they are oxidized to provide cellular energy or are used in biosynthesis of cell constituents.

Aromatic degradation genes may be either plasmid or chromosomally encoded; many have been characterized in detail, including the NAH plasmid that carries genes for naphthalene degradation and the TOL plasmid that codes for toluene degradation. Although we are unaware of any detailed studies to characterize aromatic degradation genes specifically in psychrotolerant aromatic degraders, homologues of archetypal aromatic degradation genes originally described in mesophiles have been detected in psychrotolerant bacteria and community DNA extracted from polar soils. *Pseudomonas* BI7, for example, contains a NAH plasmid carrying genes for naphthalene degradation (Whyte *et al.* 1997). The catabolic genes *ndoB* (encoding naphthalene dioxygenase) and *xylE* (encoding 2,3-catechol dioxygenase) have been detected in cold-adapted bacterial isolates and DNA from hydrocarbon-contaminated, but not pristine, Arctic soils (Whyte *et al.* 1999a). The *phnAc* gene encoding an aromatic dioxygenase subunit has been detected in pristine Antarctic soil following enrichment with naphthalene (Laurie and Lloyd-Jones 2000).

4.2.1.6 *How significant are the culturable bacteria?*

A commonly cited review (Amann *et al.* 1995) indicates that only a small percentage of microbes in an environment can be cultivated in the laboratory, and these differ from the genera detected by culture-independent methods. These observations question the significance of studying pure cultures from hydrocarbon-contaminated soils. However, some evidence to the contrary has been collected from cold soils. A culture-independent phylogenetic survey of bacterial diversity of hydrocarbon-impacted Antarctic soils identified the prevalence of *Rhodococcus*, *Sphingomonas*, *Pseudomonas*, and *Variovorax* populations *in situ* (Saul *et al.* 2005). Some of the clones were either identical to or most closely related to hydrocarbon-degrading bacteria cultured from the same location, including the alkane-degrading *Rhodococcus* 5/1 and *Sphingomonas* spp. Ant 17 and Ant 20.

Similarly, *Rhodococcus*, *Pseudomonas*, and *Sphingomonas* were the three dominant phylotypes in an enrichment culture inoculated with hydrocarbon-contaminated soil from the high Arctic and provided with Jet A-1 jet fuel as sole source of carbon and energy (Thomassin-Lacroix *et al.* 2001). Cold soils used to enrich cultures on a mixture of polycyclic aromatic hydrocarbons (PAHs) under aerobic or nitrate-reducing conditions at 7 °C and 20 °C yielded a few predominant bacterial types, including members of the genera *Acidovorax*, *Bordetella*, *Pseudomonas*, *Sphingomonas*, and *Variovorax* (Eriksson *et al.* 2003).

These results suggest that, in these soils, there is reasonable concordance between the culture-based and culture-independent detection of indigenous bacteria, and that the dominant hydrocarbon-associated genera are limited to a few culturable genera. This observation is encouraging, as it suggests that laboratory-cultivated strains are suitable candidates for further study of cold-adapted hydrocarbon degradation.

4.2.2 *Adaptations of hydrocarbon degraders to cold soil conditions*

The frequent isolation of *Rhodococcus*, *Pseudomonas*, and *Sphingomonas* spp. from contaminated cold soils suggests that these genera have physiological adaptations promoting survival and activity in these inhospitable soils. As there are few studies describing these adaptations specifically in hydrocarbon-degrading bacteria from polar regions, the following discussion includes references to other heterotrophic bacteria.

4.2.2.1 *Cold and fluctuating temperatures*

Cold-adapted indigenous microorganisms play a significant role *in situ* in degradation of hydrocarbons in cold environments, where ambient temperatures often coincide with their growth temperature range. In fine summer weather, surface polar soils may approach 20 °C, but drop to −50 °C in winter (Campbell *et al.* 1998). Interestingly, few true psychrophiles (maximum growth temperatures ≤15 °C) are isolated from polar soils. It is possible that the relatively high summer soil temperatures are responsible for the prevalence of psychrotolerant bacteria (which are active at low temperatures (4–8 °C), but have higher optimal temperatures (>15 °C)). *Sphingomonas* Ant 17, for example, mineralized phenanthrene at 4 °C, 10 °C, and 28 °C (Baraniecki *et al.* 2002) and although the rate of mineralization was highest at 28 °C, the extent of mineralization achieved at each temperature was similar.

In addition to surviving a wide range of temperatures, hydrocarbon-degrading bacteria are also subject to short-term fluctuating temperatures and freeze-thaw cycles. *Rhodococcus* was shown to predominate in an Arctic soil community following a freeze-thaw regime (Eriksson *et al.* 2001) and *Sphingomonas* Ant 17 was more tolerant of freeze-thaw than the mesophile *Sphingomonas* WPO-1 (Baraniecki *et al.* 2002). As the lower temperature threshold for significant hydrocarbon biodegradation is considered to be around 0 °C, hydrocarbon degradation activity in polar soils is likely limited to summer when the soils are thawed (Atlas 1981). However, it should be noted that recent reports suggest that microbial activity is possible at sub-zero temperatures (e.g., Rivkina *et al.* 2000; Rike *et al.* 2003a); therefore we cannot discount the possibility that hydrocarbon degradation can also occur in frozen soils.

Temperature also affects rates of microbial hydrocarbon degrading activity by its effect on the physical nature and chemical composition of oil. At low temperatures, the viscosity of oil increases, the volatilization of toxic short-chain alkanes is reduced, and their water solubility is increased. Whyte *et al.* (1999a) proposed that the decreased rates of degradation of alkanes by psychrotolerant *Rhodococcus* Q15 at low temperature was due to decreased availability of substrates.

As reviewed by Gounot and Russell (1999), Deming (2002), and also discussed in Chapter 5, psychrophiles have key adaptations for survival in constantly cold environments, including cold-active enzymes and adaptations ensuring maintenance of membrane fluidity and transport of substrates and nutrients. In addition to adjusting the chemical composition of membrane fatty acids and lipopolysaccharides (Kumar *et al.* 2002), it has been postulated that some pigmented Antarctic bacteria incorporate increasing amounts of carotenoids into membranes to adjust fluidity (Chattopadhyay and Jagannadham 2001). This adaptation may explain the frequent cultivation of pigmented colonies from polar soils.

Cold tolerance mechanisms have been investigated in *Rhodococcus* spp. strains 5/14 and 7/1 (Bej *et al.* 2000) and *Pseudomonas* 30-3 (Panicker *et al.* 2002). Temperature survival studies of *Rhodococcus* spp. 5/14 and 7/1 suggest that these bacteria can adapt rapidly to temperatures of −2 °C and retain metabolic activity for growth and survival at this temperature, whereas numbers of viable cells declined at 37 °C. *Rhodococcus* spp. strains 5/14 and 7/1 encode a homologue of the major cold shock gene *cspA* that codes for CS7.4 in *Escherichia coli* (Goldstein *et al.* 1990). While the protein CS7.4 functions as an RNA chaperone, its exact role as a survival strategy in microorganisms has still to be determined. *Pseudomonas* 30-3 grew at temperatures ranging from 0–35 °C. In a freeze-thaw study, its survival was enhanced by exposure to 4 °C prior to freezing. Exposure to 4 °C enhanced expression of a cold acclimation protein, CapB, which has significant similarity to *E. coli* CS7.4. CapB may possess antifreeze activity; however the mechanism by which CapB protein confers its adaptive role is not known.

Temperature and growth relationships can also be influenced by hydrocarbons. For example, Chablain *et al.* (1997) found that two psychrotolerant *P. putida* strains demonstrated slightly lower optimum growth temperatures when toluene or benzoate (a common intermediate of aromatic hydrocarbon metabolism) were provided as sole carbon source, rather than citrate. Similarly, the optimum growth rate of *Sphingomonas* sp. Ant 17 was shifted to a lower and less well-defined temperature optimum in minimal medium with jet fuel as carbon source (Baraniecki *et al.* 2002). This strain was able to grow on jet fuel at 1 °C, and low temperature had less of an effect on mineralization of phenanthrene than would have been predicted.

4.2.2.2 *Available nutrients*

It is reasonable to assume that microbes indigenous to nutrient-poor cold soils are capable of surviving oligotrophic conditions. However, little has been done to demonstrate this adaptation with hydrocarbon degraders indigenous to cold soils.

As in temperate soils, degradation of hydrocarbons in cold soils is generally enhanced following the addition of nitrogen and /or phosphorus (Braddock *et al.* 1997; Aislabie *et al.* 1998; Mohn and Stewart 2000) (see also Chapter 8, Section 2). However, pure culture studies with *Sphingomonas* Ant 17 demonstrated that this bacterium degraded aromatic components of fuels without the addition of nutrients (Baraniecki *et al.* 2002). It is possible that low-nutrient polar soils may contain microbes like *Sphingomonas* Ant 17 that are adapted to low nutrient *in situ* conditions.

The recent isolation of heterotrophic nitrogen-fixing bacteria from hydrocarbon-contaminated Antarctic soils (Eckford *et al.* 2002) was attributed to the *in situ* selective pressure of high soil carbon:nitrogen ratios resulting from the hydrocarbon spills. The nitrogen fixers were identified as *Pseudomonas* and *Azospirillum* spp. In addition to fixing nitrogen, some of the isolates also degraded hydrocarbons, but not concomitantly. Although non-photosynthetic *in situ* nitrogen fixation has been demonstrated in Antarctic soils (Line 1988), it has not yet been measured *in situ* in hydrocarbon-contaminated Antarctic soils.

In some polar soils, wet conditions may limit oxygen availability. Under such conditions degradation of PAHs may proceed under nitrate-reducing conditions, as recently reported in laboratory studies (Eriksson *et al.* 2003).

4.2.2.3 *Soil pH*

In temperate soils, bacterial hydrocarbon degradation has an optimum around pH 7.0 to 7.8. Polar soils, however, can be highly alkaline (Aislabie *et al.* 1998; Whyte *et al.* 1999a). Soils in coastal areas of the Ross Sea region, Antarctica, for example, typically have a bulk soil pH > 9. While hydrocarbon degradation in soil has been enhanced following lime addition to acid soils, the mineralization of hexadecane and naphthalene in an Antarctic hydrocarbon-contaminated soil with a pH of 9.4 was not enhanced following adjustment of the pH to 7.4 (Aislabie, unpublished). It should be noted that soil microsites can differ from the bulk soil and may be more conducive to growth than bulk soil pH measurements may indicate.

Sphingomonas Ant 17 grew optimally in well-buffered medium between pH 6.0 and 7.8, despite being isolated from Scott Base soil with pH >9 (Baraniecki *et al.* 2002). Lag times preceding exponential growth increased with pH, from 1 day at pH 6.4 to 6 days at pH 7.8 at an incubation temperature of 22 °C. However, at lower temperatures (4 °C and 8 °C) these differences were not detected and cultures had similar growth rates at pH 6.4 and pH 7.6.

4.2.2.4 *Salinity and desiccation*

Polar soils can be saline and dry and, furthermore, disturbance to soils may enhance salt deposition on, and water evaporation from, the surface soil. Disaccharides like trehalose and other polyols such as hydroxyectoine have been shown to be effective stabilizers of biological membranes under dry conditions. Some bacteria accumulate high levels of such compounds in response to desiccation (Potts 1994). Exopolysaccharide production has also been linked to desiccation tolerance. The sigma factor AlgU has recently been shown to be a crucial determinant in adaptation of *Pseudomonas fluorescens* to desiccation stress and high concentrations of NaCl (Schinder-Keel *et al.* 2001). In the Antarctic cyanobacterium *Chroocoidiopsis*, efficient repair of DNA damage that accumulates during dehydration has been linked to ability of the organism to survive prolonged desiccation and ionizing radiation (Billi *et al.* 2000). These same mechanisms may also play a role in the survival of hydrocarbon-degrading bacteria in polar soils.

4.3 Recommendations

Prior to advocating bioremediation at a particular site, it is important to determine that the contaminants of concern are biodegradable. Initially this information can be obtained by conducting a literature search. In the absence of published information, *in vitro* biodegradation tests can be performed. For these tests the soils to be remediated can be analyzed in the laboratory to determine whether they contain an indigenous microbial population that has potential to degrade the contaminant(s) of concern.

Methods employed include enumeration of hydrocarbon degraders, and the use of radiorespirometry to measure the potential to mineralize specific hydrocarbons (Braddock and McCarthy 1996; Braddock *et al.* 1997; Foght and Aislabie 2005). Increasingly, cultivation-independent molecular methods are used to detect bacterial degradation genes in soil to confirm the presence of a microbial population capable of degrading the contaminant(s) of concern (Whyte *et al.* 1999a; 2002a; Margesin *et al.* 2003). Here we discuss methods for detecting hydrocarbon degraders or their degradative genes in cold soils. Soil mineralization methods (e.g. use of microcosms) are described in Chapter 7.

4.3.1 *Methods for examining biodegradative microbial populations*

4.3.1.1 *Sampling methods and sample handling*

Samples for analysis should be representative and homogenized. All samples should be collected into sterile containers (e.g. Whirl-Pak bags, Nasco) with clean implements and swabbed with 70% ethanol (w/v). Samples should be stored at low temperature (e.g. 4 °C) in the dark. If they are not analyzed immediately the samples should be stored frozen at −20 °C (assuming that the soil freezes regularly), until analysis.

4.3.1.2 *Enumeration of hydrocarbon degraders*

Methods for enumeration of hydrocarbon degraders have been reviewed by Rosenberg (1992). Typically, most probable number (MPN) techniques are employed (Braddock and McCarthy 1996; Aislabie *et al.* 1998). The MPN method counts viable hydrocarbon-degrading organisms and is a statistical method based on probability theory. A suspension of cells in buffer is serially diluted ten-fold in buffer to reach a point of extinction. The dilutions are used to inoculate multiple tubes containing growth media, typically Bushnell Haas minimal media, and sterile hydrocarbon(s) are added to each tube. It is important that the hydrocarbon(s) provided as growth substrate be representative of site contamination. For example, if remediating a site contaminated with jet fuel, then jet fuel should be used as growth substrate rather than a long chain alkane such as hexadecane. It is crucial to include sufficient controls, both positive (containing known hydrocarbon-degrading microbes) and negative (lacking viable microbes). To determine the "end points" for MPN tubes, a number of parameters may be used either singly or in combination, including turbidity due to cell growth and disruption of hydrocarbon films (Brown and Braddock 1990). Sometimes a redox dye is used to indicate substrate oxidation (Johnsen *et al.* 2002), or the growth substrate is spiked with an appropriate ^{14}C-labeled compound and the endpoint is determined by trapping $^{14}CO_2$ (Atlas 1979). Incubation times and temperatures used should reflect the environment. Commonly, temperatures ranging from 4–20 °C are used to enumerate hydrocarbon degraders in cold soils, with incubation times from 2–8 weeks. Incubation times should be longer than those commonly used for enumerations at moderate temperatures. The pattern of positive and negative growth results is then used to estimate the concentration of hydrocarbon-degrading microbes in the original sample (the MPN of hydrocarbon-degrading microbes) by comparing the observed pattern of results with a table of the statistical probabilities of obtaining those results.

Liquid from MPN tubes exhibiting growth can be plated onto solid media for the isolation of hydrocarbon bacteria for subsequent investigations.

4.3.1.3 *Detection of hydrocarbon degradation genes in soil*

Molecular tools used for detection of hydrocarbon degradation genes in cold soils have been described by Whyte *et al.* (1999a; 2002a) and Margesin *et al.* (2003). DNA is extracted from the soil and the genes of interest are amplified using PCR. Catabolic genes targeted include those that encode alkane hydroxylase (*alkB*), catechol 2,3-dioxygenase (*C23DO*) and naphthalene dioxygenase (*ndoB*).

The principle of PCR is the generation of copies of two strands of DNA by a repeated sequence of denaturation of double-stranded DNA, synthesis of complementary strands, followed by the next round of denaturation and synthesis. Complementary oligonucleotides (primers) are required for synthesis, designed from knowledge of the sequence flanking or within the genes of interest. Since each newly synthesized complementary strand can serve as a template for the next round of PCR, the amount of DNA synthesized increases exponentially, and from a few starting DNA molecules, 20 cycles can generate over a million copies. The PCR products are visualized by agarose gel electrophoresis. To verify amplification of the correct PCR fragment, they are analyzed by Southern blotting and hybridization with DNA probes specific for hydrocarbon degradation genes or by sequencing. The PCR detection limits for these analyses are generally 100–1000 colony forming units per g of soil (A. K. Bej, personal communication).

To enumerate gene copy number, quantitative PCR methods such as competitive and real-time PCR assay are now being developed for investigations of hydrocarbon degrading bacteria (Mesarch *et al.* 2000; Laurie and Lloyd-Jones 2000). Competitive PCR (cPCR) includes a competitive sequence that serves as an internal control in each PCR reaction. In contrast, real-time PCR uses either a fluorogenic probe to detect a specific PCR product, or a DNA binding dye that detects double-stranded DNA, and measures fluorescence emitted continuously during the amplification reaction. cPCR requires post PCR analyses whereas real-time PCR does not.

The main advantage of a molecular approach is that it is rapid. Information on the genetic potential of the indigenous microbial community to degrade the contaminant(s) of concern can be obtained within days rather than weeks. However, a major limitation is dependence on existing catabolic gene sequences (for design of probes and primers) that are probably only a small fraction of the extant sequences in nature. Hence, inability to detect known degradative genes in a sample may indicate that the known genes are present in numbers below detection limits or alternatively that other "unknown" catabolic genes are involved in degradation at the site under investigation.

4.4 Future research

Bioremediation technology is underpinned by fundamental knowledge of pure microbial cultures, typically bacteria, able to degrade contaminants of concern. Further work that will enhance application of bioremediation to cold soils includes the development of specific microbial isolation methods that could result in the cultivation of new hydrocarbon degraders, and subsequent identification of hydrocarbon degradation genes as yet undescribed. Deposition of these new degradative gene sequences in public databases such as GenBank will lead to the design of additional gene probes for culture-independent *in situ* study. Further elucidation of cold/stress adaptations (desiccation; freeze-thaw; cold shock proteins) in combination with growth on or exposure to hydrocarbons could reveal metabolic strategies unique to cold-adapted hydrocarbon degraders.

Some activities that are currently known only from laboratory experiments must be demonstrated in cold soils. For example, the role of heterotrophic nitrogen fixers in cold soil hydrocarbon degradation and nutrient balance remains to be quantified *in situ*. The importance of hydrocarbon co-metabolism by microbial consortia may be more important in cold soils than in temperate soils, but this possibility has not yet been tested. Neither hydrocarbon degradation at sub-zero temperatures nor cold temperature anaerobic hydrocarbon degradation has been examined extensively *in situ*.

Recent advances include the development of DNA microarrays to describe the composition and function of microbial communities (Denef *et al.* 2003). This method will eventually enable simultaneous monitoring of multiple members of a community. Gene chip technology is already being developed for detecting hydrocarbon catabolism genes in contaminated environments (L. Whyte and C. W. Greer, personal communication). In the near future it should be possible to detect and evaluate, within hours, the known genetic potential for hydrocarbon degradation residing in a natural population. This technology will be a powerful adjunct to the traditional culture-based methods and laboratory studies described above. It would be constructive to perform bioremediation trials incorporating molecular techniques to promote and test new methods of determining community composition and diversity in hydrocarbon-impacted cold soils. The succession of phylotypes during hydrocarbon degradation would add to our understanding of bioremediation processes in cold soils.

5

Temperature effects on biodegradation of petroleum contaminants in cold soils

ANNE GUNN RIKE, SILKE SCHIEWER, AND DENNIS M. FILLER

5.1 Introduction

Bioremediation in cold climates is frequently regarded with skepticism. Owners of polluted sites and regulatory agencies may doubt the effectiveness of biological degradation at near freezing temperatures. While it is true that biodegradation rates decrease with decreasing temperatures, this does not mean that bioremediation is inappropriate for cold regions. Microbial degradation of hydrocarbons occurs even around 0 °C (Chapter 4). In remote alpine, Arctic, and Antarctic locations, excavation and shipping of contaminated soil may be prohibitively expensive. Bioremediation may be the most cost-effective alternative. This chapter discusses microbial adaptation to cold temperatures as well as results of laboratory and field studies of bioremediation at low temperatures.

Microorganisms can grow at temperatures ranging from subzero to more than 100 °C. Microbes are divided into four groups based on the range of temperature at which they can grow. The *psychrophiles* grows at temperatures below 20 °C, the *mesophiles* between 20 °C and 44 °C, the *thermophiles* between 45 °C and 70 °C, and the *hyperthermophiles* require growth temperatures above 70 °C to over 110 °C. The term "cold-adapted microorganisms" (CAMs) is frequently used for describing bacteria growing at or close to zero degrees Celsius. Depending on the cardinal temperatures (the minimal, the optimal, and the maximum growth temperature), CAMs can be classified as *psychrophiles* or *psychrotrophs*. Morita's (1975) definition, which holds that *psychrophiles* have a maximum growth temperature of less than 20 °C and an optimal growth temperature of less than

Bioremediation of Petroleum Hydrocarbons in Cold Regions, ed. Dennis M. Filler, Ian Snape, and David L. Barnes. Published by Cambridge University Press.

15 °C, while *psychrotrophs* have a maximum temperature of 40 °C and an optimal growth temperature higher than 15 °C, is widely accepted. *Psychrophiles* tend to have a narrower growth temperature range than *psychrotrophs*, and some strains grow faster below 10 °C than above. Adaptation to low temperatures can be caused by both genotypic and phenotypic modifications in the bacteria, and the ecological significance of the *psychroptrophic* and *psychrophilic* classification is not obvious. CAMs are ubiquitous in cold climate environments. In cold climate soils, the annual temperature variation in the surface layer often spans 30–40 °C. This is probably the reason why *psychrotrophs* seem to be more abundant than *psychrophiles* in the upper soil layer. Less is known about the growth temperature range of bacterial strains and consortia inhabiting deeper soil layers, where the temperature is less variable. Still we have to deal with the fact that cultivable bacteria frequently represent less than 1% of the total bacteria present in an ecosystem (Amann *et al.* 1995; Torsvik and Øvreås 2002), and Morita's temperature classification pertains only to cultivable bacteria.

There is no preponderance of evidence demonstrating that the microbial potential for degrading hydrocarbons is lower in cold region soils than in soils from warmer climates (Chapter 4). This is to say that although numbers of heterotrophic and hydrocarbon-degrading bacteria are not dissimilar in cold and warm climate soils (Aislabie 1997; Eckford *et al.* 2002), the annual window of opportunity for effective biodegradation is shorter in cold regions than that for temperate regions. Scientists generally recognize that biodegradation rates are a function of temperature and decrease with decreasing soil temperatures. However, biodegradation efficacy can be similar at both cold regions and temperate sites during comparable treatment periods (Braddock *et al.* 2001; Filler *et al.* 2001; Mohn *et al.* 2001). Our challenge is to understand and optimize biodegradation in cold climate soils and to determine the temperature range at which remediation can be performed with reasonably good effect. It is therefore essential to know site-specific climate conditions, especially air and soil temperatures that can be expected at a contaminated site.

5.2 Review

"Cold regions" is not a precise term; it is used to describe geographical areas with seasonal frost only for a few weeks in the winter, highly vegetated tundra, and polar deserts in the Arctic and Antarctica. The air and soil temperatures at these areas may differ greatly, and the only common characteristic is that the temperatures drop below zero degrees for some time during the year. Concerning hydrocarbon degradation at low temperatures, it is relevant to focus on the most extreme cold-climate regions where hydrocarbon contaminations and oil-spills are found, since bioremediation here represents the greatest challenge.

5.2.1 Air and soil temperature relationship

Polar regions are characterized by long, dark, and cold winters followed by a few light summer months. The temperature and climate conditions are harsh in both summer and winter. The temperature regime at a site, both in ambient air and soil, is important for planning and performing remedial actions (see Chapter 10, Section 10.3.1.3). With proper soil temperature data, it is possible to provide some estimates about the hydrocarbon degradation rate and the remediation time, provided nutrients, hydrocarbons, and oxygen (O_2) are present at non-limiting concentrations (see Chapter 4, Section 4.2.2).

Soil temperature varies from air temperature as a function of soil thermal diffusivity and a temperature phase lag. This temperature variability is often characterized by the mean annual air temperature (m.a.a.t.) $\pm$ a sinusoidal function with a surface temperature amplitude of variation from that of the m.a.a.t (see Chapter 2, Sections 2.2.2 and 2.2.3). The m.a.a.t. in high-Arctic locations varies from about $-19\,^{\circ}\mathrm{C}$ to $-5\,^{\circ}\mathrm{C}$, and is highly influenced by local climate and the ocean. In Antarctica, the m.a.a.t. is generally lower than in the Arctic; where m.a.a.t. of about $-20\,^{\circ}\mathrm{C}$ applies to large regions (Bockheim and Tarnocai 1998). In general, soil temperature varies with depth, with decreasing amplitude of variation (see Section 5.2.3 this chapter). At a depth where the soil temperature variation becomes negligible, isothermal conditions prevail. At shallow permafrost sites with little snow cover, or high latitude sites, the prevailing isothermal temperature can approach the sub-zero m.a.a.t. At sites with seasonal frost and substantial snow cover (providing insulation), the isothermal soil temperature may exceed the m.a.a.t. and be above zero degrees.

Of course, the amount of snow and/or vegetative cover (surface insulation) greatly influences subsurface temperatures, and soil thermal regimes must be evaluated based on site-specific conditions (e.g. ground-slope aspect and cloud cover influence on solar radiation). At locations with similar air temperatures, the yearly soil temperature variation at highly vegetated sites is often lower than those at sites with sparse vegetation. Changes in the ground's natural thermal regime have been recorded at contaminated sites in cold regions where petroleum has destroyed or damaged vegetation. For example, at Caribou-Poker Creek, Alaska, the depth to permafrost increased from 0.7 m to 1.9 m as a result of an oil spill 25 years earlier (Braddock *et al.* 2003).

5.2.2 *Implications of low temperature on microbial kinetics*

The implication of low temperature on microbial activity is very complex. In general, both abiotic and biotic processes are influenced by temperature. Thermodynamic mechanisms and microbial adaptations seem to be interrelated in their roles for sustaining microbial life and activity at low temperatures. Presently we have a vague understanding of their interactions, and even less is known about their relevance to biodegradation of hydrocarbons.

Microbial life and activity exists in cold-climate soil exposed to extreme temperature variations and in soil where the temperature never exceeds zero degrees for thousands of years (Rivkina *et al.* 2000). Although a number of studies of microbial mechanisms and adaptation to low-temperature conditions have been performed, very few deal with hydrocarbon-degrading bacteria. It is nevertheless important to be aware of the mechanisms and adaptations reported, since similar mechanisms may be involved in hydrocarbon-degrading strains in cold-climate soil.

5.2.2.1 *Reaction rate and temperature*

The thermodynamic relationship between temperature and chemical reaction rate is fundamental in our understanding of temperature effects, although it does not explain why some microbial strains are active at freezing to sub-zero temperatures. In the 1880s, Svante August Arrhenius, a physical chemist, observed that the rate constant of chemical reactions (k) varies with temperature (T) in a manner similar to the variation of an equilibrium constant with temperature:

$$k = Ae^{-E_a/RT} \tag{5.1}$$

The quantity E_a is known as the Arrhenius activation energy, constant A as the "pre-exponential factor," R is the universal gas constant, and T is absolute temperature. This expression, and studies on the quantitative correlations between reaction rates and chemical equilibrium constants, suggests that the reaction rate constant k can be expressed as a constant term that is proportional to temperature and a constant $K^{\pm}$, which has the properties of an equilibrium constant for formation of the transition state of a reactant. Based on the proposal that all transition states break down with the same rate constant ($k_B T/h$), the reaction rate is formulated as:

$$k = \left[\frac{k_B T}{h}\right] K^{\pm} \tag{5.2}$$

In Eq. 5.2, h is Planck's constant and k_B is Boltzmann's constant, and the reaction equation can be rewritten as:

$$k = \left[\frac{k_B T}{h}\right] e^{-\Delta G^{\pm}/RT} \tag{5.3}$$

where $\Delta G^{\pm}$ is the free energy of activation (Stumm and Morgan 1996). The relation between reaction rate and temperature in Eq. (5.3) is approximate. Nevertheless, this thermodynamic summary illustrates what researchers have known for more than a century that the rate of chemical reactions is reduced at lower temperatures. Eq. (5.3) also shows that anything that tends to stabilize the transition state of the reactant (decrease of $\Delta G^{\pm}$) increases the reaction rate.

The Arrhenius temperature-reaction rate correlation for a single enzyme/reactant provides a general understanding of temperature effects on biochemistry. However, microbial growth responses to temperature are more complex. In chemical reactions, catalysts permit formation of a transition state in which the reactant has a lower energy level (i.e. higher stability). Enzymes are proteins that function as catalysts in biochemical reactions. Their greater affinity for the transition state rather than substrate results in reduced activation energy and an increase in reaction rate. Indirectly, this affinity for the transition state is dependent on the amino acid sequence and the enzyme polypeptide structure. Thus, an enzyme reaction rate is dependent on both temperature and the enzyme structure. This idea is the theoretical basis for cold adaptation of microorganisms: their enzymes achieve higher reaction rates at low temperatures, in contrast to corresponding reactions catalyzed by enzymes of *mesophilic* organisms. Today, it is widely accepted that thermodynamics alone cannot explain why CAMs can grow at and below 0 °C. Cold adaptation of enzymes is probably essential for microbial activity and metabolism at near freezing temperatures.

5.3 Recent advances

In this section we summarize recent advances in our understanding of low-temperature impact on microbial processes, and results of biodegradation experiments and remediation trials performed with temperature controls or monitoring.

5.3.1 The effect of water at sub-zero temperatures

The lower temperature limit to growth of microorganisms seems to be determined by properties of the aqueous solvent system inside and outside the microbial cell. Russell (1990) suggested that the lower growth temperature of *psychrophiles* is restricted to around −10 °C to −15 °C. He proposed that microorganisms remain unfrozen at the lower growth temperature limit and that cytoplasmic water is super-cooled. In this state, water has higher vapor pressure than the extra-cellular ice. As a result, water leaves the cell and the cytoplasm concentrates, thereby reducing the freezing point. In this process intracellular toxic conditions caused by impaired ionic balance, pH, and water activity may result.

Microbial metabolic activity in response to temperatures ranging from −20 °C to 5 °C, was studied in permafrost soil samples by Rivkina *et al.* (2000). Incorporation of ^{14}C-acetate into lipids occurred down to −20 °C, although the activity was considerably reduced compared to 5 °C (on the order of magnitude of 10^{-5}). Rivkina *et al.* (2000) compared the correlation between ^{14}C-acetate incorporation in the lipid fraction initially in stationary phase, the percent of unfrozen soil water, and the thickness of the unfrozen water film on soil particles at different temperatures. They concluded that microorganism growth at the lower temperature in their experimental system was limited by ice acting as a diffusion barrier. The thickness of the unfrozen water film surrounding the soil particles at sub-zero temperatures regulated the diffusion of nutrients and substrate to the cells as well as elimination of waste products from the cells.

5.3.2 *Microbial adaptations to low temperature activity*

The ability to grow at low, but not at moderate, temperatures is proposed to be dependent on adaptive changes in the microorganisms. It is supposed that the main adaptations appear in proteins and lipids. Adaptation to cold environments can be genotypic or phenotypic. Genotypic adaptations are changes that have occurred over an evolutionary time scale, while phenotypic adaptation occurs within the lifetime of an organism (Russell 1990). Cold-adapted changes in enzymes and protein structures are fixed in the genome. Furthermore, although phenotypic changes in microbial lipid composition are not genetic, the ability to adjust lipid composition (e.g. genes encoding desaturases) is.

5.3.2.1 *Cold-adapted proteins*

In general, enzymes from CAMs have higher specific activity and catalytic efficiency at lower temperatures than their counterparts from mesophiles and thermophiles. This is generally achieved by synthesizing enzymes that are more flexible at low temperatures, which may be associated with reduced stability at higher temperatures. The thermostability of a protein is dependent on many non-covalent interactions (e.g. hydrogen bonds, salt bridges, van der Waals interactions, and hydrophobic bonds) between the peptide backbone and the amino acid side groups. Only recently have the structural changes in cold-active enzymes been studied (Russell 2000). The present knowledge on structural data comes from studies of soluble cytoplasmic enzymes. Different adaptations are used to achieve conformational flexibility in the active site at low temperatures, and the changes are not necessarily the opposite of those that confer thermostability. Hydrogen bonds and electrostatic interactions are formed exothermically, so they are stronger at low temperatures. In contrast, hydrophobic bonds are formed endothermically and will be weaker at low temperatures. According to Russell (2000), some general adaptations that confer the necessary conformational flexibility to the active site in cold-active enzymes include:

- residues with greater polarity and less hydrophobicity
- additional glycine and low arginine/lysine ratio in the polypeptides
- fewer hydrogen bonds, aromatic interactions, and ion pairs
- lack of salt bridges
- additional surface loop(s) with increased polar residues and decreased proline content
- modified α-helix dipole interactions
- reduced hydrophobic interactions between enzyme sub-units.

Berchet *et al.* (2000) performed structural analysis of elongating factor G (EF-G), an essential protein involved in protein synthesis. EF-G was isolated from the bacterium *Arthrobacter globiformis* SI55, which grows between −5 °C and 32 °C. Several structural and conformational changes were supposed to be important for low-temperature flexibility and activity when the EF-G from the cold-adapted strain was compared to EF-Gs from two related mesophilic bacterial strains.

Presently our knowledge about membrane-bound cold-active enzymes and their mechanisms is lacking. Enzymes involved in redox reactions, such as in oxidation of hydrocarbons and electron transfer, are located in lipid membrane structures. Russell (2002) suggests that enzymes embedded in the cytoplasmic membrane also need to be cold adapted in order to function at low temperature, and that this adaptation will depend not only on the intrinsic protein structure of the enzyme, but also on the physical properties of the surrounding membrane lipids. Interactions between the membrane lipids and embedded proteins are probably also important in microbial temperature sensing.

5.3.2.2 *Lipids and membranes at low temperature*

The cytoplasmic membrane in bacteria is predominately proteins and lipids. The lipid fraction accounts for about 25% of the total dry weight of the membrane in *E. coli*, and phospholipids dominate the lipid fraction (>90%). The fatty acids in the phospholipids vary in chain length, and in number and type of double bonds and alkyl side groups. The chemical structure of the fatty acids determines to a large extent the thermostability and flexibility of the cytoplasmic membrane.

Since mesophilic bacteria often have very similar lipid composition to psychrotolerant microorganisms, the membrane composition does not correlate with the microbial classification in growth temperature ranges. Phenotypic changes in membrane lipids are observed as responses to changes in growth temperature. This mechanism is not restricted to CAM but is found in mesophiles and thermophiles as well. Bacteria adjust the fatty acid composition of membrane phospholipids in response to changes in the growth temperatures. To maintain a normal fluid state of the membrane at low temperatures, modification of the fatty acids occurs. Membrane fluidity is necessary for survival and growth, and when the growth temperature falls the membrane fluidity will decrease, and membrane-associated metabolic processes mediated by enzymes, cytochromes, and permeases will be slower. Several lipid changes, such as shortening of fatty acid chains, trans- and cis-desaturation, and branching, increase the membrane fluidity at low temperatures. The rate of lipid modifications in response to transfer to and from cold temperatures will depend on the biosynthetic mechanisms involved. Changes in the degree of desaturation are normally fast, since the modification occurs directly on the lipids in the existing membrane. Changes in methyl-branching normally take more time, since *de novo* synthesis of the fatty acid molecules is necessary (Gounot and Russell 1999).

5.3.2.3 *Microbial response to temperature reduction*

Sudden changes of temperature may induce the synthesis of stress proteins in bacteria. If the temperature is shifted to just beyond the lower or upper growth limits, then cold shock or heat shock will occur. Cold-shock response (CSR) is observed in many bacterial strains. In mesophilic bacteria, cold shock results in a growth lag or acclimation phase. In this period, the number of polysomes decreases while 70 S monosomes accumulate. Genes for several different cold-shock proteins (CSP) are also expressed in mesophiles (Thieringer *et al.* 1998). He suggested that CSP are synthesized to enable gene expression and as noted by Cavicchioli *et al.* (2000) to continue protein synthesis at low temperatures. Many of these proteins seem to be involved in stabilizing ribosomes and protein synthesis at low temperatures. CSP probably contribute to ensuring that balanced microbial growth can continue. At present, we do not have a complete picture of CSP in different microorganisms and little is known about their specific function.

One of the most significant differences between mesophilic and cold-adapted bacteria is that ribosomes retain their ability to form polysomes at temperatures as low as 5 °C (Broeze *et al.* 1987). The relative rate of synthesis of most cellular proteins is also maintained after cold shock (Whyte and Inniss 1992; Michel *et al.* 1997). The findings show that cold-adapted bacteria sustain their activity at low temperatures, and that temperature-mediated inhibition in mesophiles might be responsible for their inability to grow at low temperatures (Gounot and Russell 1999).

Another bacterial response to low temperature exposure is the onset of cryoprotective mechanisms. Trehalose is found at high concentrations in many organisms that naturally survive dehydration. Accumulation of this disaccharide up to 500 mM in response to heat shock, osmotic stress, and during the stationary phase is found in bacteria and yeasts. Recent studies indicate that this sugar plays a major role in cell protection against harsh environmental conditions (Panicker *et al.* 2002; Bej *et al.* 2000). Kandror *et al.* (2002) demonstrated that synthesis and accumulation of trehalose is essential for viability at low temperatures. The enzymes for trehalose synthesis are induced in wild-type *E. coli* under cold-shock conditions, and the resulting trehalose accumulation increased viability when temperatures fell to near freezing.

Although at present we do not know much about these temperature-induced effects on hydrocarbon-degrading bacteria, it is nevertheless important to consider temperature relative to nutrient uptake when laboratory experiments with soil samples and field bioremediation schemes are planned, performed, and explained. Better understanding of the response of hydrocarbon-degrading soil bacteria to temperature reduction will help us in future efforts to optimize biodegradation rates at low temperatures.

5.3.3 *Temperature effects on bioremediation*

It is well documented that petroleum hydrocarbons are persistent in cold climate soils (Chapter 4). In our attempt to optimize hydrocarbon degradation at low temperatures, a crucial task is to identify and overcome the limitations of low temperature, oxygen depletion, nutrient and water deficiencies, and hydrocarbon availability (Leahy and Colwell 1990; Colwell and Waker 1977; Westerlake *et al.* 1973). It is also difficult to compare degradation rates at different temperatures, since temperature directly influences other soil parameters as well. Furthermore, soil heterogeneities and a high degree of variability between experimental and field trial methodologies make it difficult, if not impossible, to compare results.

A further complication relates to differentiating biotic and abiotic contributions to hydrocarbon removal. Especially light hydrocarbons can volatilize into the atmosphere. Similar to biological degradation, volatilization also increases with temperature (Snape *et al.* 2005). To estimate the relative contributions of biotic and abiotic processes, it can be useful to measure changes in the chemical composition. While branched molecules have similar volatilization rates as

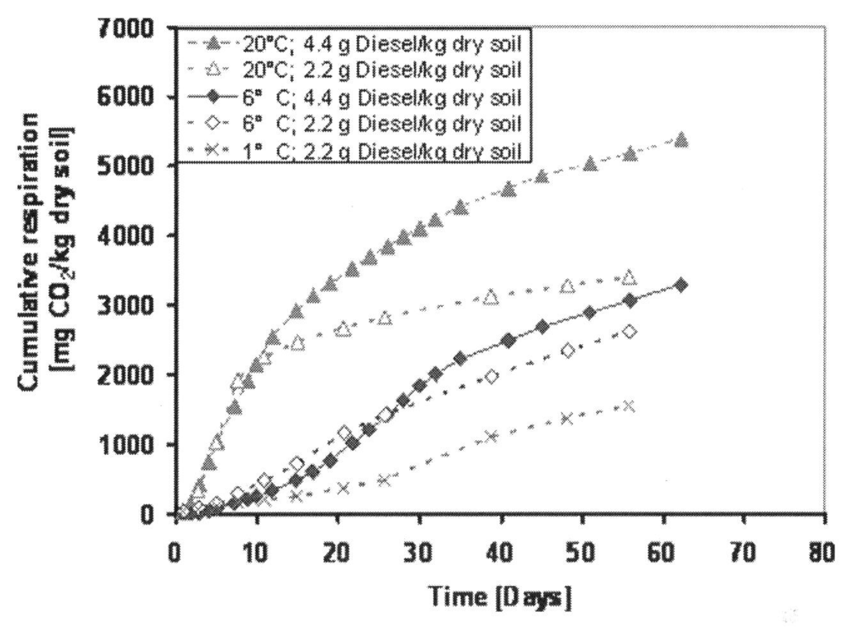

Figure 5.1. Diesel degradation in contaminated soils from Dalton Highway maintenance facility, Alaska (Schiewer and Niemeyer, 2006).

straight alkanes of similar mass, bioremediation of branched compounds is lower due to steric hindrance of enzymes involved in the degradation process. Therefore, the relative rate of change for recalcitrant branched hydrocarbons such as pristane and phytane compared to more easily biodegradable linear hydrocarbons can provide an important indication of the relative contributions of biotic and abiotic removal (Reimer *et al.* 2003; Snape *et al.* 2005).

5.3.4 Laboratory experiments

A number of laboratory studies investigated the effects of temperature on hydrocarbon bioremediation rates in microcosms. While higher temperatures typically lead to higher conversion of hydrocarbons, the differences are not as pronounced as one might expect. Several studies showed that higher temperatures increase the biodegradation rate during an initial period, but similar rates were observed for higher and lower temperatures shortly thereafter, indicating that factors other than temperature are limiting (Schiewer and Niemeyer 2006; Wong *et al.* 2003). For example, Figure 5.1 results depict the total amount of CO_2 production over two months from diesel-contaminated soil sampled at a maintenance facility located on the Dalton Highway, on the North Slope of Alaska (Schiewer and Niemeyer 2006). The slope of the curves corresponds to respiration rates. The results indicate that for an initial diesel concentration of 2.2 g kg^{-1} soil, biodegradation rates in the first ten days were much higher at 20 °C than at 6 °C and 1 °C. From day 15 to 56, however, biodegradation rates were actually higher at 6 °C than at 20 °C, and after 56 days, the respective total CO_2 productions at 1 °C and 6 °C were 47% and 76% that of soil at 20 °C.

To investigate whether substrate availability may have been limiting at 20 °C, leading to the reduced rates after ten days, a second set of experiments was performed on soil samples from the same site with a higher initial diesel concentration of 4.4 g kg^{-1} soil (Schiewer and Niemeyer 2006). This time, degradation at 20 °C and 6 °C continued to increase through the sixty-day trial period, indicating that substrate (diesel) availability was the limiting factor for the comparative trials with lower substrate concentrations in gravel.

Table 5.1 summarizes experimental conditions and results from recently reported laboratory experiments on biodegradation of petroleum hydrocarbons in cold soils, primarily in soil microcosms taken from cold-climate sites. The table reports endpoint or highest degradation rates achieved and extent of degradation. In all the experiments, nutrients or other amendments were added to the microcosms. References are cited for more specific information about the experiments.

5.3.5 Field trials

For field applications of bioremediation in cold climates factors other than temperature must be considered (Rowsell 2003). Often cold sites are remote, present logistical challenges, and experience a short summer season. Therefore, careful planning is of the utmost importance. At remote locations, where no facilities and electricity are available, passive systems may be employed. Wind-powered aeration was used at a former Defense Early Warning (DEW)-line site at 69° N latitude, in Northwest Canada (Pouliot *et al.* 2003). Within three treatment seasons, the TPH concentration was lowered by 85%, achieving an average concentration of 649 mg kg^{-1} soil to meet the applicable cleanup criterion. Heat generated by microbial degradation processes contributed to achieving temperatures high enough for efficient treatment. Other techniques to improve bioremediation rates in field applications include addition of fertilizer, soil amendments (e.g. addition of bulking agents for improved aeration), aeration, and soil warming in combination with insulation. In a field study by Reimer *et al.* (2003), biopiles with blanket-type insulation were actively heated and aerated by hot air. It was possible to maintain a core temperature of 15 °C even at ambient air temperatures of −42 °C.

Table 5.1 *Laboratory experiments of low-temperature biodegradation of petroleum hydrocarbons in soil from cold climate sites*

	Site Characteristics			Experimental conditions			Result		
Site	Characteristic site temp.	Petroleum concentration in soil (mg kg^{-1})	Depth of soil samples (m)	Experimental system Analytical method	Hydrocarbons degraded	Temp (°C)	Degradation rate	Extent of degradation	References
Naval Arctic Research Laboratory, Barrow, Alaska	a.d.t. in July: 2.9 °C	250–860	0.5	Microcosms Respirometry	JP-5 and diesel	10	5 mg C $kg^{-1} day^{-1}$	Not reported	Braddock *et al.* 1997
Naval Arctic Research Laboratory, Barrow, Alaska	a.d.t. in July: 2.9 °C	100–7100	0.5	Soil slurries Radio-respirometry	^{14}C-Hexadecane ^{14}C-Naphthalene	10 10	1.3 mg CO_2 kg^{-1} day^{-1} 4.5 mg CO_2 kg^{-1} day^{-1}	Not reported	Braddock *et al.* 1997
Naval Arctic Research Laboratory, Barrow, Alaska	Soil August: 7.3 °C (0.1 m) Permafrost at 1.3 m	250–860	0.5	Microcosms Respirometry	Diesel	5 10 15 20	7 mg C kg^{-1} day^{-1} 12 mg C kg^{-1} day^{-1} 21 mg C kg^{-1} day^{-1} 26 mg C kg^{-1} day^{-1}	Not reported	Walworth *et al.* 2001
Canadian Force Station–Alert, Ellesmere Island, Canada	a.d.t. in July: 3.6 °C	200–26 900	Not reported	Microcosms Radio-respirometry TPH determination	^{14}C-Hexadecane TPH	5	Not reported	32 days: 40% 45 days: 30%	Whyte *et al.* 1999
Alpine subsoil from 1 700–2 000 m a.s.l.	Not reported	Not reported	Not reported	Microcosms TPH determination	Diesel (5000 mg/kg^{-1} added)	10	Not reported	20 days: 53%	Margesin 2000

(cont.)

Table 5.1 *(cont.)*

Site Characteristics				Experimental conditions			Result		
Site	Characteristic site temp.	Petroleum concentration in soil (mg kg^{-1})	Depth of soil samples (m)	Experimental system Analytical method	Hydrocarbons degraded	Temp (°C)	Degradation rate	Extent of degradation	References
Eureka, Ellesmere Island, Canada	Not reported	4257–5166	Not reported	Microcosms Radio-respironetry Radio-respirometry TPH determination	^{14}C-Hexadecane ^{14}C- Naphthalene TPH	5	Not reported	45 days: 20% 45 days: 70% 16 weeks: 53%	Whyte *et al.* 2001
Eureka, Ellesmere Island, Canada	Not reported	4257–5166	Not reported	Microcosms TPH determination	TPH	5	Not reported	16 weeks: 53%	Whyte *et al.* 2001
Ft. Wainwright, Fairbanks, Alaska	Soil August: 8 °C (1 m) October: 3 °C (3 m)	8100	2–6	Microcosms Respirometry	Diesel	1 11 21	1.7 mg HC kg^{-1} day^{-1} 8.2 mg HC $kg^{-1}day^{-1}$ 15.1 mg HC kg^{-1} day^{-1}	Not reported	Walworth *et al.* 2001
Greenhouse ground, University of Calgary, Canada	Soil Constant 5 °C at 5 m	20 000	Not reported	Soil columns Respirometry and TPH determination	Crude oil	5 21 5 21	*Growth phase* 64 mg HC kg^{-1} day^{-1} 100 mg HC $kg^{-1}day^{-1}$ *Stationary phase* 11 mg HC kg^{-1} day^{-1} 11 mg HC kg^{-1} day^{-1}	Not reported	Gibb *et al.* 2001
Ellesmere Island, Canada	a.d.t. in July: 6.2 °C	900–1000	Not reported	Microcosms Respirometry and TPH determination	Weathered arctic diesel	−5 0 7 7/-5	*Max* (C_{11}-C_{15}): 0.71 mg kg^{-1} day^{-1} 0.95 mg − kg^{-1} day 1 1.8 mg kg^{-1} day^{-1} 3.5 mg kg^{-1} day^{-1}	48 days: 0 300 mg kg^{-1} 450 mg kg^{-1} 600 mg kg^{-1}	Eriksson *et al.* 2001

Ellesmere Island, Canada	a.d.t. in July: 6.2 °C	2400	Not reported	Microcosms TPH determination	Weathered diesel	7	50 mg TPH $kg^{-1} day^{-1}$	92 days: 1800 mg kg^{-1}	Thomassin-Lacroix *et al.* 2002
Longyearbyen, Spitsbergen,	m.a.a.t.: −6 °C Soil July: <5 °C at 1 m Permafrost at 2 m	3600–21 500	0.5	Microcosms Radio-respirometry	^{14}C-Hexadecane	5	20 mg $kg^{-1} day^{-1}$	42 days: 586 mg kg^{-1}	Børresen *et al.* 2003a
Longyearbyen, Spitsbergen,	m.a.a.t. −6 °C	5224	0.2–0.4	Microcosm	^{14}C-Hexadecane	5	40 mg $kg^{-1} day^{-1}$	200 days: 14%	Børresen *et al.* 2003b
Sag River, Dalton Hwy, Alaska, USA	Not reported	1180	0.3	Microcosms Respirometry and DRO determination	Diesel	1 6 20	2 mg CO_2 $kg^{-1} day^{-1}$ 4 mg CO_2 $kg^{-1} day^{-1}$ 15 mg CO_2 $kg^{-1} day^{-1}$	Not reported 56 days: 73% 56 days: 90.3%	Niemeyer & Schiewer, 2003
Cape Dyer, Nunavut, Canada,	Not reported	3330–8970	Not reported	Microcosms Radiorespirometry TPH determination	^{14}C Dodecane Weathered diesel, Jet fuel	7	Not reported	54 days: 27% 54 days: 78%	Reimer *et al.* 2003
Mackenzie River	Not reported	10 000	Not reported	Slurry	#1 arctic diesel	10	Oxygen uptake 115.2 mg $L^{-1} day^{-1}$	44 days: 53% for C_{10}-C_{16}	Wilson *et al.* 2003
Calgary	Not reported	10 000 (spike)	1.5	Slurry	Crude oil	5 20	Not reported	60 days: 40% 60 days: 60%	Wong *et al.* 2003

a.d.t. – Average daily temperature (air)
m.a.a.t. – Mean annual air temperature
a.s.l. – Above sea level

Table 5.2 summarizes some of the key conditions and results reported from field studies and remedial trials at cold-climate sites. Site locations, site characteristics, and endpoint results and/or highest degradation rates achieved are included. Petroleum hydrocarbon concentrations cited are typically initial concentrations before remediation.

One study, the Longyearbyen, Spitsbergen (78° N latitude) trial highlighted in Table 5.2, was a good attempt to correlate temperature and bioactivity with depth *in situ* at a permafrost site contaminated with a mix of fuels (Rike *et al.* 2003a). Soil temperatures, oxygen (O_2) and carbon dioxide (CO_2) in soil gas at 0.7 m, 2.0 m, and 3.5 m depths were continuously measured and compared to data from a nearby uncontaminated control site in assessing biodegradation rates.

The hydrocarbon degradation rate at 0.7 m depth varied between 3 to 7 mg TPH kg soil^{-1} day^{-1} from October to mid-November, and biodegradation continued at this rate for nearly 12 days after the soil at this depth froze. By comparing empirical O_2 concentrations with data from a model that calculates O_2 variability with soil depth over time (under the assumption that microbial O_2 consumption ceased after soil freezing), the researchers concluded that hydrocarbon biodegradation occurred at subzero temperatures during winter periods at this Arctic site. The data gave the first indication that microorganisms were metabolizing hydrocarbons at subzero temperatures at an Arctic site (Rike *et al.* 2003).

Beginning the next spring, site bioactivity (as O_2 consumption versus CO_2 production) was monitored in the active layer through the thaw season and through summer (Rike *et al.* 2005). At the 0.7 m depth, measurable O_2 consumption initiated during the middle of April (day 109), about 45 days before the soil thawed (day 154) at this depth (Figure 5.2). A theoretical model was developed and used to predict O_2 concentrations as a function of soil depth and time at different microbial activation temperatures in spring until steady-state O_2 conditions were achieved in summer (Figure 5.3). Figure 5.3 data does not evidence correlation between the microbial activation time and soil thawing. The CAM became active at temperatures of about $-6\,°C$, yet significant biodegradation activity did not occur until soil temperatures warmed above $-3\,°C$; hydrocarbon degradation was probably limited by O_2 availability early on. Later in summer, bioactivity again became limited with O_2 depletion. Figure 5.3 illustrates the good correlation between empirical and modeled data for O_2 evolution with depth (i.e. at the different activation temperatures).

Results from the Longyearbyen field trial indicate that some microbial strains inhabiting this Arctic site can degrade petroleum hydrocarbons down to $-6\,°C$, and a presumption that the active strains possess a kind of cryo-protective mechanism (e.g. extracellular polymer or surface compound) that remains active in

Table 5.2 *Field biodegradation trials of petroleum hydrocarbons in soil at low temperatures*

Site characteristics				Experimental conditions			Result		
Site and location	Site temp.	Petroleum type	Petroleum concentration in soil (mg kg^{-1})	Experimental conditions / treatment	Nutrients or additives	Temperature regime in study	Degradation rate	Extent of degradation (final concentration)	References
Alpine ski resort 3 000 m a.s.l., Austria	Not reported	Biodiesel	2600	Surface soil treatment	Fertilizer	Natural fluctuation	Not reported	12 months: 54% 27 months: 72%	Margesin 2000
Deadhorse Airport, Prudhoe Bay, Alaska	August: approx. 5 °C November: approx. −25 °C	Arctic diesel	800–11 000	Biopile (49 m × 40 m × 2.4 m) Thermal insulation Electrical heating Temperature and soil gas control	Fertilizer	Above freezing 0.5 to 7.8 °C with seasonal variations	5.65 mg kg^{-1} day^{-1}	329 days: 93% (142.1 mg kg^{-1})	Filler *et al.* 2001
Cambridge Bay, Northwest Territories, Canada (69° N, 105° W)	Not reported 4 months with air temp. above freezing July: 7.6 °C	No. 1 Arctic diesel	196	Biopiles 0.25 m^3 Aeration pipes	Nutrients, peat, inoculum	July–August: 2.7–10 °C in covered biopiles 1.7–6.7 °C in uncovered biopiles	Not reported	12 months: 95% (10 mg kg^{-1}) 12 months: 91% (195 mg kg^{-1})	Mohn *et al.* 2001
Komakuk Beach, Yukon Territory, Canada (70° N, 140° W)		No. 1 Arctic diesel	2109						

(cont.)

Table 5.2 *(cont.)*

Site characteristics				Experimental conditions			Result		
Site and location	Site temp.	Petroleum type	Petroleum concentration in soil (mg kg^{-1})	Experimental conditions / treatment	Nutrients or additives	Temperature regime in study	Degradation rate	Extent of degradation (final concentration)	References
Caribou-Poker Creeks Research Watershed, Fairbanks, Alaska	Natural soil temperatures 8–13 °C at 0.1 m in summer Permafrost depth increased from 0.7 m to 1.9 m in contaminated soil	Crude oil (Prudhoe Bay) Spilled 25 years ago	1000–659 000	Natural weathering in soils from 0–18 cm depth based on hopane determination in GC-detectable oil	Not for 25 years	Natural fluctuation	Not reported	25 years: 10–83% Variation between oil components Biodegradation is the main contributor to the observed weathering	Braddock *et al.* 2002
Ellesmere Island, Canada (82° N, 62° W)	a.d.t. in July: 6.4 °C	Weathered arctic diesel or jet fuel	2900	Biopiles 0.5 m^3 Passive aeration Plastic cover	Nutrients Surfac-tants Soil structure amend-ments	a.d.t. in experiment: 10 °C to −14 °C	90 mg kg^{-1} day^{-1} (for 14 days)	65 days: 83% (500 mg kg^{-1})	Thomassin-Lacroix *et al.* 2002
Saviktok Pt, Tuktoyaktuk, NT, Canada	a.d.t. in summer: ~10 °C	Diesel	4136	Biopiles Wind powered aeration	Not specified	Not specified	Not reported	1 season: 34% 3 seasons 85%	Pouliot *et al.* 2003

Longyearbyen, Spitsbergen (78° N)	m.a.a.t. −6 °C Permafrost at 2 m depth	Weathered diesel and fuel oil	3600–21 000	*In situ* attenuation On-line temperature and soil gas measurements at 0.7, 2.0 and 3.5 m	No additives	Frozen soil profile −2 °C to < 0 °C in the period where degradation was measured	3–7 mg HC kg^{-1} day^{-1} (for 12 days)	Site still active	Rike *et al.* 2003a
CFS Alert, Ellesmere Island, Nunavut, Canada (82° N)	a.d.t. 5 °C in summer ~ −20 °C in winter min. −45 °C	Not specified, diesel?	Not reported	Biopiles, Forced aeration, Heating by warm air	Not reported	Core temp.: 15 °C even at a.d.t. −42 °C	Not reported	141 days: 63%	Reimer *et al.* 2003
Casey Station, Antarctica	a.d.t. in summer: 10 °C	Special Antarctic Blend Diesel	~ 23 000	*In situ*	Controlled release nutrients	Natural fluctuation	Not reported	3 years: ~90%	Snape *et al.* 2003
Eureka, Nunavut, Canada	a.d.t in summer ~5 °C	Diesel	Up to 32 000	*In situ*	Fertilizer	Natural fluctuation	Not reported	Strongly varied	Whyte *et al.* 2003

a.d.t. – Average daily temperature (air)
m.a.a.t. – Mean annual air temperature
a.s.l. – Above sea level

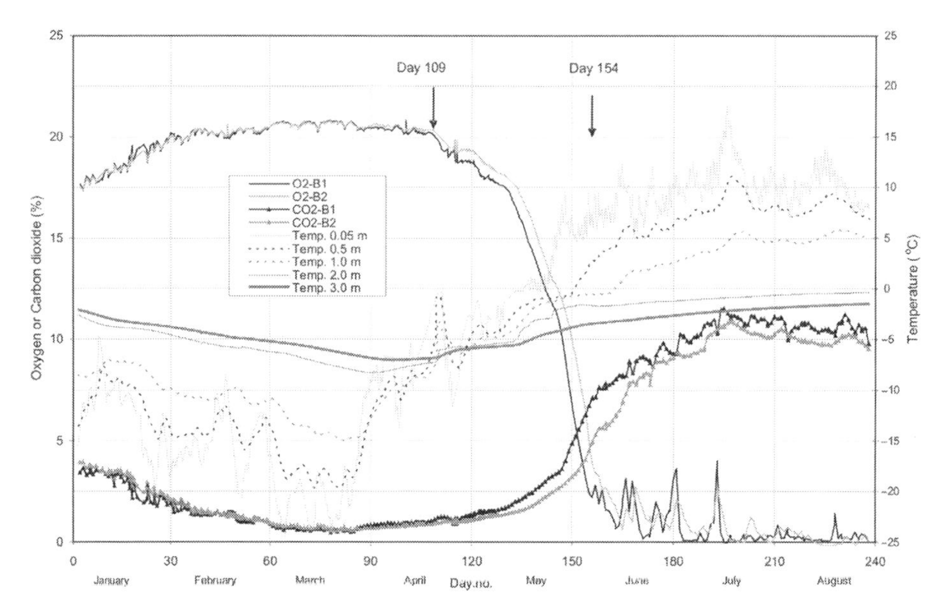

Figure 5.2. Longyearbyen (Spitsbergen) trial: O_2 and CO_2 in soil gas at B1 (0.7 m) and B2 (2.0 m) within the fuel spill area, and soil temperature 0.05 m to 3.0 m. Day 109 indicates the microbial activation time at 0.7 m, and Day 156 the thawing time at 0.7 m (reprinted from Rike *et al.* 2005).

capturing unfrozen water around the cell at subzero temperatures. Furthermore, in absence of other limiting conditions, the active biodegradation period can be extended to about six months annually at this high-Arctic site.

5.4 Guidelines and recommendations

The accurate collection of certain information from experimental and field trials would aid our understanding of temperature effects relative to remediation efforts at petroleum-contaminated sites. With future experiments it is important to carefully report the parameters and conditions summarized in Table 5.3. Periodic monitoring of these parameters and conditions through treatment duration aids our ability to optimize treatment, understand data spikes (treatment interruptions and/or discontinuities), and assess treatment progress.

Temperature-water relationships will affect bioremediation effectiveness, especially during seasonal transition periods. Soil type, moisture content, and porosity provide important information about oxygen availability. For instance, in cold-climate regions, soil moisture contents generally increase in spring when snow thaws. Poorly drained soil can be saturated in the spring and give rise to temporary anaerobic conditions as soil voids are filled with water. On the other

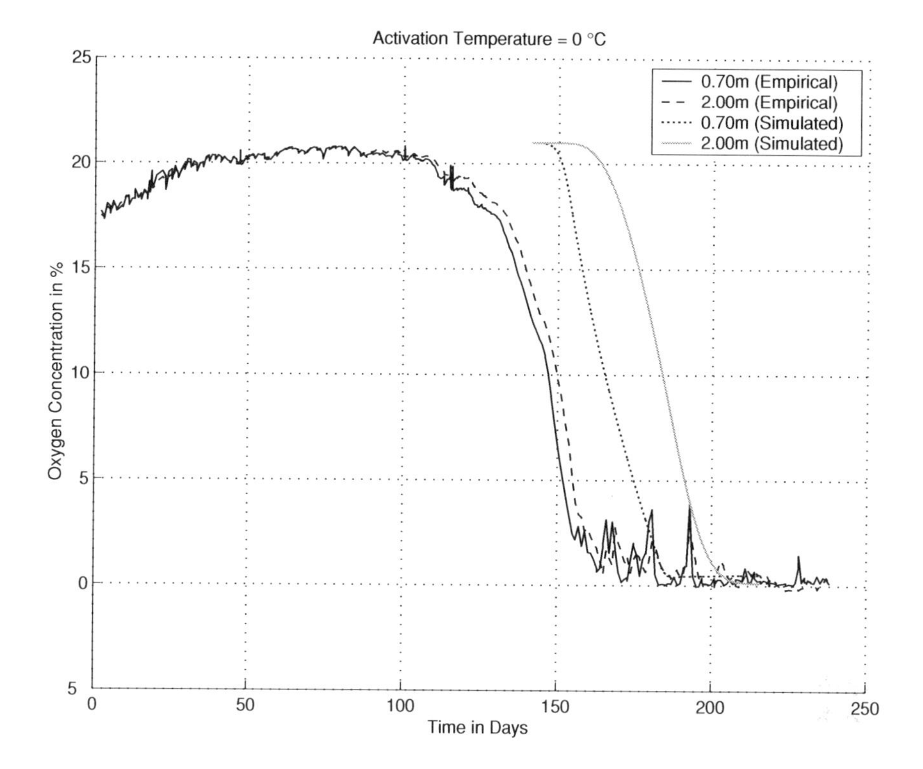

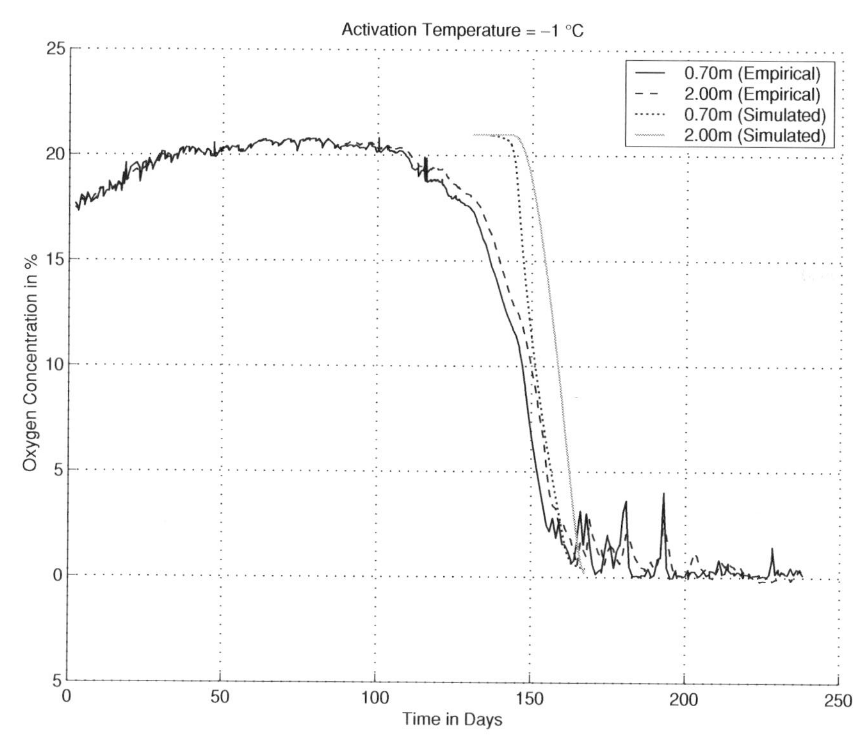

Figure 5.3. Longyearbyen (Spitsbergen) trial: A comparison between empirical and simulated oxygen concentrations at depths of 0.7 m and 2.0 m, for 0 °C, −1 °C, −2 °C, −3 °C, −4 °C, and −6 °C activation temperatures (reprinted from Rike *et al.* 2005).

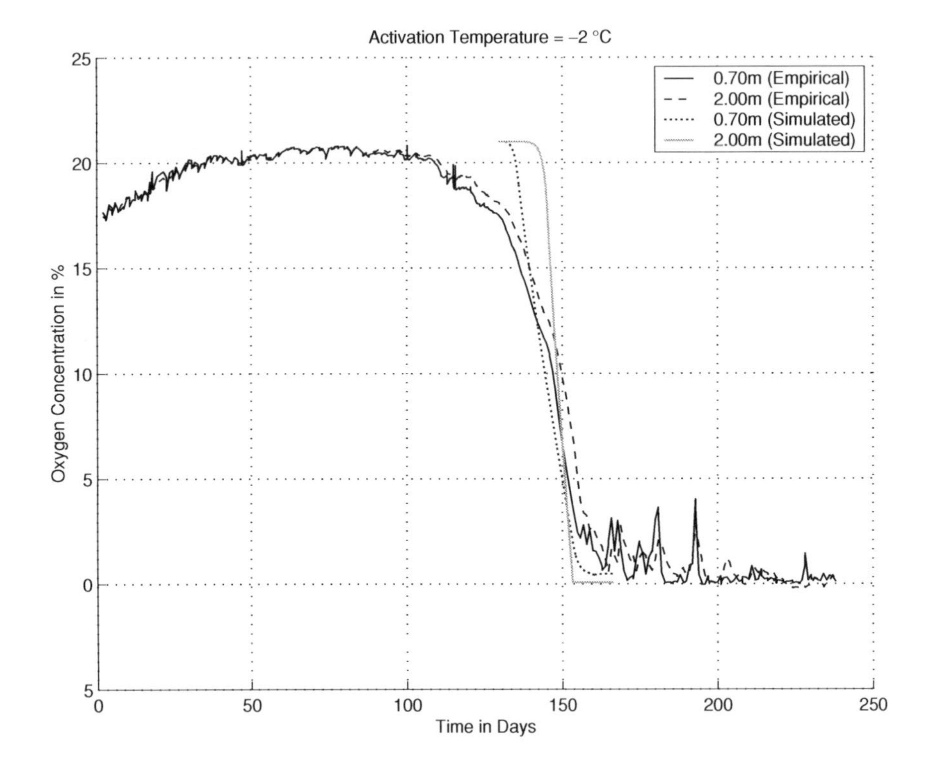

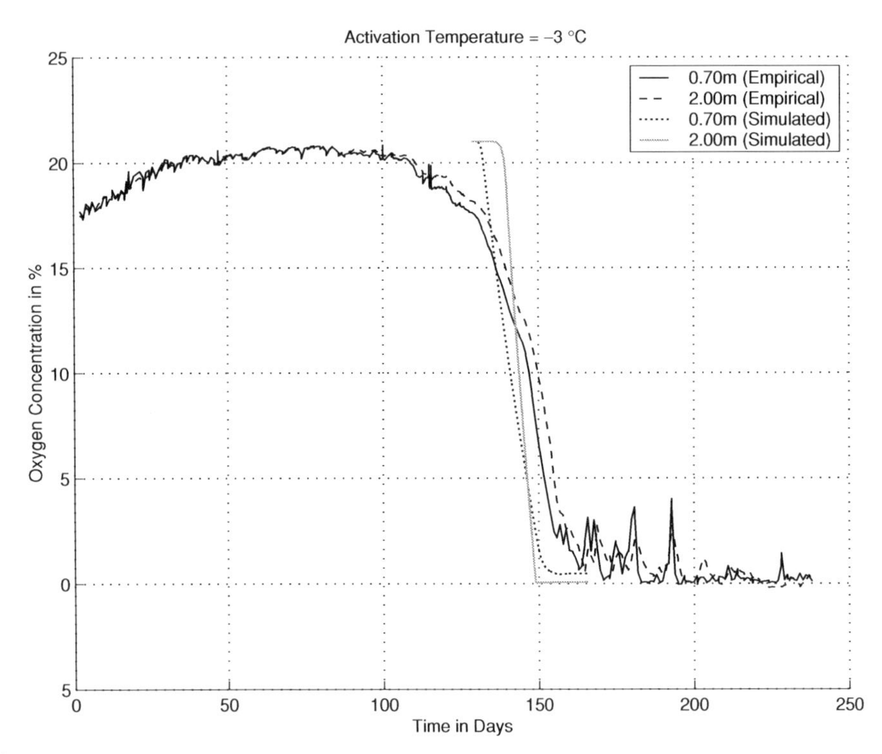

Figure 5.3. (cont.)

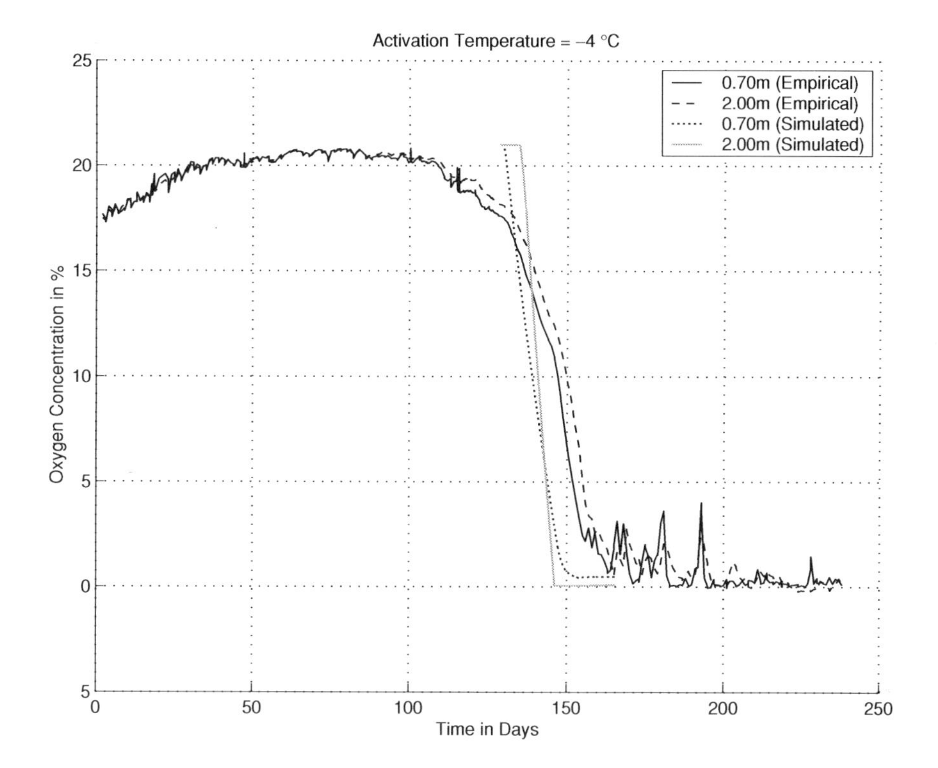

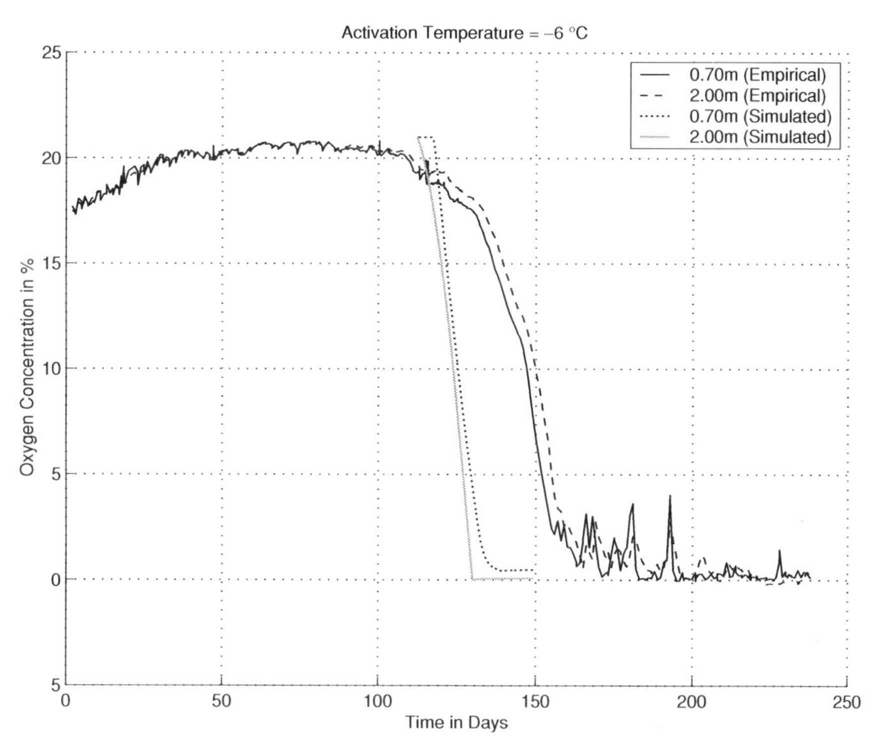

Figure 5.3. (cont.)

Table 5.3 *Conditions and parameters important for interpretation of temperature effects on hydrocarbon degradation studies*

Initial site and soil conditions	Experimental and treatment considerations	Method used in determination of degradation
Ambient air and soil temperatures	Temperature regime and temperature control	Chemical analysis of the initial and final petroleum hydrocarbon concentrations
Soil type and grain size distribution	Storage and pretreatment of soil used in the experiments	Chemical analysis of the hydrocarbons at several times during the experiment
Soil moisture	Addition of water	Chemical analysis and ratio to biomarkers (e.g. pristane, phytane, hopane)
Soil porosity	Supply of electron acceptors	Radiorespirometry of a single radio-labeled compound and $^{14}CO_2$ determination
Natural organic matter	Alternate substrate	Pyrolysis GC/MS
Nutrient concentration and pH	Amendments (e.g. nutrients, emulsifiers, or co-substrates) and pH adjustment	Respirometry with measurement of oxygen consumption
Population size of hydrocarbon-degrading and heterotrophic microorganisms	Experimental scale and system-scale examinations (e.g. *in situ*, on site, laboratory experiments)	Respirometry with measurement of carbon dioxide formation
Chemical composition and concentration of the petroleum contaminant	Addition of external petroleum hydrocarbons (type and concentration) in experiments	IR spectroscopy, GC/MS or GC/FID
Weathering degree of the petroleum contaminant	Sampling strategy and sampling time	GC/MS chromatograph interpretation

hand, mobilization of water-soluble nutrients coincides with the snow thawing in the spring, and availability of fresh nutrients may increase microbial proliferation in the soil at this time. Furthermore, since the content of unfrozen water in soil at subzero temperatures is dependent on the soil type and the grain size, grain size distribution curves may provide valuable information about potential diffusion barriers.

The extent of hydrocarbon biodegradation is frequently reported as percent of initial concentration of a hydrocarbon compound (e.g. benzene) or group (e.g. TPH, DRO, GRO) degraded over a certain period. Precise information about the initial and final hydrocarbon concentrations, degradation time, and temperature regime is recommended. Biodegradation rates are reported in various ways: as carbon degradation (in mg C $kg^{-1}day^{-1}$) or carbon fraction degradation (in mg C_{11}-C_{15} $kg^{-1}day^{-1}$), as analyte specific or total hydrocarbon degradation (mg benzene kg^{-1} day^{-1} or mg TPH kg^{-1} day^{-1}), or in terms of carbon dioxide evolution (mg CO_2 kg^{-1} day^{-1}). In some studies degradation rates have been estimated simplistically from initial and final hydrocarbon concentrations, under the assumption of a constant and linear biodegradation through the degradation period. Other studies differentiate between degradation rates during the acclimation, logarithmic growth, and stationary phases of bioactivity. In general, these parameters represent how we evaluate treatment effectiveness today. However, more accurate measurements for rate are necessary that account for contaminant or analyte-specific degradation rates relative to the contamination mix that may be present in a soil, and differentiation between volatilization and biodegradation contributions. A more thorough discussion on assessing biodegradation rates is presented in Chapter 7.

5.5 Future research

It may not be feasible to remediate contaminated soil at subzero temperatures from a practical and regulatory perspective, if remediation time is a governing criterion. However, bioremediation during the polar summer months with temperatures around 5 or 10 °C appears very feasible and has been achieved in numerous field trials. Biodegradation rates and the annual period of effective treatment can be increased by active or passive soil warming. However, since contaminated sites in polar regions are often situated in remote locations, energy availability may limit the options for aeration, irrigation, and soil warming. Hydrogen fuel cells and other alternative energy sources are being considered for cold regions environmental applications (discussed in Chapter 11). Passive methods using wind energy for aeration and solar energy for soil heating may become feasible tools for remote contaminated sites in the future.

A requirement for efficient remediation in cold climates is that all potentially limiting factors be addressed so that effective bioremediation can be achieved under the existing temperature regimes. For the most effective use of the short Arctic and Antarctic summers, additional research is necessary to investigate interrelationships between temperature and mass transfer, contaminant partitioning, and abiotic processes relative to biological degradation of petroleum hydrocarbons. As with the Longyearbyen (Spitsbergen) field study, *in situ* relationships between biomass and soil parameters, and their spatial variability with depth, will provide valuable site-specific information to enhance our understanding and ability to achieve successful remediation of hydrocarbon-contaminated sites in cold regions. The more site-specific information that can be synthesized the better our ability to develop and apply generic remedial methods in extreme cold environments.

6

Analytical methods for petroleum in cold region soils

DANIEL M. WHITE, D. SARAH GARLAND, AND
CRAIG R. WOOLARD

6.1 Introduction

In order to demonstrate the effectiveness of a bioremediation project, one must have an accurate measure of the contaminants, both at the start of the project and throughout the treatment process. The measurement of the contaminants throughout the process is important to demonstrate that the treatment is successful and to identify advances or set-backs quickly and effectively.

Proving the disappearance of hydrocarbons is important to the success of a bioremediation project. An accurate measurement of hydrocarbons and their biodegradation products is needed to confirm that petroleum was actually consumed by bacteria (discussed in Chapter 7, Section 7.3). One method of confirming biodegradation of petroleum is the coupled measurement of biodegradation rates by proxy methods and the disappearance of the contaminant. Biodegradation rates do not, in and of themselves, prove the decomposition of contaminants. Measurement of biodegradation rates, however, can be an easy way to demonstrate that the potential exists for contaminant removal. While measures of biodegradation rates are often used to estimate time to closure for a site, or proof of technology, biodegradation rates can be unreliable. Common measures of aerobic biodegradation are loss of contaminants, oxygen (O_2) consumption, and carbon dioxide (CO_2) evolution. Unfortunately, the CO_2 can result from non-biological sources (see Chapter 7, Section 7.2.2.2 for additional discussion). Particularly in low pH groundwater, pH adjustment made during bioremediation could result in CO_2 off-gassing from groundwater. Oxygen depletion in the subsurface is also not proof of biodegradation. Particularly in highly reduced

Bioremediation of Petroleum Hydrocarbons in Cold Regions, ed. Dennis M. Filler, Ian Snape, and David L. Barnes. Published by Cambridge University Press.

water, oxygen scavenging by reduced chemicals, such as iron and manganese, can appear to be biodegradation.

Many applications of biodegradation, particularly of hydrocarbons, rely on the microbial requirement for carbon (C) and energy. Ideally, microbes consume organic contaminants to remediate a site. This process is complicated in many cold regions soils by the presence of high concentrations of organic matter. Organic matter accumulates in soil when primary production exceeds microbial decomposition. In the cold soils of northern climates, organic matter has accumulated for millions of years. The result is that up to 30% of the world's soil carbon is stored in the Arctic (Oechel and Vourlitis 1995). For microorganisms, a contaminant is another organic chemical which may or may not be an attractive source of C and energy.

Thus the accurate evaluation of the effectiveness of bioremediation and the accurate application of site closure criteria require accurate contaminant measurements. Unfortunately, many contaminant measurement methods are prone to significant error, particularly in the rich organic soils found in northern cold regions. This chapter provides an overview of the measurement techniques used to determine contaminant concentration in soil and a review of the biogenic interferences common in organic soils. Section 6.3 presents one method proposed for solving the complication to analytical methods posed by organic soils. The chapter concludes with a discussion of future research needs.

6.2 Review

6.2.1 Measuring hydrocarbons in soil

Petroleum products such as crude oil, diesel fuel, and gasoline are complex mixtures that contain both non-polar and polar compounds (Dragun 1988). For example, non-polar compounds such as aromatics and straight, branched, and cyclic alkanes, alkenes, and alkynes comprise 82–97% of crude oil (Potter and Simmons 1998a). The polar fraction, which is frequently referred to as the nitrogen, sulfur, and oxygen containing fraction, is generally high in molecular weight and heterocyclic in nature (Gill and Robotham 1989).

6.2.1.1 Extraction

The methods commonly used to quantify petroleum in soils rely on extraction of the contaminant from the soil matrix. The sparingly water soluble solvents are non-polar and are commonly used as petroleum solvents. Common solvents used to extract petroleum from soil are methylene chloride, hexane, and hexane-isopropanol mixtures. Solvents used in the past have included benzene, carbon disulfide, chloroform, and 1,1,2-trichloro-1,2,2-trifluoroethane (freon-113). Because crude oil includes some polar compounds, the very non-polar solvents such as hexane are sometimes avoided.

Many of the northern European countries use the Dutch Guidelines to analyze soils contaminated with petroleum. Other countries have modified the Dutch Guidelines or use them in combination with the USEPA guidelines (MFE 2003). In the United States, each state specifies the solvent used in hydrocarbon analysis. For example, Alaska's methods for determining the concentration of diesel range organics (DRO) and residual range organics (RRO), specify the use of methylene chloride. In some cases, the solvent choice is limited by the analytical method used to quantify the hydrocarbon. For example, USEPA Method 418.1, Total Petroleum Hydrocarbons (TPH), uses infrared spectroscopy to measure the CH bond stretch of hydrocarbons in a solvent extract. This analytical method requires use of a solvent, such as Freon-113, that does not itself have a CH bond. A review of the solvents used in different hydrocarbon measurement methods is available in Potter and Simmons (1998b).

Using the solvent required by the appropriate regulatory agency, a contaminated soil is extracted by one of a number of mechanisms. The extraction techniques most commonly used for extraction of petroleum from soil are reflux apparati (e.g. soxhlet extraction), bath or horn sonicators, shake flasks, microwaves, end-over-end tumblers, and supercritical fluids. Each extraction method has a different recovery efficiency depending on the properties of the contaminant. The use and efficiency of different extraction methods was thoroughly reviewed in Dean (1998). The data in Figure 6.1, modified after Dean (1998), show that methods for extraction can recover greater than 100% and less than 60% of two similar polynuclear aromatic hydrocarbons (PAH). The graph further shows that no one method consistently performed better than any of the others.

6.2.1.2 *Concentration and cleanup*

Solvent extracts are often too dilute to get an accurate measurement using many analytical techniques. When this is the case, the solvent is concentrated, typically by evaporating the solvent. Evaporation techniques, such as the Kuderna-Danish method rely on the fact that most semi-volatile hydrocarbons can be condensed from a volatilized solvent matrix while allowing at least a portion of the solvent to escape. By selectively evaporating the solvent, the semi-volatiles of interest can be concentrated (Dean 1998). Depending on the molecular weight of the target analyte, some portion will be lost during solvent concentration (Potter and Simmons 1998b). This is particularly a problem with semi-volatiles. In addition, by incidentally removing the light fraction of

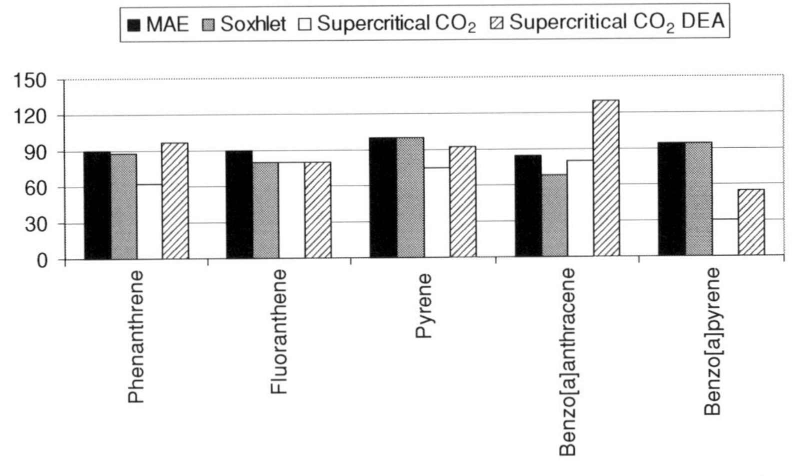

Figure 6.1. Extraction efficiency for different techniques and different polynuclear aromatics. MAE refers to microwave assisted extraction and CO_2 DEA refers to a CO_2 + 10% diethylamine extraction solvent (modified from Dean 1998).

the contaminant, detailed analysis of the contaminant chemistry could be misleading.

Solvents can also extract a portion of the natural organic matrix of the soil along with the hydrocarbons. In order to remove natural interferences, semi-polar gels, such as silica gel, are used as a cleanup step. Silica gel is specified as the cleanup sorbent in USEPA Method 418.1. In Table 6.1, modified from Potter and Simmons (1998b), the Freon-113 extract from different organic matrices is presented. Silica gel removed only a portion of the natural hydrocarbon extracted from uncontaminated grass, nuts, and needles after one pass through the cleanup column. For pine needles, three passes over silica gel were required to remove the natural hydrocarbons. For grass and gall nut extracts, three passes of the extract over silica gel did not completely eliminate the natural hydrocarbons. A discussion of cleanup methods as applied to extracts from cold regions soils is discussed in the section on biogenic interference (Section 6.2.2).

6.2.1.3 *Quantification*

Once a contaminant is extracted from the soil it must be quantified using gravimetric, chromatographic, spectroscopic, immunoassay, or spectrometric techniques. These techniques will quantify the contaminant concentration in the soil extract with varying degrees of accuracy and precision. White and Irvine (1996) compared gravimetry, gas chromatography/mass spectrometry, and infrared spectrometry for the quantification of petroleum in cold regions soils, particularly in the presence of natural organic matter (NOM). Other studies

Table 6.1 *Total petroleum hydrocarbons by USEPA Method 418.1 in natural materials (modified after Potter and Simmons 1998b)*

Material	TPH (mg kg^{-1})	TPH (mg kg^{-1}) after 1 pass through silica gel	TPH (mg kg^{-1}) after 2 passes through silica gel	TPH (mg kg^{-1}) after 3 passes through silica gel
Grass	14 000	4500	2700	2600
Gall nuts	9700	4500	1300	1200
Pine needles	16 000	1700	1400	0

have described and compared advantages and disadvantages of different detection methods (Potter and Simmons 1998a, 1998b). For additional detail on quantification methods, the reader is encouraged to review these references. Each method has advantages and disadvantages but no one method is effective for multiple contaminants in multiple matrices.

6.2.2 *Biogenic interference in analyses of petroleum hydrocarbons*

Biogenic interference is the term given to the fraction of NOM that is counted as petroleum in a standard petroleum analysis. The amount of biogenic interference for a particular soil is a function of both the quantity and the character of the NOM in soil.

6.2.2.1 *NOM in soil*

Northern soils contain an estimated 25–33% of the world's soil carbon (Oechel and Vourlitis 1995). Cold temperatures reduce the rate of both aerobic and anaerobic decomposition of NOM in most northern soils, resulting in an accumulation of NOM. Until recently, the net carbon balance in northern ecosystems tended toward carbon preservation since the primary production rate exceeded the decomposition rate (Oechel and Billings 1992).

Soil NOM is divided into two categories, humic and non-humic substances. Humic substances are refractory, mainly hydrophilic, high molecular weight, chemically complex, polyelectrolyte-like substances (Schnitzer 1991). Molecular weights of humic substances range from 300 to more than 300 000 daltons (Stevenson 1985). Humic substances are further subdivided into three operationally defined categories: fulvic acids which are soluble in water at all pH values; humic acids which are water-soluble in alkaline conditions but precipitate at $pH < 2$; and humin which is not water-soluble at any pH (Aiken *et al.* 1985). In mineral soils, nearly all of the humic substances are bound in

Table 6.2 *Some chemicals that are present in both natural organic matter and petroleum (Modified from Jorgenson 1990)*

Acetic acid	Heptacosanoic acid	Propanoic acid	Alkanes n-C_{14} to n-C_{33}
Benzene	Hexadecanoic acid	Tetradecanoic acid	Octanoic acid
1,2-benzofluorene	Methane	Toluene	Ethyl benzene
Benzoic acid	Methanethiol	Pentanoic acid	Nonanoic acid
Butanoic acid	Methanol	Perylene	Ethanol
2,6-dimethylundecane	Xylenes	Phenanthrene	Formic acid
Eicosanoic acid	Naphthalene		

clay metal-humus complexes. In organic soils, insoluble macromolecular complexes of humic substances predominate (Stevenson 1985). Nonhumic substances are low molecular weight, recognizable compounds including carbohydrates, proteins, peptides, amino acids, nucleic acids, purines, pyrimidines, fatty acids, waxes, resins, and pigments (Schnitzer 1991). Nonhumic substances are normally readily biodegraded. The NOM in organic soils is composed, on average, of 65% humic substances, 10% carbohydrates, 10% nitrogen-containing compounds, and 15% lipids (Schnitzer 1991). In general, biogenic interference arises from the lipid fraction of NOM (White and Irvine 1998b).

Lipids, also called bitumens, include hydrocarbons, alcohols, acids, sterols, terpenes, pigments, heterocyclics, and phospholipids (Braids and Miller 1975). The compounds listed in Table 6.2 were reported by Jorgenson (1990) to be present in both crude oil and NOM. A review of the origin, nature, and abundance of soil bitumens as they influence petroleum analyses was published in White and Irvine (1996, 1998a).

6.2.2.2 *Biogenic interference in analytical methods for petroleum*

The mass and composition of soil bitumens detected in soil varies according to the type of solvent used to extract the soil and the character of the NOM. Figure 6.2 represents the amount of NOM that was extracted from clean cold regions soil by different solvents and could be quantified as biogenic interference (White 1995). White (1995) used USEPA Method 413.2, Oil and Grease for this comparison. USEPA Method 413.2 is a gravimetric determination. In a soil that contains 50% by weight organic matter, 10% of the NOM constitutes 50 g kg^{-1}, or 50 000 mg kg^{-1}. If this material were quantified as petroleum, the soil would not meet cleanup limits whether it was contaminated by petroleum or not. In most cases, acidification increased the amount of NOM extracted by a solvent. Acidification is specified by most methods, including the Alaska methods

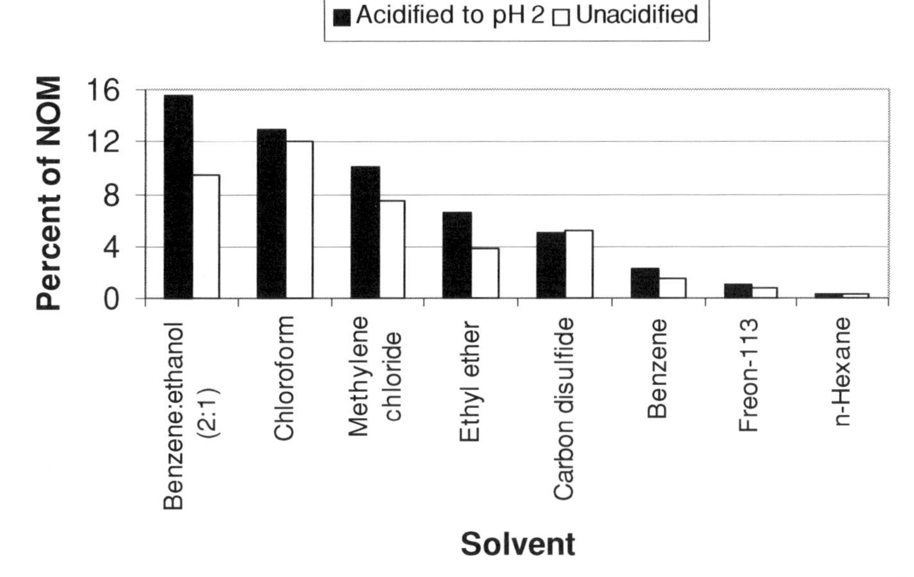

Figure 6.2. Percent of NOM extracted and quantified as petroleum hydrocarbons using various solvents (modified after White 1995).

in order to capture any slightly polar molecules in petroleum. The consequence is that in most cases, NOM capture has also increased.

6.2.3 Some approaches to the biogenic interference problem

6.2.3.1 Background sampling

A common method for estimating biogenic interference in contaminated soil is to use a comparable but uncontaminated soil sample to establish the background level. Standard petroleum analyses are performed on both the contaminated and uncontaminated samples. The "petroleum" levels of the uncontaminated background samples are subtracted from those of the contaminated samples to estimate the amount of actual petroleum in the contaminated samples.

The inherent problems of this method are obvious, and it may be impossible to prove that a background sample truly is uncontaminated as the boundaries of spill-affected areas are often unknown. High concentrations of measured "petroleum" in the background sample could indicate a high level of biogenic interference, or it could indicate that the "background" sample was affected by the spill. Finally, even if a background sample is known to be uncontaminated and appears to have an organic matrix similar to that of the contaminated sample, high variability in biogenic interference makes it difficult to determine the background level. Biogenic DRO and RRO each varied by an order of magnitude at one small site on the North Slope of Alaska (Woolard *et al.* 1999a). The cost of collecting and analyzing enough samples to get useful data in the face of such high variability would be highly impractical. The 95% confidence limits for biogenic interference in a collection of background samples could be greater than the actual level of petroleum contamination in a contaminated sample.

6.2.3.2 *Solvent cleanup and biogenic interference*

Most bituminous material in soil has a slight degree of polarity, usually provided by a carboxylic acid or alcohol group at the end of a long carbon chain. To exclude NOM from contaminant quantification, some methods for hydrocarbon analysis, such as USEPA Method 418.1, require that the soil extract be passed through a column containing silica gel. Silica gel selectively removes compounds with oxygen containing functional groups. Other cleanup methods, such as USEPA Method 3620 or 3610/3611, stipulate the use of fluorisil or alumina gel. The disadvantage of removing semi-polar compounds derived from NOM from the soil extract is the coincident removal of similar compounds from the contaminant matrix. White (1995) found that approximately 14% of crude oil from a soil extract was lost on silica gel while only 94% of the biogenic interference was removed. In a soil that is 50% by weight organic matter, if 10% of the NOM is extracted by the solvent, and only 94% of this material is removed from the solvent, this leaves 3000 mg kg^{-1} of natural matter reported as "hydrocarbons." A concentration of this magnitude may be in excess of a country's cleanup standards (Chapter 1).

6.2.3.3 *Other methods for identifying biogenic interference*

Other methods, such as analyzing stable isotopes, odd to even chain alkanes, the n-C_{17}/pristane to n-C_{18}/phytane ratios for solving the problem with biogenic interference have been explored elsewhere (White 1995; Garland 1999). While all of the methods have provided some guidance, no method has been able to address the problem of identifying biogenic interference in a previously contaminated sample.

6.3 Recent advances in solving the NOM problem

An alternative approach to identifying biogenic interference in petroleum analyses uses pyrolysis-gas chromatography/mass spectrometry

(GC/MS) (White and Irvine 1998; White *et al.* 2004). Garland *et al.* (1999) developed the method for using pyrolysis-GC/flame ionization detection (FID) to differentiate between petroleum hydrocarbons and biogenic interference. The pyrolysis-GC/FID method was used to quantitatively characterize soil NOM in terms of biogenic interference in the Alaska methods for contaminant analysis. The pyrolysis-GC/FID method was based on a pilot study by White *et al.* (1998) that indicated pyrolysis-GC/MS could be used to predict biogenic interference for USEPA Method 413.2, Oil and Grease. The study showed that for organic soil samples, the amount of biogenic interference could be estimated in the presence of high concentrations of petroleum in the sample. This method was the first to quantify biogenic interference in a contaminated soil matrix, eliminating the need for an uncontaminated "background" sample.

The pyrolysis-GC/FID method was developed specifically to quantify biogenic interference in the Alaska Department of Environmental Conservation (ADEC) methods, AK 102 for DRO and AK 103 for RRO (www.dec.state.ak.us/spar/csp/process.htm). With the appropriate calibration, the method could be adapted to any national or international cleanup standard. Since the method development was based on cleanup standards in Alaska, the AK methods are referred to in this section. FID was chosen in place of the mass spectrometry that was used in White *et al.* (1998) to make the method less expensive and better suited for use in contract laboratories.

In the pyrolysis-GC/FID method, specific compounds in chromatograms of whole, contaminated soil are used as "biogenic indicators." Mathematical models relating biogenic DRO and RRO to the biogenic indicators were developed using regression analysis. New acronyms were needed to describe the estimates of biogenic interference calculated using the models. The acronyms used were "diesel range *biogenic* organics" (DRBO) and "residual range *biogenic* organics" (RRBO).

The pyrolysis-GC/FID method was developed using samples from 57 sites from across Alaska. The samples included tundra, boreal forest, southeastern muskeg, buried peats, as well as less organic silt loam and loess samples. These samples had no known history of contamination. The method was based on the principle that an analysis of complex fatty acid-like materials in soil could be used to estimate the mass of biogenic interference in the sample. The complex fatty acid-like materials could be analyzed independently and in the presence of petroleum. For the 54 samples, biogenic indicators were identified and tested in models until the predicted biogenic interference (measured as DRO or RRO) was equal to the measured biogenic interference (see Figure 6.3). Since these samples were uncontaminated, the measured DRO and RRO was entirely composed of biogenic interference. The development of the methods was described previously (Garland *et al.* 1999).

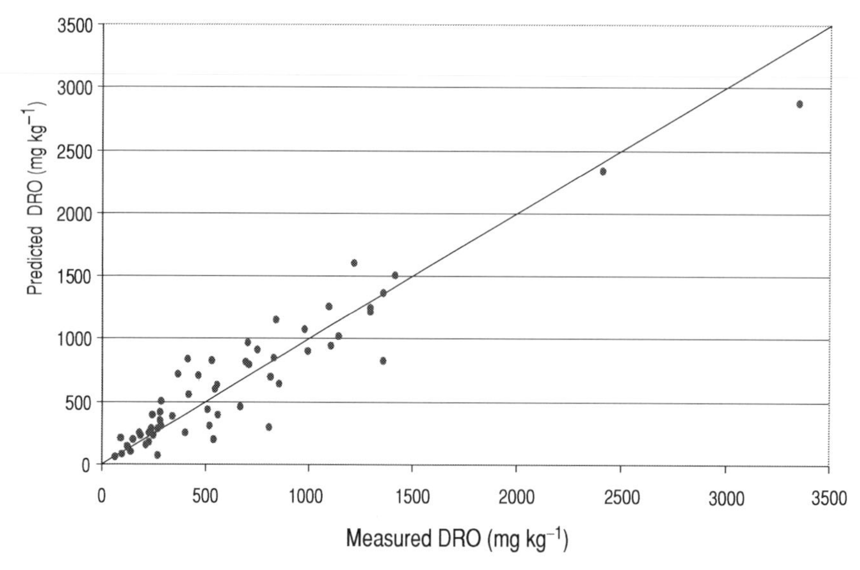

Figure 6.3. Data showing predicted vs. measured DRO for 57 sampling sites. Since these samples were uncontaminated, the "DRO" is natural organic matter that would show up as biogenic interference in a petroleum analysis. The line does not represent the best fit, but rather the theoretical perfect fit. Since it is not a "best fit" line, there is not an associated r^2 value.

6.3.1 *Example use of the pyrolysis-GC/FID method for one set of analytical methods*

6.3.1.1 *Collection of samples*

Conduct sample collection according to standard methods (e.g. Alaska Methods for DRO and RRO). For each sample, collect approximately 500 g of soil. This will allow for 250 g of soil for the DRO/RRO analysis and 250 g for the DRBO/RRBO analysis. If both sets of analyses will be conducted in the same laboratory, send the full 500 g sample to the lab for splitting. If the DRO/RRO and DRBO/RRBO will be conducted in separate laboratories, split the samples in the field. Prepare, package, and ship samples according to standard methods.

6.3.1.2 *Laboratory preparation of samples*

Split the sample into two equal, representative portions if the sample was not split in the field. Perform DRO and RRO analyses of one portion, using standard methods (e.g. AK102 and AK103 for Alaska). Oven-dry the sample for DRBO/RRBO following American Standards and Testing Methods D 2974. Record the oven-dry mass. If DRO/RRO analysis is to be performed at the same laboratory, the oven-dried DRBO/RRBO sample may be used to calculate moisture content for the DRO/RRO sample. Grind the dried sample with a mortar and pestle and shake through a #30 sieve. Visually inspect the material not passing the sieve to ensure that it contains only mineral matter. Repeat the grinding process as needed until all vegetable material is reduced to a fine powder. A ball mill may be used to grind tough fibrous organic matter. Weigh and set aside the coarse fraction. The fraction passing the sieve is referred to as the dry fines. Stir the dry fines, pour into a vial or jar of volume at least double that of the fines, and shake thoroughly to homogenize. Store the dry fines in a desiccator until use. Long-term storage of the fines is possible providing the samples remain dry.

Calculate the coarse adjustment factor (CAF) as follows:

$$\text{CAF} = (\text{total oven} - \text{dry mass}) \div [(\text{total oven} - \text{dry mass}) - (\text{mass of coarse fraction})] \quad (6.1)$$

Conduct a loss on ignition test on a subsample of the dry fines as follows. Weigh an aluminum sample dish. Add approximately 10 g of dry fines and weigh again; subtract the weight of the dish to determine the original sample mass. For samples with high organic content and high moisture content, less than 10 g of dry fines may be available; in this case, it is acceptable to use as little as 1 g of dry fines for the LOI test. Place the aluminum dish and sample in a muffle furnace at 550 °C for 1 hour. Remove, cool in a desiccator, and weigh again. Subtract the weight of the dish to determine the combusted sample mass. Discard the combusted fines. Calculate loss on ignition (LOI) by:

$$\text{LOI} = [(\text{original sample mass}) - (\text{combusted sample mass})] \div (\text{original sample mass}) \quad (6.2)$$

Calculate the target mass: Target mass $= (2\,\text{mg}) \div \text{LOI}$. Place a plug of quartz wool in the bottom of a 2 mm ID quartz sample tube for pyrolysis. Add dry fines to the sample tube to achieve the target mass of fines, $\pm 10\%$, and weigh on the microbalance. Use of a static ionizing unit, such as the Staticmaster® model number 2U500, is recommended to minimize the effect of static on the microbalance. Add a small plug of quartz wool at the top of the tube to hold the sample in place. Inject the sample with 5 μl of a 0.8 mg ml^{-1} solution of polyalphamethylstyrene dissolved in methylene chloride.

6.3.1.3 *Sample Analysis*

Carrier gas and flow rates

Pyrolysis results are sensitive to changes in the flow conditions through the pyrolysis-GC apparatus. Use of a fine pressure regulator in the line between the nitrogen tank and the pyrolysis-GC is recommended. Maintain the GC column head pressure at 103 kPa, flow through the FID at 1 ml min^{-1}, split vent flow at approximately 50 ml min^{-1}, and GC purge vent flow at approximately 15 ml min^{-1}. Adjust the inlet pressure of the carrier gas to obtain the desired flow rate.

Pyrolyzer

In the development of this method, pyrolysis was performed on a CDS Analytical Pyroprobe 2000 with an autosampler AS2500. Maintain the pyrolysis interface chamber at 280 °C. Before pyrolyzing the sample, allow the sample to sit in the 280 °C interface for 10 minutes while the nitrogen carrier gas flowing through it is purged to vent. At the end of the purge, switch the nitrogen gas flow online to the GC and begin pyrolysis. To pyrolyze, heat the sample at 5 °C min^{-1} from 280 °C to 700 °C and then hold at 700 °C for 9.9 seconds. Keep the pyrolyzer online to the GC for two minutes following the start of pyrolysis, then discharge to vent and supply clean carrier gas to the GC for the remainder of the run.

Gas Chromatograph

The pyrolyzer was connected to an HP 5890 GC with a flame ionization detector. Maintain the injection port at 270 °C and the detector at 300 °C. Use a cross-linked methyl siloxane, 25 m × 0.2 mm × 0.33 μm film thickness capillary column, such as the HP-1, and a FID. Run the GC in splitless mode for the first minute of each run, then return to split mode. Use an 80-min temperature program: hold at 40 °C for 10 minutes, ramp up at 3 °C min^{-1} to 85 °C, ramp up at 5 °C min^{-1} to 270 °C, and hold at 270 °C for 18 minutes.

6.3.1.4 *Interpretation of data*

Peak identification and integration

Run a blank of quartz wool injected with the surrogate, polyalphamethylstyrene, to determine the time at which the surrogate elutes, approximately 20 minutes. Retention times vary slightly from sample to sample and also shorten over time as the column ages. Retention times shorten more over time for the surrogate than for the compounds of interest, and more for earlier-eluting compounds. Pyrograms of organic soils contain a distinctive series of double peaks (Figures 6.4 and 6.5). These double peaks are referred to with letters from "a" to "l." The double peaks are further split into a first and second as, for example, e1, e2, f1, f2, etc., with 1, the first, and 2, the second peak in each pair, and retention times increasing with alphabetical order. The double peaks are usually much larger than nearby single peaks in the pyrogram. Lag times between the first and second peaks of each pair, and between the first peak of a pair and the first peak of the next pair, decrease with increasing retention time.

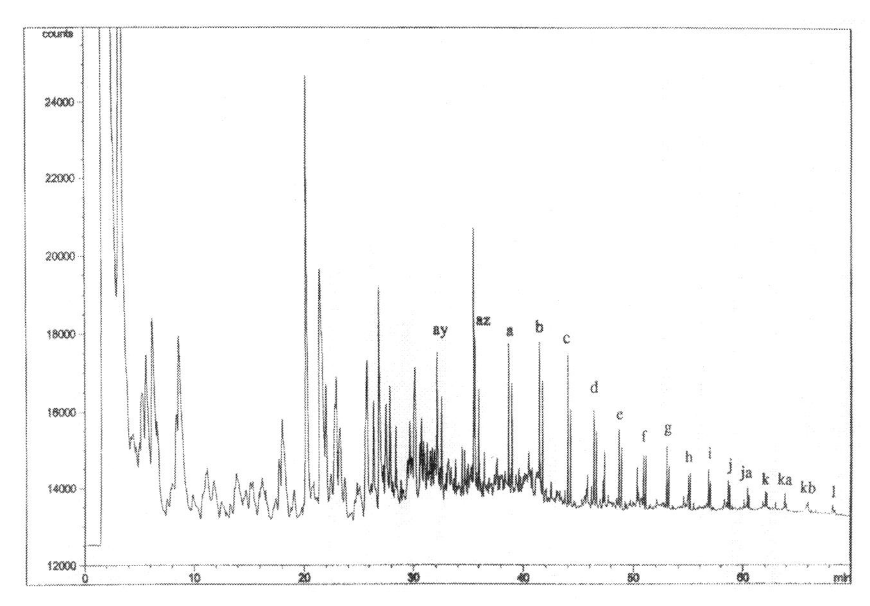

Figure 6.4. Chromatogram showing the double peak series with peaks labeled.

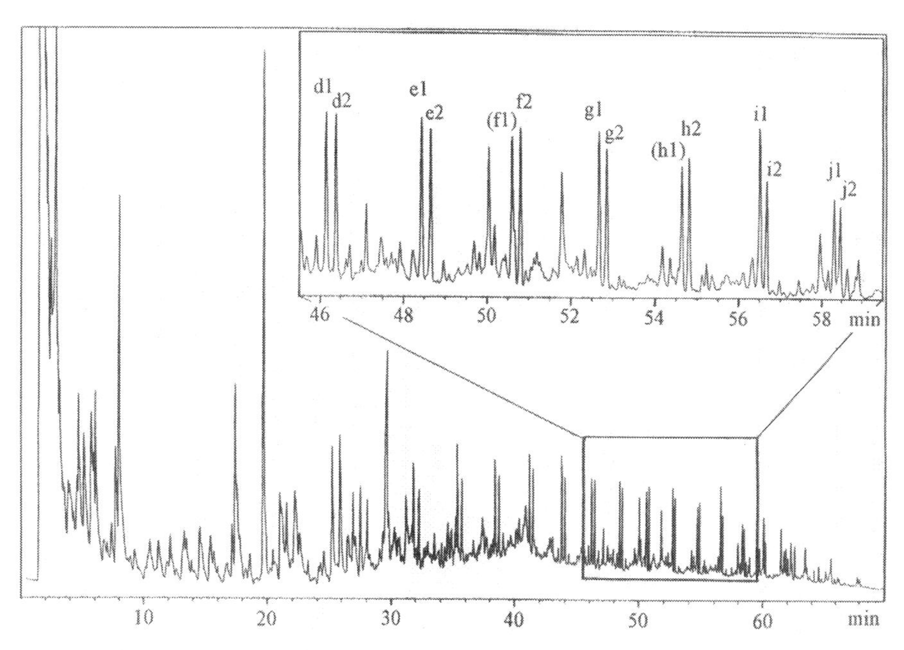

Figure 6.5. Expanded view of the double peak series used for calculation of biogenic interference.

Peaks a1 and a2 elute approximately 18.5 minutes after the surrogate. Peaks b1 and b2 are the next pair after a; c1 and c2 are after b, etc. If the pyrogram does not contain the distinctive double-peak pattern, DRBO and RRBO are non-detectable (ND) by the pyrolysis-GC/FID method.

Identify peak pairs d through j. The peaks of interest for the Alaska standard methods were d1, e2, f2, g2, h2, i1, i2, and j1. These were selected by trial and error from the suite of double-peak compounds. Visually examine a close-up view of the integrated chromatogram to ensure that the peaks of interest are integrated properly. Adjust integration parameters if necessary. For each peak of interest in the sample, calculate the numerical value of the associated biogenic indicator as, e.g., d1 = (area of peak d1 in pyrogram) ÷ [(mass of sample pyrolyzed) × (CAF)]. For each sample, calculate the weighted area sums x_D and x_R:

$$x_D = 0.0200\text{d1} + 0.0864\text{g2} - 0.1086\text{h2} - 0.0265\text{i1} + 0.0527\text{j1} \tag{6.3}$$

$$x_R = 0.0200\text{d1} - 0.0287\text{e2} + 0.0156\text{f2} + 0.0605\text{g2} - 0.0648\text{i2} \tag{6.4}$$

Equations 6.3 and 6.4 were derived from a statistical analysis of the pyrogram with respect to biogenic DRO and RRO. A complete accounting of the statistics and origin of the equations can be found in Garland (1999). The unresolved fraction of the chromatogram includes peaks associated with non-lipophilic compounds and is important to the analysis.

Calibration

Three uncontaminated soils of known DRBO and RRBO are used as standards to create a calibration curve. Standards are run according to the same conditions as the sample. The DRBO calibration curve is created by plotting known DRBO of the standards on the vertical axis and calculated x_D on the horizontal axis. The $y = mx + b$ form of the equation of the best-fit line is the calibration curve, with $x_D = x$ and DRBO $= y$. Similarly, the known values of RRBO and calculated x_R are used to create the RRBO calibration curve. DRBO and RRBO of unknown samples are found by inserting the calculated values of x_D and x_R for the sample into the equations of the calibration curves. It is important to note that the calibration equations actually convert the pyrolysis-GC/FID data to the relevant standard method. In this case, the result is the equivalent of the Alaska Methods for DRO and RRO. Full detail relative to the development of the DRBO and RRBO calibration method is presented in Garland (1999).

6.3.1.5 *Summary of the pyrolysis-GC/FID method*

The pyrolysis-GC/FID method could serve as one tool for quantifying biogenic interferences in samples. At this time, there are very few laboratories that have pyrolysis interfaces for GCs. In order for this method to become widely available, commercial laboratories would need to invest in the technology. This would likely require regulatory agency certification of the method. Pyrolysis-GC/FID is not without its limitations. In particular, pyrolysis can be subject to matrix effects. The study described herein attempted to overcome matrix effects by analyzing large numbers of samples of differing origin and mineral content. Any one sample, however, particularly those with unusual matrices (e.g. marine sediments) could have unexpected matrix effects leading to inaccurate quantification of DRBO or RRBO. In the course of the method development, however, no samples were discovered for which inaccurate DRBO or RRBO was quantified.

It is important to note that this method is quite detailed and requires significant development. It is unlikely that this would be a routine method for biogenic interference where inexpensive, and easy methods, such as silica gel cleanup or background sampling are practical. In particularly difficult cases, it may be worth developing and applying the pyrolysis-GC/FID method.

6.4 Guidelines and recommendations

It is important to remember that a solvent extracts a different portion of the organic matter in the contaminant and in the organic soil. In cold region soils where the organic content can be over 50% of the soil mass, large errors in contaminant concentration can result. A very non-polar solvent such as hexane will extract the least amount of NOM but also sacrifices the semi-polar fraction of petroleum products. Therefore, more refined products, such as gasoline and diesel fuel, are better suited to very non-polar solvents. The more complex, asphaltene-containing oils require a slightly more polar solvent to capture all of the petroleum. In addition, weathering generally makes an oil more polar as microbes add oxygen to some of the molecules, or the non-polar light fraction evaporates.

Whole soil techniques such as the pyrolysis-GC/FID method can eliminate some of the error inherent in the soil extraction process. Without this method, use of a background (if available) and solvent cleanup will improve one's ability to identify biogenic interference. The limitations to background and cleanup methods are well documented, however. It is critical to understand the limitations of any petroleum quantification method prior to use.

6.5 Future research

The need persists to better understand the role of NOM in contaminant analyses. As contaminants weather they are chemically transformed and indeed become more like NOM. Future research must address the transition from when a contaminant is a contaminant to when it is part of the natural environment.

Future research must continue to address the interaction between NOM and contaminants, and the impact of that interaction on contaminant extraction and analysis. Since the chemical and physical properties of NOM are unique to a given environment, it is important to characterize NOM in a way that allows one to predict contaminant behavior in any NOM matrix.

7

Treatability studies: microcosms, mesocosms, and field trials

IAN SNAPE, C. MIKE REYNOLDS, JAMES L. WALWORTH, AND SUSAN FERGUSON

7.1 Introduction

Treatability assessments are used to identify limitations to the rate or endpoint of bioremediation for a specific soil-contaminant combination. For treatability studies, the degradation pathways for the contaminant are generally known (see Chapter 4, Section 4.2.1), but the limitations in a particular soil or at a specific site are less well understood. The tremendous utility of treatability studies is in evaluating practical treatment regimes prior to full-scale implementation. The goal is to demonstrate practicability, optimize treatment design, and provide information for project planning. Sometimes this is an essential proving step for clients or regulators because choice of treatment depends primarily on urgency of remediation and cost. The cost-time relationship for different treatment types is illustrated in Chapter 1, Figure 1.1. The ability to predict the rate of bioremediation progress for a treatment scheme is particularly important in cold regions where costs are higher and treatment times are longer than in temperate regions.

In an effort to understand and improve the bioremediation process in cold regions, researchers have used treatability experiments to:

- identify the presence or absence of microbial activity for a particular contaminant or group of contaminants;
- determine optimum requirements, such as temperature, nutrients, oxygen, and water, for bacteria and fungi to metabolize contaminants in the soil regime;

Bioremediation of Petroleum Hydrocarbons in Cold Regions, ed. Dennis M. Filler, Ian Snape, and David L. Barnes. Published by Cambridge University Press.

- examine the effects that natural cycles, such as freezing-thawing and wetting-drying, have on microbial activity and degradation rate;
- estimate achievable endpoints;
- predict and compare treatment times and costs.

Treatability studies can involve *in vitro* microcosms with individual bacterial species or consortia from the soil incubated in liquid or slurry media, mesocosm studies with soils and natural microfauna, or field trials. There is a continuum between laboratory microcosms, mesocosms, and field trials; the terms are used interchangeably by many and the distinction between them is subjective. For the purposes of this critical review, we have adopted a distinction here whereby:

- microcosms are small-scale relatively simple experimental designs that do not attempt to simulate soil ecosystems;
- mesocosms attempt to simulate natural soil conditions, and they incorporate some indirect effects but attempt to control others. It is possible to conduct a mesocosm study in the field;
- field trials are larger-scale open-system experiments.

In this chapter we consider how well experimental designs address the various goals of a treatability evaluation and, most importantly, whether the results and interpretations can be related back to bioremediation in soils at low temperatures or that undergo seasonal freezing. The primary limitations to modeling the real world in treatability studies stems from necessary simplification of the interactions that occur in complex real systems, and the imposition of artificial treatment conditions. For example, the difference in oxygen diffusion and hydrocarbon volatilization in small volumes of soil in mesocosms will be significantly different than a larger-scale trial. Similarly, field changes in soil water content or temperature can have a profound effect on bioremediation rate. Most mesocosm studies use static temperature and soil water conditions and few, if any, consider oxygen diffusion at all.

An additional limitation that potentially affects all treatability studies concerns extrapolating data beyond the period when observed relations actually hold true. In cold regions, bioremediation by any low-cost scheme is likely to take several years, whereas most treatability evaluations often run for three months or less.

7.2 Review

A summary of the advantages and disadvantages of the most common treatability studies is provided in Table 7.1, and a review of the key experiments and findings is given in Table 7.2.

7.2.1 Microcosms

Simple *in vitro* microcosms using liquid or slurry systems that are incubated can be used to investigate processes, rates, or simple interactions. The main advantage of undertaking relatively simple microcosm experiments is that process-oriented hypotheses can be rigorously tested cost effectively. Individual environmental manipulations can be isolated from secondary effects and causal relationships can be explored with confidence. Simple system approaches have been used to demonstrate the presence of hydrocarbon-degrading bacteria in a particular soil or region (Margesin and Schinner 1998; Powell *et al.* 2004), to measure the effect of low temperatures on microbial activity (Ferguson *et al.* 2003b; Ferguson *et al.* 2006; Margesin *et al.* 2003; Margesin and Schinner 1998), and to elucidate particular hydrocarbon degradation pathways (Whyte *et al.* 1998).

The main limitation of *in vitro* liquid or slurry systems is that few petroleum-contaminated sites can be modeled this simply – soils are complex ecosystems comprising particulates, soil organic matter, water, gas, and micro- and macrofauna and flora. For example, in a biotreatability study undertaken by Wilson *et al.* (2003) on a contaminated silty soil, lower than expected degradation rates were found in a slurry experiment that was thought to be optimized for the fastest possible degradation. In this case oxygen diffusion and volatilization were recognized as important processes that were limited in the slurry, and faster rates were observed in an open soil-pan experiment. In another simple *in vitro* experiment, Margesin and Schinner (1998) determined the diesel-degrading potential of 29 enrichment cultures from different alpine habitats. The conclusion from the study was that efficient hydrocarbon degraders could be enriched from both pristine and contaminated sites in liquid cultures. Extrapolating such findings as scientific evidence to support bioaugmentation, however, would have been erroneous. As Margesin and Schinner (1997b) demonstrated in their other studies where inocula were added to soil in an effort to improve biodegradation efficacy, the reality of biodegradation in soil systems is much more complex. They demonstrated that diesel oil degradation by a cold-adapted inoculum in an experimentally contaminated alpine soil decreased with increasing temperature and incubation time, whereas degradation activity by the indigenous soil microorganisms increased. Comparable results were obtained in a soil with aged contamination (Margesin and Schinner 1997c), and their studies show the inefficiency of bioaugmentation in alpine soils. Several other studies have also demonstrated the complex interactions and ecological dynamics that can occur during progressive stages of hydrocarbon degradation. Individual species or strains are often not capable of degrading the full range of compounds in petroleum products. In many cases, indigenous flora and fauna appear to be best adapted to petroleum degradation at that particular site, and bioaugmentation has proven to be without effect or counterproductive over longer time periods in many instances (Dott *et al.* 1989; Margesin and Schinner 1997c; Moller *et al.* 1997; Venosa *et al.* 1992).

7.2.2 Mesocosms

Mesocosm studies that use soils and their autochthonous microbial consortia from the contaminated site more accurately reflect the substrate to be treated than do simple liquid or slurry microcosms. Four main types of systems are commonly used to measure biodegradation progress in mesocosms (see Table 7.1). The simplest systems follow the concentrations of petroleum hydrocarbons through time by repeated sacrificial sampling. Hydrocarbon losses measured in this way are a reliable and relatively unambiguous way of documenting the most important aspect of petroleum bioremediation, which is a reduction in contaminant concentration. The other three experimental set-ups use a tracer or surrogate, such as bacterial abundance, gas evolution, O_2/CO_2 respiration (e.g. Ferguson *et al.* 2003a; Ferguson *et al.* 2003b; Walworth *et al.* 1997a; Walworth *et al.* 1997b), RNA/DNA changes or increases in intermediate degradation products (e.g. Eriksson *et al.* 2001; Eriksson *et al.* 2003; Whyte *et al.* 1998), as estimates of biodegradation progress. It is perhaps surprising that in some of the more advanced experimental designs, little attempt is made to relate observed tracers or surrogates back to hydrocarbon concentrations and rates of breakdown (e.g. Stallwood *et al.* 2005).

It can be difficult to isolate individual cause and effect or treatment-response relationships in such complex systems. For example, petroleum-degrading soil consortia will respond to external pressures or treatments dynamically as part of an ecosystem. There may be shifts in competitive advantage of different groups of organisms as conditions change, synergistic or antagonistic relationships between variables and processes, second-order or more complex interactions, or stochastic biological feedback mechanisms. The question of how closely such mesocosms are analogs of site conditions is also important. In particular, most polar, alpine, and some sub-polar low altitude soils naturally experience a periodically freezing temperature regime, and spatially and temporally variable frozen water content. Contaminant distribution and partitioning, nutrient bioavailability, and aeration or dissolved oxygen levels could in turn be strongly influenced by seasonal conditions.

7.2.2.1 Sacrificial flask experiments

The simplest experimental mesocosms use a repeated sacrificial sampling strategy to obtain rate data. Sacrificial flask experiments are perhaps most commonly employed to follow the breakdown of petroleum hydrocarbons by repeated analysis of soil chemistry, sometimes augmented by other parameters, such as microbial enumeration (e.g. Ferguson *et al.* 2006). Slightly more sophisticated approaches use chemical indices to unequivocally attribute decreases in hydrocarbon concentration to biological, rather than abiotic, processes. This involves a comparison of variably bioavailable compounds to relatively recalcitrant compounds (Snape *et al.* 2005). Alternatively, abiotic flasks can be established by treating soil with Na azide, $HgCl_2$, gamma radiation, or by autoclaving (see Section 7.3.1 this chapter). The primary weakness is the destructive nature of the sacrificial approach. This leads to a need for suitable replication, combined with repeated measurements over meaningful time frames, which tends to be labor intensive and costly.

Although some studies have undertaken repeated sampling of soil in each mesocosm flask, the mesocosm becomes open to air. Evaporation of volatiles in this situation is inevitable, and there is a potential for contamination by allochthonous bacteria. Where biodegradation with indigenous consortia is being studied, cross contamination is a potential concern that is difficult to overcome when sampling the soil in the bottom of the flask.

Providing sufficient replication for a time-integrated study or development of a predictive model using sacrificial flasks becomes a significant undertaking.

Case study 1

An example of one of the limitations inherent in mesocosm studies is illustrated by the sacrificial flask study of oxygen release kinetics at low temperatures undertaken by Schmidtke *et al.* (1999). As part of a remedial action plan for a contaminated gravel pad, a series of simple gas displacement experiments were undertaken to quantify oxygen release rates from magnesium peroxide at low temperatures under a range of conditions (abiotic, with and without catalase, no soil) to determine if this method is appropriate for supplying oxygen. They predicted that release rates were two to three times less at 7 °C than 21 °C, but that sustained oxygen release was possible using their oxygen release compound. However, the short-term release of oxygen observed in the laboratory did not continue for as long as predicted in the field. Other factors such as oxygen loss to the atmosphere were difficult to account for and, most importantly, oxygen release was significantly reduced by a crust that developed on the magnesium peroxide under field conditions. This was not something that could have been easily predicted before or during the treatability evaluation.

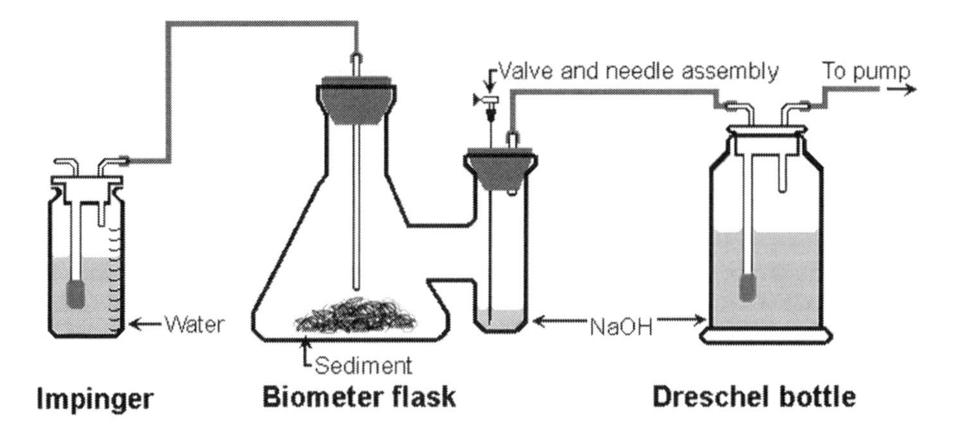

Figure 7.1. Schematic respiration system typical of the type that uses ^{14}C-labeled compounds as a surrogate tracer. Reprinted from Ferguson *et al.* (2003a), with permission from Elsevier.

7.2.2.2 *Monitoring CO_2 evolution with and without radiolabeled ^{14}C mineralization*

These systems use respired CO_2 produced during aerobic biological mineralization of petroleum hydrocarbons. During biodegradation, some portion of contaminant carbon is respired as CO_2 (see Chapter 4, Section 4.3.1.2 for additional discussion). The rest is assimilated into biomass and stable organic products. The fraction of carbon not directly converted to CO_2 varies depending on microbial species, contaminant properties, contaminant concentration, and temperature, and may range from a few percent to more than 90% (Alexander 1999). Graham *et al.* (1999) also reported that the fraction of carbon incorporated into soil and biomass varied by more than a factor of two, depending on the availability of nitrogen and phosphorus.

Notwithstanding carbon-cycling in the soil, the remaining carbon respired as CO_2 can be measured by placing a strong base trap into the mesocosm, usually as a separate sidearm in a biometer arrangement (Figure 7.1). A simple measure of CO_2 production, and hence respiration, can be made by measuring the dissolved CO_2 in the base solution. Some studies spike the soil with an isotopically labeled hydrocarbon, such as ^{14}C-labeled hexadecane (see Table 7.1). The evolved ^{14}C can then be quantified via scintillation counter. Degradation of ^{14}C-labeled hydrocarbon measured in this way precisely quantifies the biological mineralization *of the tracer*. The advantage of the system is that a surrogate radiolabeled hydrocarbon is monitored through time as biodegradation proceeds and background respiration of organic soil constituents does not contribute to the measured respiration. A disadvantage is that measured loss is restricted to that of a surrogate tracer, or more exactly, one carbon of a surrogate tracer, rather than the hydrocarbon contaminant, and the bioavailability, and hence mineralization of the tracer may be different than that of petroleum fractions in the soil. This leads to the important question of how closely mineralization of the tracer relates to the biodegradation of the full suite of hydrocarbon contaminants in the soil. Petroleum is a complex mixture of compounds that invariably have quite different bioavailabilities, toxicities, and resistance to biodegradation. If a surrogate radiogenic tracer is used, it is important that it degrades in a similar fashion to the contaminant of concern.

Case study 2

Treatability studies undertaken at Cape Dyer in Nunavut (Canada) by Reimer and co-workers (Reimer *et al.* 2003) found that nutrient amendment greatly stimulated ^{14}C-dodecane mineralization, but there was no correlation with TPH removal. In this study, mesocosms with fresh and weathered contaminants showed similar TPH removal over 54 days, but ^{14}C-dodecane mineralization was much faster in aged and weathered soils that were (presumably) partly already degraded, with a microbial consortia already selected for hydrocarbon degradation. Reimer *et al.* (2003) concluded that ^{14}C-dodecane did not allow estimates of degradation kinetics for weathered hydrocarbon spills. Alternative tracers or even a suite of tracers could potentially provide such information. However, for some classes of compounds commonly found in fuel spills, such as isoprenoids or aromatics, the cost of even small amounts of labeled tracers (50 μCi) is prohibitively expensive.

The amount of material used in a typical mesocosm can also have a major influence on the interpretation of results. Most economically competitive bioremediation operations usually involve treating tens to thousands of tons of soil, and landfarms (Chapter 9), biopiles (Chapter 10), or *in situ* treatment regimes are usually a half to several meters deep. By comparison, mesocosms typically only use 10–100 g soil. An important consequence of this is that oxygen diffusion, one of the major limitations to field-scale biodegradation, is not generally a limitation in a mesocosm.

However, to maintain an aerobic soil, the headspace is usually periodically flushed with fresh air to replenish the oxygen. This means that volatilization in a mesocosm is much more than would be predicted for deeper soil systems and the abiotic transfer of volatiles from the mesocosm soil to the headspace is very important. Although the evolved ^{14}C method is (mostly) not directly affected by abiotic losses when a relatively heavy tracer is chosen, Ferguson *et al.* (2003a; 2003b) found that volatilization was the dominant process operating in their mesocosms, and abiotic loss of the petroleum hydrocarbons was possibly greater than the changes produced by the biological processes that were being investigated. The biotic-abiotic question is a function of the contaminant and its volatility, and treatability estimates from microcosms may overestimate field losses of lighter more toxic components of the fuel. Volatilization can also be an important factor that influences toxicity, bioavailability, and other related parameters.

Case study 3

By using cultures of *Rhodococus sp.* strain Q15 in diesel-amended liquid microcosms, Whyte *et al.* (1998) were able to demonstrate that the psychrotrophic Q15 strain was very effective at degrading a range of hydrocarbons at low temperatures. To test the hypothesis that bioaugmentation with Q15 strain was beneficial in the degradation of fuel in a soil substrate with bacterial consortia, they used soil mesocosms with uncontaminated soil and soil contaminated with 5000 mg crude oil kg^{-1} soil. Seeding the soil with Q15 strain significantly reduced the lag time in the mineralization of ^{14}C-labeled hexadecane. They cautiously concluded that such a reduction in lag time, through bioaugmentation, could be crucial for cold regions where short summer seasons do not permit long acclimatization periods.

However, other interpretations of the mesocosm data are possible. The unequivocal mineralization of the ^{14}C-labeled hexadecane indicates little about the degradation of the bulk of the crude oil. The short lag phase in the Q15-augmented uncontaminated treatment is offset to some extent by the early onset of a reduced degradation rate or plateau phase. In the unamended uncontaminated soil, there is a longer lag period but the rate of degradation in the plateau phase is still faster than the Q15 treatment at the end of the experiment. It is possible that after a longer period than their 23 day study, the natural consortia might have been more effective at removing the tracer. Similarly it would have been interesting to compare the ^{14}C-labeled tracer with analysis of TPH and chemical indices. The results presented by Whyte *et al.* (1998) reflect a thorough experimental design. They do, however, also illustrate some of the uncertainties in extrapolating to a remedial scheme. Their observations of diesel degradation by Q15 in a simple system are convincing, but the extrapolation to complex systems and field applications are less certain, and other bioaugmentation studies have reached different conclusions.

7.2.2.3 *Monitoring CO_2 evolution through closed-system O_2 delivery*

Automated monitoring of CO_2 evolution is possible by using a closed system reactor coupled to a sensitive pressure transducer that replaces small amounts of O_2 to maintain constant pressure as gaseous CO_2 is partitioned into a strong liquid base. A data-logging system, such as the N-CON COMPUT-OX, detects this drop in pressure and delivers O_2 to the reactor in measured increments. The system records the volume of O_2 delivered to allow time-integrated monitoring of respiration (Figure 7.2).

The main advantage of this system over other respirometers is that aerobic headspace can be maintained without opening the mesocosm to the atmosphere. This prevents evaporative loss of volatile hydrocarbons because the equilibrium partitioning of volatile hydrocarbons into a relatively small headspace is negligible. The closed system also avoids the possibility of cross contamination. Additionally, moisture is not lost, so soil moisture can be closely regulated. These strengths coupled with the ability to take continual automated measurements for the duration of an experiment make the closed-system respirometer an attractive proposition.

The main disadvantage is that respiration measurements do not differentiate between natural soil respiration and respiration directly involving petroleum contaminant degradation. Controls can be used to estimate the background respiration. However this does require uncontaminated soil with the same properties as the contaminated substrate. This necessity introduces quantitative uncertainty into the estimation of the respiration that is directly involved with contaminant degradation. Another potential problem concerns reactions other than O_2-CO_2 respiration that could cause pressure increases or decreases. For example, nitrification or denitrification reactions could potentially trigger or suppress O_2 delivery by changing the partial pressure of nitrogen gas species in addition to consuming or liberating O_2 directly. One final difficulty concerns the ability to operate the system in a cyclic freeze-thaw regime. Because gas pressure is temperature related for a fixed volume, comparable temperatures are required in both mesocosm and reference chamber to be able to detect the pressure drop associated with CO_2 evolution and dissolution in the base trap (Figure 7.2).

7.2.2.4 *Monitoring respiration with flow-through system gas analysis*

Biodegradation progress can be monitored by measuring headspace gas concentrations in a flow-through (open) system. One popular system is the Columbus Micro-Oxymax respirometer which is capable of measuring oxygen, carbon dioxide, methane, carbon monoxide, hydrogen sulphide, and hydrogen in the headspace of the reactors (adapted Schott bottles). The system monitors composition of exhausted headspace gas via infrared absorbance, electrochemically,

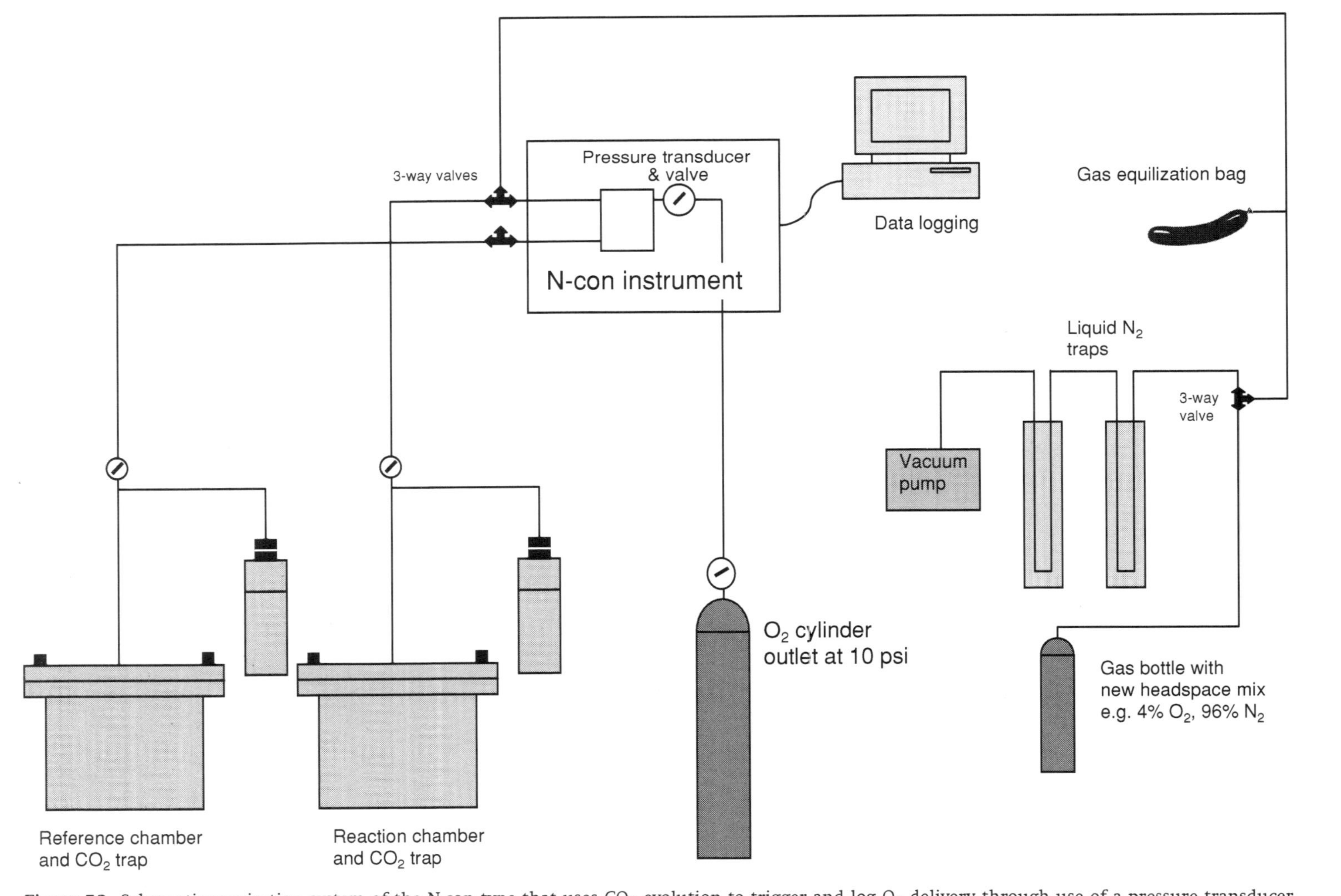

Figure 7.2. Schematic respiration system of the N-con type that uses CO_2 evolution to trigger and log O_2 delivery through use of a pressure transducer. The system depicted is designed for use at a range of temperatures, with variable gas mixtures, or fluctuating freeze-thaw conditions.

or with other suitable methods. While the Micro-Oxymax is a closed circuit, the mesocosm arrangement is an open system in terms of gas exchange in the headspace. The primary advantage of the system is that it offers simultaneous analysis of several gases, which can be important in understanding the importance of reactions such as nitrification. However, the system is relatively complex and maintaining a gas-tight seal is known to be a challenge. There is also potential for substantial volatilization of petroleum hydrocarbons and water, although these losses can be minimized by condensing volatiles in the exhaust stream and returning them to the reaction vessel.

7.2.3 Field trials

Properly designed field trials should largely avoid the scale-up issues associated with microcosm and mesocosm approaches. Field trials overcome the problem of identifying and incorporating the many environmental variables that could potentially affect biodegradation. Natural variation or heterogeneities are often large in contaminated soils; therefore field trials need appropriate replication to be able to confidently distinguish between treatments or accurately model biodegradation rates. Scientific information gained from field trials is often at considerable financial cost because of the scale of pilot trials and the logistic cost of undertaking bulk earth works is expensive, especially in remote high latitudes. Microcosm and mesocosm studies, by contrast, involve a relatively small number of treatment variables and variable combinations can be more readily studied than in field trials. Because many environmental variables influence field trials, it is also difficult to isolate key processes or rate limitations. The open-system nature of field trials also needs careful consideration when designing experiments to evaluate treatment options, and especially where determination of processes and rates are required. Early bioremediation field trials in cold regions involved artificially contaminating plots with hydrocarbons, and then applying various treatments (e.g. Delille *et al.* 2004a; Delille *et al.* 2004b; Delille *et al.* 2002). This is a reasonable analogy for spills where remediation response is prompt and identifying the degradation mechanism is not a goal of the evaluation. However, many spills currently scheduled for remedial action in cold regions are old and weathered (e.g. Reimer *et al.* 2003; Snape *et al.* 2006a; Snape *et al.* 2001), and measured treatment rates might be significantly slower than those measured in a field trial involving an artificial spill. Such trials also often fail to achieve mass balance within the plot in terms of attributing losses to a specific process. Biodegradation is implicated by changes in microbial abundance, species composition, or gene expression, but quantities of hydrocarbons lost through subsurface runoff or evaporation are often neglected or difficult to quantify.

Case studies 4 and 5

The first Antarctic bioremediation field trial was conducted in 1990 by Kerry (1990), who investigated the effect of nutrient and water addition on biodegradation rates of an artificial fuel spill at Davis Station. Kerry found that (1) significant hydrocarbon losses occurred in surface soils in one year, (2) an alternative carbon source inhibited degradation, and (3) significant evaporative losses occurred in the surface soils.

This was a landmark study for Antarctic remediation research because it demonstrated for the first time that biodegradation was possible at low ambient temperatures using indigenous microorganisms. The experimental approach, however, also opened up many scientific questions. There was considerable variability in TPH measurements, and because the plots were open, infiltration of hydrocarbons and an associated decrease in TPH in the soil surface could not be reconciled. Long-term rates could not be predicted from the trial, and it was not possible to apportion losses to either abiotic or biotic processes (cf. Snape *et al.* 2006a). Anaerobic biodegradation did not appear important from the results (cf. Powell *et al.* 2006b), extrapolation to degradation of older, partly weathered spills was not possible, and the issue of residual toxicity and determination of a suitable endpoint were not investigated. This important study clearly posed many questions.

Some of these issues were investigated by Delille and co-workers through a number of Antarctic and sub-Antarctic field trials. In their 1996–9 study of crude oil degradation on sub-Antarctic Anse Sabloneuse (Kerguelen Archipelago), Delille *et al.* (2002) found that the aliphatic fraction readily degraded, but that the soil toxicity remained high. They attributed this to a toxic residue, probably containing PAH; noting also that there is often a poor correlation between measured microbial parameters and indices of biodegradation or hydrocarbon loss because of confounding environmental factors such as photooxidation and alternative carbon sources. Delille *et al.* (2002) concluded that environmental factors strongly influenced the ability of microbial populations to degrade PAH and aliphatic compounds, and that the conclusions from their study could not necessarily be extrapolated to other sub-Antarctic beaches.

7.3 Recent advances

A number of recent technological advances and scientific approaches have been applied to bioremediation treatability studies for cold regions.

7.3.1 Distinguishing between abiotic losses and biodegradation

The importance of distinguishing between abiotic and biodegradation losses is important when attempting to optimize hydrocarbon degradation schemes that aim to use *ex situ* techniques. Simple attempts to determine abiotic losses have been undertaken by comparisons with chemically inhibited controls (e.g. Ferguson *et al.* 2003b). The limitation of this approach is that between-sample heterogeneity in hydrocarbon concentrations in soil is often quite large, making quantification of evaporation or biodegradation losses difficult. More recent approaches have used a number of biomarkers to follow biodegradation progress. For example, studies have documented fractionation of highly degradable *n*-alkanes relative to conservative biomarkers such as hopanes (e.g. Prince *et al.* 2002). Similarly, Snape *et al.* (2005) related changes in a range of indices that are sensitive to evaporation (e.g. i-C_{13}/pristane) and biodegradation (e.g. n-C_{10}/R+UCM) to the proportion of mass lost through evaporation and biodegradation in fuels adapted for use in cold regions that do not contain hopanes.

Efforts to compare biotic and abiotic systems by sterilizing soil treatments are also fraught with difficulty. It is notoriously difficult to kill all the microbes in a soil system without volatilizing fuel. To avoid high temperature treatment, chemically inhibited controls are often employed. However, these methods generally do not render the soil abiotic and are often more of a way to attenuate the biotic component, as microorganisms generally re-establish in the soil. A recent method that could become more widely adopted involves gamma irradiation, which sterilizes soil whilst avoiding excessive heating and the associated risks of volatilization or changes in soil physico-chemical properties (McNamara *et al.* 2003). Using this method, Ferguson and co-workers (unpublished) found discoloration of their experimental equipment during irradiation that they attributed to enhanced oxidation caused by the irradiation procedure.

7.3.2 Accommodating the effects of repeated freeze-thaw

Despite the distinction from better studied warm soil systems that do not freeze each year, few cold-region treatability studies have attempted to incorporate repeated freeze-thaw as a variable. Several recent studies have explored what effect freeze-thaw might have on biodegradation, although a clear pattern of cause and effect is yet to emerge (see Chapter 3, Section 3.3 and Chapter 5, Sections 5.2.2 and 5.3.2 for related discussions).

Eschenbach *et al.* (1998) studied the effects that periodic freezing had on ^{14}C-labeled PAH mineralization in a mesocosm study, and found that the process had essentially no effect on biodegradation. The unpublished PhD study by Leszkiewicz (2001), however, reported an inhibitory effect. He used a sacrificial mesocosm system to simulate the effect that repeated freeze-thaw cycles had on biodegradation of petroleum contaminated soils from McMurdo Station, Antarctica. In contrast, Eriksson *et al.* (2001) found that repeated freeze-thaw cycling stimulated microbial activity in mesocosms using contaminated soil from Ellesmere Is. (Canadian Arctic). They used an open-system approach that involved repeated sampling of the same microcosm flasks at various times, and used a multi-parametric monitoring approach. Reynolds (2004a) demonstrated that diurnal temperature cycles of 20 °C had little effect, relative to temperature held constant at the daily mean, as long as the temperatures were above freezing. When the diurnal cycle included a freeze-thaw event, the soils undergoing freeze-thaw cycling had greater respiration.

7.3.3 Multi-parametric monitoring

Technological developments in analytical instrumentation for quantifying chemical and microbiological parameters have increased significantly in the last five years. It is now possible to integrate sophisticated approaches that measure and relate contaminant changes in the soil substrate to microbial processes and degradation pathways. Such approaches offer the potential to take identification of rate limitations to a new level for improved biodegradation efficiency. For example, identifying bacteria and specific genes that are associated with degradation of a range of fuel and oil fractions is the first step towards developing environmental conditions optimal for their growth. The study by Whyte *et al.* (1998) is perhaps the most comprehensive treatability evaluation for a cold regions soil/isolate. They used both liquid microcosms and soil mesocosms, and monitored a wide range of parameters including mineralization, using four isotopically labelled compounds, CO_2 and TPH, and catabolic pathways using solid phase micro-extraction gas chromatography – mass spectrometry and polymerase chain reaction genetic approaches. This study was published in 1998, but the monitoring techniques used were more comprehensive than many later treatability evaluations.

7.3.4 Modeling and predictive studies

Modeling biodegradation data such as oxygen consumption, carbon dioxide production, or isotope recovery from treatability studies is often attempted to assist in the quantification of rates or processes, to compare treatments, and facilitate extrapolation of results. Fitting empirical kinetic models to data is usually undertaken to assist *a posteriori* analysis rather than to describe from first principles the type of reactions involved within microcosms. Empirical models provide a means to relate various microbial and physical variables to hydrocarbon degradation rates, which in turn can sometimes be used to predict likely rates or times to closure at a field site.

The simplest mathematical models describe treatability results through zero- and first-order rate equations; the first-order rate equation is shown below (Eq. 7.1). This mineralization rate equation has been used to model contaminants such as benzene (Franzmann *et al.* 1999), and diesel and crude oil (Aichberger *et al.* 2005). Unfortunately, it has been noted that first-order kinetics are only valid when the contaminant concentrations are low (Bekins *et al.* 1998).

$$k_1 = \ln\left(\frac{C_o - C_t}{C_o}\right) \tag{7.1}$$

where k_1 is the mineralization rate, C_o is the initial concentration of a reactant added to the mesocosms (such as a radiolabeled hydrocarbon) and C_t is the concentration of a product produced (e.g. $^{14}CO_2$) at time t.

When microbial growth is non-linear, which is typically the case in treatability studies, the Monod equation (Eq. (7.2)) sometimes provides a better description of the mineralization data as it incorporates limiting growth conditions. The Monod equation was used by Bury and Miller (1993) to describe the effect of surfactants on solubilization of two *n*-alkanes and the effect this had on microbial growth in a single substrate system. Reardon *et al.* (2000) found that the Monod equation was able to describe the growth of *P. putuda* F1 on toluene, phenol, and benzene. However, for growth on multiple substrates, Reardon *et al.* (2000) found that adaptations to the equation involving a sum of kinetic terms and interaction parameters was needed.

$$\mu = \frac{\mu_{\max} S}{(K_S + S)} \tag{7.2}$$

where μ is the specific growth rate, $\mu_{\max}$ is the maximum specific growth, S is the substrate concentration and K_S is the saturation constant for the substrate.

Other attempts to model non-linear mineralization have used the Arrhenius equation, which was originally developed to describe the effect of temperature on the rates of chemical reactions (Eq. (7.3)). The Arrhenius equation assumes that temperature affects biological behavior and chemical reactions in the same manner. Chemical reactions and biological mineralization (up to a point) increase with increasing temperature (also discussed in Chapters 5 and 11). Baker (1974) successfully described the temperature characteristics for five psychrophilic bacteria isolated from Signy Island (Antarctica) by modeling their growth rates between ~5 and 30 °C with the Arrhenius equation (Eq. 7.3):

$$\ln k_1 = -\frac{E_a}{RT} + \ln A \tag{7.3}$$

where k_1 is the first-order rate coefficient (from Eq. 7.1), E_a is the activation energy and can be considered as a measure of the effect of temperature on biological activity in mesocosms, R is the universal gas constant, T is temperature (K), and A is the frequency factor which can be assumed constant over the small temperatures used in treatability studies.

However, it has been noted that this sort of relationship does not hold well for microbial communities that respond dynamically to environmental changes such as temperature and nutrient regime (e.g. Ferguson *et al.* 2003a; Ferguson *et al.* 2006).

In a development of the Arrhenius equation, Ratkowsky (1983; 1982) also considered the importance of temperature, but whereas the Arrhenius model assumes that the activation energy is constant at different temperatures, the Ratkowsky model recognized that activation energy decreases as temperature increases (Eq. (7.4)). The model has been successfully used on mixed culture experiments, for example a study by Pelletire *et al.* (1999) that examined the effect of temperature on denitrification in a Quebec cropped soil.

$$\sqrt{r} = b(T - T_{min})(1 - e^{c(T - T_{max})}) \tag{7.4}$$

where r is the growth rate or the reciprocal of time required to achieve a level of growth (or a surrogate measure of growth), b is the regression coefficient (described by the slope of the regression line below the optimal growth temperature), c is a fitting parameter for data above the optimal growth temperature, T is temperature (K), and T_{min} and T_{max} are the minimum and maximum growth temperatures, respectively (conceptual temperatures of no metabolic significance but at which the equation predicts that the generation time is infinite).

Mohn and Stewart (2000) used a similar logarithmic equation to model their mineralization data from Arctic Canada (Eq. (7.5)). However, in the Mohn and Stewart (2000) and Pelletire *et al.* (1999) studies, mineralization rates did not exhibit plateau or long lag phases, whereas many treatability evaluations that involve a significant shift in microbial community have pronounced lags and a plateau phase of degradation.

$$Y = \frac{Y_{max}(-Y_{max} - Y_{init})}{(1 + e^{k(t - t_{mid})})} \tag{7.5}$$

where Y is the extent of mineralization at time t, Y_{init} and Y_{max} are the initial and maximal extents of mineralization, t_{mid} is the time of half-maximal mineralization, and k is a rate constant.

Brunner and Focht (1984) addressed the common features of mineralization in their three-half-order model. The two forms of the model, involving either linear or exponential microbial growth rates, have been used to describe the mineralization kinetics of a variety of chemicals, such as di (2-ethylthexyl) phthalate and *n*-dodecylbenzene sulfonate (Dörfler *et al.* 1996), pyrimidine 2-c-14 rimsulfuron (Metzger *et al.* 1999), mecoprop and isoproturon (Helweg *et al.* 1998), atrazine (Grigg *et al.* 1997), and octadecane (Ferguson *et al.* 2003a; Ferguson *et al.* 2003b). Non-logical negative values of the fitting variable describing the initial lag phase of mineralization resulted when the Brunner and Focht (1984) model was applied to mineralization studies using octadecane (Ferguson *et al.* 2003b), phenol, aniline, and nitrilotriacetic acid (Scow *et al.* 1986). Trefry and Franzmann (2003) addressed this issue by extending the Brunner and Focht three-half-order model to include terms that further characterize the initial lag phase. They found that lagged production curves were best described by a delayed onset of the mineralization process, whilst the biomass growth phase conformed well to the Brunner and Focht (1984) exponential model. The linear and exponential forms of the Brunner and Focht (1984) model with Trefry and Franzmann (2003) extensions are shown in Equations (7.6) and (7.7) respectively.

$$P(t) = S_o\left(1 - e^{-k_1 t - \left(\frac{k_2 t^2}{2}\right)}\right) + k_0\left(t - \frac{\sqrt{\pi}}{2} v\ \mathrm{erf}(t/v)\right) \tag{7.6}$$

$$P(t) = S_o\left(1 - e^{-k_1 t - \frac{E_0}{\mu}\left(e^{\mu t} - 1\right)}\right) + k_0\left(t - \frac{\sqrt{\pi}}{2} v\ \mathrm{erf}(t/v)\right) \tag{7.7}$$

where P is the concentration of the product, S_o is the initial substrate concentration, k_1 is the first-order rate coefficient, t is time, k_2 is a constant, k_o is the zero-order rate coefficient, E_o is the initial biomass growth rate, μ is the non-linear exponent of growth, and erf is an error function.

Although the three-half-order models are powerful in that they can accurately describe treatability evaluations that exhibit lag, exponential growth, and plateau phases of mineralization, a major disadvantage is their inherent complexity. The equations are difficult to fit to data and the various parameters do not relate in an intuitive way to the properties that can be measured or manipulated in the mesocosms. Also, they can not be used in mineralization studies when all three phases of mineralization do not occur. For example, Ferguson *et al.* (2003a) found that low-temperature water-nutrient interactions did not reach a plateau phase of mineralization in their 90-day treatability evaluation of Antarctic soil from Casey Station.

In conclusion, there is no one unified mineralization model that can be used to mathematically describe the various responses that are possible for treatability studies that range from simple to complex analog systems, such as closed-system monoculture microcosms and open-system mesocosms. This review of models is not exhaustive and with many possible models to choose from it is often difficult to compare between different treatments, rates, and studies. Nevertheless, future initiatives aimed at comparing treatability evaluations, or relating laboratory experiments to field trials, will benefit greatly from a more quantitative approach.

7.4 Guidelines and recommendations

7.4.1 Choose a suitable treatability technique

Treatability studies should be a precursor to any large-scale remediation scheme. Unfortunately there is currently no definitive treatability system that meets all requirements. Table 7.1 summarizes the relative merits and disadvantages of different systems. Attempts to reproduce the effects of freeze-thaw and accurately monitor petroleum hydrocarbon losses over time are currently limited to simple sacrificial approaches. These have the advantage of simplicity and few uncertainties concerning flow rate, leakage, and sorption of the ^{14}C-labeled materials (Semple *et al.* 2003). The disadvantage is that they are prone to volatilization as the mesocosms are re-oxygenated, and they are expensive and time consuming where an accurate measure of biodegradation is needed over time. Where detailed time-integrated studies are needed, surrogate systems that monitor respiration through CO_2 evolution are better. They also more readily provide data that can be used to develop predictive testable models. However, the difficulties inherent in accurately reproducing the many environmental variables that could affect biodegradation mean that on-site pilot trials are still the most reliable method of predicting the efficacy and likely cost of a large remediation project.

7.4.2 Monitor appropriate variables

When used in conjunction, chromatographic and radiometric tracers or systems that utilize gas-flux approaches clearly do elucidate those parameters that influence the rate of petroleum degradation at low temperatures under a variety of treatment regimes. The most important chemical parameter to measure, however, is the concentration of the contaminant in the soil. Even if this can only be done in the pre-treatment and post-incubation times, this measure is critical for a full and meaningful evaluation of treatability test variables. Chemical indices of biodegradation (e.g. Ferguson *et al.* 2003a; Ferguson *et al.* 2003b; Stallwood *et al.* 2005), evaporation (Snape *et al.* 2006a; Snape *et al.* 2005), and dissolution (Arey *et al.* 2005) are useful, but need relating back to concentration losses of contaminant (usually TPH). Advanced chemical methods for apportioning mass transfer to different processes have recently been demonstrated; it is likely that they will be of considerable benefit to cold regions treatability evaluations in the near future. For example, Wang *et al.* (1998; 2003; 2005) outlined an approach to differentiating abiotic and biotic losses using a number of different indices. Similarly, Arey *et al.* (2005) illustrate how two-dimensional gas chromatography can be used to estimate environmental partitioning properties for a complete set of diesel fuel hydrocarbons (see Chapter 6, Sections 6.2 and 6.3 for additional discussions).

Where the purpose of the treatability evaluation is identification of microbial processes and pathways, enumeration of viable consortia or identification of specific genes may be the most important information (see Chapter 4). Unfortunately, the analysis of microbial community structure and density in treatability studies is challenging; there are serious limitations in the techniques currently used to describe the shifts in microbial community (Head *et al.* 1998; Powell *et al.* 2003). Culture-based methods, such as most probable number (MPN) estimations or the Biolog system only detect between 0.1–10% of the total microbial community (Alexander 1999; Garland and Mills 1991). Furthermore, these methods tend to be highly variable; it is often difficult to detect significant differences between treatments (Ferguson *et al.* 2006) and there is generally poor correlation between hydrocarbon-degrading populations and changes in chemical methods used to track hydrocarbon degradation (Delille *et al.* 2002). Emerging molecular technologies, such as microarrays (Zhou 2003), functional gene expression (Watanabe and Hamamura 2003), and stable isotope (Manefield *et al.* 2002; Pelz *et al.* 2001; Pombo *et al.* 2002) techniques allude to the possibility of linking members of a microbial community to particular metabolic functions, and thus the isolation and characterization of the microbe(s) responsible for contaminant removal in treatability studies. Powell *et al.* (2006a) were able to demonstrate significant changes in the alkane-degrading population in a field trial by analyzing for the *AlkB* gene; it was not possible to detect these changes using MPN methods. As new techniques develop and existing ones continue to improve, the identification and characterization of the key microbial elements responsible for the bulk of hydrocarbon degradation may be possible with a single technique. Unfortunately we are currently reliant on a combination of culture-dependent, molecular, chemical, and physiological approaches.

7.4.3 Consider possible scale-up issues

A degree of caution is required when extrapolating the results from analog experiments to the conditions at a field site, or for predicting the outcomes of remediation actions. The issue of scale-up differences is particularly important. Leszkiewicz (2001) considered the issue of scale-up in his treatability assessment. He noted that the batch-reactor design possibly influenced variables such as mass transfer of O_2 and water, and that bioavailability might have been different in full-scale trials. It was also difficult to predict end-points from relatively short-lived experiments, with recalcitrant compounds possibly extending out treatment times considerably. The overall conclusion from the Leszkiewicz (2001) study was that a larger-scale treatability assessment was needed before developing a full-scale treatment plan. The difficulty of extrapolating to long time frames and for recalcitrant compounds was also of concern to Walworth *et al.* (2007). In their study of various nutrient amendments to increase respiration in biodegradation of sub-Antarctic Macquarie Island, there was considerable uncertainty how long it would take remediation operations to reach target concentrations of 200 mg fuel kg^{-1} soil. As a result, Walworth *et al.* (2007) faced considerable uncertainty in project planning and associated costs. Their final evaluation of treatment time and effectiveness will more likely be derived from a full-scale pilot microsparge remediation scheme, with design parameters only loosely guided by their treatability evaluation.

In summary, the main areas of concern regarding scale-up are:

- unrepresentative water potentials, nutrient concentrations, and bioavailabilities in slurry or liquid microcosms
- different oxygenation regimes in soil mesocosms compared with large-scale biopiles or landfarms
- the influence of freeze-thaw events
- extensive role of volatilization in shallow mesocosm soils
- suitability of surrogates as tracers for biomineralization of contaminants in mesocosms
- extrapolating short-term experimental data to long-term field applications
- adequate replication and suitable monitoring in pilot-scale field trials to draw scientifically robust conclusions.

7.5 Future research

More intricate methods and apparatus are needed to define contaminant degradation rates and pathways using a range of tracers. It is possible that integrating a number of tracer techniques will allow better elucidation of the various parameters and interactions that limit biodegradation in cold regions. It is also possible that that in turn will help to develop or underpin more usable predictive models. Building on this mesocosm approach, low-cost quantitative microbial ecology tools are needed to examine microbial dynamics, soil ecosystem function, and ultimately to identify rate limitations to those organisms that are specifically needed to degrade target contaminants. Modern molecular techniques such as denaturing gradient gel electrophoresis and real-time polymerase chain reaction have gone some way to addressing this issue, but considerable additional work remains to be done (Powell *et al.* 2006b).

Another exciting area of future research concerns developing systems or mesocosm techniques that will allow quantitative study of anaerobic degradation rates. Recent work by Powell *et al.* (2006b) demonstrated the importance of anaerobic processes in the degradation of nutrient-amended field trials undertaken in Antarctica. This observation, that anaerobic biodegradation with microbes utilizing nitrogen as an alternative electron acceptor to yield environmentally significant rates of degradation, offers considerable scope for low-cost nutrient-amended *in situ* treatment for some remote sites. Ironically, Eriksson *et al.* (2003) raise the possibility that such anaerobic processes could actually be faster at relatively low temperatures. This would necessitate a considerable change in our model paradigms concerning the rate limitations of temperature on mineralization reactions.

Table 7.1 *Summary of the different types of treatability evaluation with some of the features, advantages, and limitations of each system*

Experimental design	Experimental characteristic	Advantages	Disadvantages
Microcosms	Closed or open simple systems with few variables designed to examine first-order relationships (treatments and responses).	• Can provide well-constrained insight into specific processes, rates, pathways, or treatments without influence by second-order interactions.	• Poor analogs for soil or soil water systems. • Experiments involving nutrient amendments may have unrealistic water potentials.
Mesocosms – sacrificial flask experiments	Open system that uses soil samples with indigenous flora and fauna. Direct measurement of target variables.	• Most directly measure hydrocarbon changes over time.	• Usually prohibitively expensive to undertake replicated successive sampling over protected periods. • Not possible to maintain headspace O_2 for long periods without air flushing and associated volatilization. • Can be difficult to isolate abiotic and biotic effects in aerobic systems.
Mesocosms – ^{14}C-labeled surrogate hydrocarbon tracer. Biological mineralization of ^{14}C to trace hydrocarbon breakdown via CO_2 respiration.	Open system that uses soil samples with indigenous flora and fauna. Indirect measurement of variables.	• Continual measurements possible over long periods. • Precisely measures surrogate hydrocarbon mineralization. • Not labour intensive.	• OHS concerns regarding handling of radioactive material. • Not possible to maintain headspace O_2 for long periods without air flushing and associated volatilization.

Mesocosms – Gas analysis (Columbus type)	Open system that uses soil samples with indigenous flora and fauna. Indirect measurement of variables.	• Provides information on the total gas exchange. • Continual respiration measurements possible.	• Prone to high volatilization through flushing of the headspace.
Mesocosms – pressure change (CO_2) evolution (N-Con type) Uses CO_2 respiration and associated O_2 consumption to trace microbial activity by metered O_2 delivery.	Closed system that uses soil samples with indigenous flora and fauna. Indirect measurement of variables.	• Continual measurements possible over long periods. • Does not have problems with volatile losses. • Cost-effective and relatively simple experimental set up. • Can be used for anaerobic systems.	• Does not distinguish between natural soil respiration and respiration associated with contaminant mineralization. • Not readily amenable to modeling mineralization in a freeze-thaw regime. • Adversely affected by gas exchanges other than CO_2 respiration/O_2 consumption, such as nitrification.
Field Trials – Field experiments.	Open system that uses soils on site with all natural environmental parameters and associated interactions. Monitoring can include direct or measurement of variables.	• Most accurate way of evaluating potential full-scale treatment efficacy. • Most realistic accommodation of site-specific factors.	• Expensive and can be technically difficult. • Monitoring is expensive and time consuming • Difficult to isolate the primary rate-limitation at any given stage to optimize treatment. • Not often possible to identify or quantify the contribution of complex interacting variables.

Table 7.2 *Review of treatability evaluations for cold regions*

Reference Year	Design type	Region	Product(s)	Treatments	Monitoring Parameters	Findings, limitations, and uncertainties
(Delille *et al.* 2002)	Field trial Beach Landfarming (3 years)	Sub-Antarctic Kerguelen Is.	Arabian light crude oil	Vegetated soil Non-vegetated soil Nutrient (Inipol EAP-22) Nutrient (Fish compost)	Microbial enumeration Microtox TPH PAH Chemical indices	Slow release fertilizer stimulated biodegradation. Greatest benefit to desert soil. Soil heterogeneity noted as important. Treatments did not successfully degrade PAHs. High residual toxicity in soil. Caution when extrapolating to other soils/environmental conditions.
(Delille *et al.* 2004a; Delille *et al.* 2004b)	Field trial Beach Landfarming (1 year)	Sub-Antarctic Kerguelen Is.	Arabian light crude oil Diesel	Vegetated soil Non-vegetated soil Nutrient (Inipol EAP-22) Passive heating	Microbial enumeration Microtox TPH PAH Chemical indices	Fertilization had most beneficial effect on non-vegetated soil. Degradation of alkanes and light aromatics. High residual toxicity in soil. Passive heating using black plastic covers achieved a 2 °C temperature increase. The black plastic cover favored degradation of alkanes over aromatics.
(Whyte *et al.* 2001)	Mesocosm $^{14}CO_2$ (16 weeks)	Arctic Ellesmere Is. Eureka	Diesel	Sandy loam #1 Sandy loam #2 Clay loam 5 and 23 °C Nutrient (Inipol EAP-22) Nutrient (20:20:20) Nutrient + tilling Nutrient + peat moss Nutrient + water	Microbial enumeration PCR alkB, xylE, ndoB TPH ^{14}C-hexadecane ^{14}C-napthalene	Indigenous bacteria can degrade substantial amounts of hydrocarbons in nutrient amended soils at low temperatures. 20:20:20 nutrient supplements increased biodegradation rates more than Inipol EAP-22. Greater mineralization rates at 23 °C than 5 °C. In addition to nutrient amendment, tilling and water addition may be needed for some soils. Significant residual hydrocarbons after treatability evaluation.

(Walworth *et al.* 1997a)	Mesocosm CO_2-O_2	Arctic Alaska	Crude oil Diesel JP-5	Sand Sandy loam Silt loam Nutrient (var) Water (var) NaCl	TPH CO_2 production-O_2 delivery	Soil water potential and O_2 consumption were best related by concentration of N measured as mg N kg^{-1} soil H_2O. 2000 mg N kg^{-1} soil H_2O identified as an optimal nutrient addition. Coarse textured and/or dry soils are most likely to be sensitive to over fertilization.
(Walworth *et al.* 1997b)	Mesocosm CO_2-O_2	Arctic Alaska	Jet-A1	10 and 20 °C Nutrient (DAP) Nutrient (Bonemeal)	TPH CO_2 production-O_2 delivery	Fish bonemeal was a potentially useful source of nutrients for diesel bioremediation.
(Walworth *et al.* 1999)	Mesocosm CO_2-O_2	Arctic Alaska	Diesel	Soil water potential (var) NaCl Nutrient (var)	Microbial enumeration CO_2 production-O_2 delivery	Soil water potential affects microbial respiration in a negative way. Concentrations of N>~3000–5000 mg N kg^{-1} soil H_2O inhibit biodegradation.
(Mohn and Stewart 2000)	Mesocosm $^{14}CO_2$	Arctic Canada	Jet-A1	18 soils Temperature (var) Nutrient (var) Inoculation	Microbial enumeration TPH ^{14}C-dodecane	Heating can increase biodegradation rates. Inoculation increased biodegradation rates, but questionable cost–benefit. All soils were nutrient limited; type of nutrient influenced degradation rate, but was soil specific. Soil type affected biodegradation kinetics.
(Kerry 1993)	Field trial Landfarming (1 year)	Antarctic Davis St.	SAB light diesel	Nutrient Carbon	Microbial enumeration TPH Chemical indices	Nutrient addition increased degradation, but losses restricted to <3 cm. High residual hydrocarbons at depth. Alternative carbon source inhibited biodegradation. Difficult to quantify losses through infiltration and evaporation.

(cont.)

Table 7.2 (*cont.*)

Reference Year	Design type	Region	Product(s)	Treatments	Monitoring Parameters	Findings, limitations, and uncertainties
(Mohn *et al.* 2001)	Field trial Biopiles (1 year)	Arctic Canada Cambridge B	#1 Arctic diesel (~Jet-A1)	2 field sites Nutrient Nutrient, peat Nutrient, peat, inoculum Passive heating	TPH	Passive heating with clear plastic cover increased soil temperature and improved moisture retention. Nutrient additions greatly increased hydrocarbon degradation rates. Peat did not aid degradation, possibly by limiting O_2 diffusion.
(Pritchard and Costa 1991 & therein)	Field trial Landfarming Beach (var times)	Sub-Arctic Alaska	Prudhoe Bay crude oil	Nutrient (var)	Microbial enumeration TPH Chemical indices	Sustained nutrient addition increased degradation rates. Combination Inipol EAP-22 and customblen (aka MaxBac) was most effective for subsurface product. Liquid amendment was fastest but operationally more difficult to control.
(Powell *et al.* 2006b)	Field trial Landfarming (5 years) Mesocosm (sacrificial)	Antarctic Casey St.	SAB light diesel (20yrs old)	Nutrient (MaxBac) Aerobic Anaerobic	Microbial enumeration DGGE Real-time PCR *nosZ, nirS, bzdQ* TPH	Fertilization increased degradation in aerobic and anaerobic soils, but was most beneficial in anaerobic soils where evaporation was negligible. Denitrifiers actively involved in hydrocarbon degradation. When anaerobic communities are exposed to aerobic conditions, degradation is suppressed. Anaerobic remediation needs evaluation.
(Ferguson *et al.* 2003a; 2003b)	Mesocosm $^{14}CO_2$ (45/90 days)	Antarctic Casey St.	SAB light diesel (20yrs old)	Temperature (var) Nutrients (var) Water (var)	^{14}C-octadecane TPH Chemical indices	Higher temperatures led to higher degradation rates. High temperatures (37 and 42 °C) and low temperatures (−2 to 4 °C) had an initial lag phase. Moderate temperatures (10–28 °C) did not. Optimum nutrients were 1000–1600 mg N kg^{-1} soil H_2O.

(Eriksson *et al.* 2001)	Mesocosm CO_2 (48 days)	Arctic Canada Ellesmere Is.	#1 Arctic diesel	Temperature (var) Freeze-thaw Nutrients-water (fixed)	CO_2, TPH, SPME Microbial enumeration PCR, RISA RNA/DNA	Biodegradation observed at low temperatures 0 °C, but not < 0 °C. Freeze-thaw possibly stimulated microbial activity.
(Eriksson *et al.* 2003)	Micocosm Liquid culture (90 days)	Arctic Canada Ellesmere Is. & Labrador	11 PAH with 2–5 aromatic rings	Aerobic Anaerobic Inoculation (4 soils) Temperature (7 and 20 °C)	PAH quantitation Metabolites w/ SPME Microbial enumeration PCR, RNA/DNA RIS-RFLP	Low temperature inhibited aerobic PAH biodegradation, but had less of an effect on anaerobic biodegradation. Anaerobic biodegradation associated with nitrate reduction. Enriched microbial communities were more influenced by the presence of oxygen than either temperature or source of inoculum.
(Whyte *et al.* 1998)	Microcosm Liquid culture (6–52 days) Mesocosm $^{14}CO_2$ (23 days)	Sub-Arctic Canada Quebec	Diesel Crude oil	Temperature (var) Inoculation with Q15	TPH ^{14}C-dodecane ^{14}C-hexadecane ^{14}C-octadecane ^{14}C-dotriacontane CO_2 PCR, *thcA*, *alkB* Catabolic pathway with SPME GC-MS	Bioaugmentation of soil with a known hydrocarbon degrader beneficial to biodegradation of tracer (inoculation avoided the lag phase). Longer soil mesocosm incubation could have produced different results as soil microbial communities changed. Alkanes degraded in order dodecane-dotriacontane. Intermediary biodegradation products indicate Q15 degrades alkanes by both terminal subterminal oxidation pathways. PCR indicates that Q15 has aliphatic aldehyde dehydrogenase gene highly homologous to *Rhodococcus erythropolic thcA*.

(cont.)

Table 7.2 (*cont.*)

Reference Year	Design type	Region	Product(s)	Treatments	Monitoring Parameters	Findings, limitations, and uncertainties
(Reynolds *et al.* 1998)	Field trial (54 days)	Arctic Alaska	Diesel Crude oil	Nutrients (fixed) Grass enhancement	TPH	Greatest biodegradation found in treatments that had nutrient supplements and Annual Ryegrass and Articulated Fescue. Plants without nutrient supplements decreased rates, possibly through competition for nitrogen.
(Aislabie *et al.* 1998)	Mesocosm $^{14}CO_2$ (90 days)	Antarctic Ross Dependency	JP8 jet fuel	Nutrients (fixed)	TPH ^{14}C-hexadecane ^{14}C-napthalene Microbial enumeration	Demonstrated biodegradation potential of Antarctic micro-organisms.
(Stallwood *et al.* 2005)	Microcosm Sacrificial slurry (12 weeks)	Sub-Antarctic Signy Is.	Polar blend marine gas oil	Nutrients (fixed) Bioaugmentation (ST41) Water (fixed)	Chemical indices Microbial enumeration PCR-TTGE	Bioaugmentation by a proven indigenous hydrocarbon degrader in nutrient-amended microcosms yielded the fastest biodegradation rates. Stimulation by nutrient amendment achieved the same endpoint for n-alkane degradation, but took four days longer. More recalcitrant compounds not studied.
(Børresen *et al.* 2003)	Mesocosm $^{14}CO_2$ (128 days) Microcosm liquid culture	Arctic Norway Spitsbergan	Arctic diesel	Nutrients (fixed) Soil from 3 depths Inoculation	TPH ^{14}C-hexadecane Microbial enumeration	In liquid microcosms inoculated in with microbes from soil, the highest degradation rates were observed in cultures from permafrost at 3.5 m. In soil mesocosms, the highest degradation rates were observed in soil from the transition zone at 2.0 m. Degradation rate was directly proportional to contaminant concentration.

(Margesin and Schinner 2001)	Mesocosm field trial (780 days)	Alpine Austria	Diesel	Nutrients were applied at the beginning and after the first winter season	TPH Microbial enumeration Catalase, lipase	Nutrient amendment increased degradation over three seasons but was most effective during the first season. Residual hydrocarbons were very high.
(Margesin and Schinner 1997a)	Mesocosm Semi-open pan (155 days)	Alpine Austria	Diesel	Nutrients (fixed) 2 soils Inoculum (RM7/11)	TPH; Microbial enumeration Basal respiration Dehydrogenase activity	Biostimulation with nutrient amendment led to greater biodegradation than bioaugmentation with an actinomycetes inoculum. Abiotic losses accounted for ~30% of degradative losses (because of fresh contamination).
(Margesin and Schinner 1997c)	Mesocosm Sacrificial flask (20 days)	Alpine Austria	Diesel	Nutrients (fixed) 5 soils Inoculum	TPH Microbial enumeration	Biostimulation with nutrient amendment led to greater biodegradation than bioaugmentation with a psychrotrophic inoculum. Abiotic losses were as significant as biotic degradation. Residual hydrocarbons were very high.
(Margesin and Schinner 1997d)	Mesocosm Semi-open flasks (30days)	Alpine Austria	Diesel	Nutrients (var) Inoculum (var) Temperature (var)	TPH Microbial enumeration Basal respiration Dehydrogenase activity	Bioaugmentation with both psychrotrophic and mesophylic inocula, compared with biostimulation, was influenced by nutrient amendment. Abiotic losses did not contribute significantly to degradative losses (possibly because contamination was aged).
(Whyte *et al.* 2003)	Field trial	Arctic Eureka	Diesel	Nutrients (var) Tilling	TPH Microbial enumeration Microarray PCR, *alkB*, *ndoB*	Reduced TPH concentrations from ~13 700 to ~5 016 mg TPH kg^{-1} after two seasons. Abiotic losses through tilling and runoff identified as being important, though residual TPH after two seasons still very high.

8

Nutrient requirements for bioremediation

JAMES L. WALWORTH AND SUSAN FERGUSON

8.1 Introduction

Nutrients are required to support biological activity, and hence bioremediation. It is recognized that, although the microbial community requires numerous nutrients, nitrogen and phosphorus are the nutrients most often lacking, and thus limiting to biological hydrocarbon degradation in cold region soils (Mohn and Stewart 2000). Numerous studies have reported that biodegradation of hydrocarbon contaminants in cold region soils has been enhanced by the addition of one or both of these nutrients (Walworth and Reynolds 1995; Braddock *et al.* 1997; Walworth *et al.* 1997; Braddock *et al.* 1999; Mohn and Stewart 2000; Mohn *et al.* 2001; Ferguson *et al.* 2003a).

Nitrogen most often provides positive responses, although methodologies for determining application levels are not well defined. Proper nitrogen management can increase cell growth rate (Hoyle *et al.* 1995), decrease the microbial lag phase (Lewis *et al.* 1986; Ferguson *et al.* 2003a), help to maintain populations at high activity levels (Lindstrom *et al.* 1991), and increase the rate of hydrocarbon degradation (Braddock *et al.* 1997; Braddock *et al.* 1999). Whereas many studies indicate positive effects of supplemental nitrogen (Rasiah *et al.* 1991; Allen-King *et al.* 1994; Walworth and Reynolds 1995), a surprisingly large number report no benefit, or even deleterious effects when excessive levels of nitrogen are applied (Watts *et al.* 1982; Brown *et al.* 1983; Huntjens *et al.* 1986; Morgan and Watkinson 1990; Genouw *et al.* 1994; Zhou and Crawford 1995; Braddock *et al.* 1997; Walworth *et al.* 1997; Braddock *et al.* 1999; Mohn *et al.* 2001; Ferguson *et al.* 2003a). Thus proper nitrogen management is critical in bioremediation management.

Bioremediation of Petroleum Hydrocarbons in Cold Regions, ed. Dennis M. Filler, Ian Snape, and David L. Barnes. Published by Cambridge University Press.

Identification of proper application rates is crucial for maintaining optimum conditions for microbial hydrocarbon degradation, as is an understanding of the mechanisms through which excessive nitrogen inhibits bioremediation.

Lack of phosphorus does not appear to limit rates of biodegradation as frequently as does nitrogen, nor are responses as large as those to added nitrogen, however phosphorus application can stimulate contaminant biodegradation (Huntjens *et al.* 1986; Walworth and Reynolds 1995; Braddock *et al.* 1997) and decrease microbial lag time (Lewis *et al.* 1986). Phosphorus application can also increase microbial response to added nitrogen (Chang *et al.* 1996; Braddock *et al.* 1997; Graham *et al.* 1999). Excess phosphorus is less likely to inhibit biodegradation because of its very low solubility, and because phosphorus fertilizers rapidly react with soil particles. However, large applications of phosphorus can inhibit hydrocarbon degradation (Mills and Frankenberger 1994). Other nutrients may also enhance bioremediation. For example, addition of low levels of a mixture containing magnesium, sodium, calcium, iron, and sulfur increased respiration in a PAH-contaminated soil, although the increase was not attributed to any one constituent of the nutrient mix (Liebig and Cutright 1999).

8.2 Review and recent advances: nitrogen and phosphorus use in bioremediation systems

8.2.1 Nitrogen in bioremediation

Waksman (1924) found that soil bacteria contained carbon and nitrogen in a ratio of approximately 5:1. Assuming that these microorganisms assimilated between 5 to 10% of carbon substrates, he calculated that complete microbial degradation of a given amount of carbon substrate would require a carbon to nitrogen ratio (C:N ratio) of 100:1 to 50:1. Redfield *et al.* (1963) estimated that carbon, nitrogen, and phosphorus are required in a ratio of approximately 106:16:1, a ratio that has been adopted as an optimal nutrient ratio for microbial activity. Alexander (1999) suggests that a typical C:N ratio for soil microbes is about 10:1 and, assuming that 30% of substrate carbon is assimilated, the nitrogen required for complete substrate degradation is about 33:1.

Despite extensive use of C:N ratios to estimate nitrogen demand for bioremediation, there is no consensus on the optimum ratio. For example, Morgan and Watkinson (1989) reported optimal soil C:N ratios ranging from 200:1 to 9:1, Dibble and Bartha (1979) found that 60:1 was optimum, the United States Environmental Protection Agency recommends variously 12:1 (Sims *et al.* 1989), or a range of 100:1 to 10:1 (USEPA 1995), and Cookson (1995) recommends 10:1. Data of Brown *et al.* (1983) suggest that the optimum C:N ratio may depend on the level of soil contamination. In their studies, 9:1 was the optimum C:N ratio in refinery sludge with a contaminant level of 3500 mg kg^{-1} sludge, but the optimum ratio in a petrochemical sludge with 21 000 mg kg^{-1} of contaminants was 124:1 and reducing the ratio to 23:1 decreased biodegradation.

Alexander (1999) pointed out that C:N ratios can be used to calculate the amount of nitrogen required for total degradation of a given amount of substrate carbon, but not to determine the amount of nitrogen required to maximize the rate of degradation. Several factors that may affect the optimum C:N ratio should be considered. First, the relationship between cell composition and substrate composition is dependent on efficiency of substrate use. Some portion of the mineralized carbon is converted to CO_2 and not incorporated into microbial cells, although the efficiency rate is variable, depending on substrate qualities, microbial properties, and environmental conditions (McFarland and Sims 1991; Alexander 1999). Second, that portion of contaminant carbon which is not bioavailable because of sorption or other mechanisms will not decompose, and so imposes no nutrient demand. Third, nitrogen use efficiency varies widely. Nitrogen can be lost from the system when nitrous oxide is produced via denitrification, through ammonia volatilization, when nitrate is leached through the soil profile, if ammonium is complexed by soil clays, or when soluble nitrogen is incorporated into non-bioavailable complex humic compounds (Stevenson and Cole 1999). Fourth, nitrogen can be recycled, or used multiple times during biodegradation, so using the C:N approach may overestimate nitrogen demand. Last, use of the C:N ratio ignores toxic effects of excessive nitrogen application.

Nonetheless, the C:N approach correctly indicates that hydrocarbon-contaminated soils generally do not contain enough nitrogen to sustain optimal contaminant biodegradation. Petroleum products are carbon rich, and contain very small quantities of nitrogen. Therefore, supplemental nitrogen is usually required to optimize biodegradation. As expected, most studies have found that addition of nitrogen enhances biodegradation (Allen-King *et al.* 1994; Walworth and Reynolds 1995; Walworth *et al.* 1997; Mohn and Stewart 2000). On the other hand, some studies, including those by Huntjens *et al.* (1986), Morgan and Watkinson (1990), Genouw *et al.* (1994), Manilal and Alexander (1995), Walworth and Reynolds (1995), and Braddock *et al.* (1997) found that addition of nitrogen inhibited microbial hydrocarbon degradation. It is worth noting that in some studies (Morgan and Watkinson 1990; Durant *et al.* 1997), when excess nitrogen was added the rate of biodegradation or microbial activity was actually reduced below the unfertilized soil, indicating that fertilizer application was inhibiting bioremediation. Thus it is clear that, although addition of nitrogen is usually beneficial for soil bioremediation, use of excessive levels of nitrogen may negatively impact biological soil cleanup processes (mechanisms for this phenomenon will be discussed later). In addition, excess nitrogen entering surface and ground waters in the form of nitrate can constitute an environmental hazard.

8.2.2 Nitrogen in soil

Most soil microorganisms rely on the inorganic soil nitrogen forms nitrate and ammonium as a nutrient supply. However, the bulk of the nitrogen in most soil systems is contained in the organic fraction of the soil, and is not directly bioavailable. As soil organic material is oxidized by various soil fungi and bacteria, excess nitrogen may be converted to inorganic forms. If organic matter is sufficiently rich in nitrogen, degradation of the organic matter will result in the release of ammonium and nitrate nitrogen, or mineralization. On the other hand, degradation of organic matter with low levels of nitrogen will consume or immobilize ammonium and nitrate nitrogen as degrading microorganisms scavenge available nitrogen from the soil system (Stevenson and Cole 1999). In general, decomposition of organic materials with a C:N ratio of less that 35:1 will mineralize nitrogen, whereas degradation of those with a C:N ratio greater than 35:1 will immobilize nitrogen (Vigil and Kissel 1991).

Nitrogen mineralization can be divided into ammonification (Eq. (8.1)), conversion of organic nitrogen to ammonia (NH_3) which hydrolyzes in water to form ammonium (NH_4^+) (Eq. (8.2)); and nitrification, which is the oxidation of ammonium nitrogen to the nitrate form (NO_3^-). In ammonification, organic nitrogen compounds such as proteins, amino sugars, and nucleic acids are biologically degraded into ammonium by a wide range of aerobic and anaerobic heterotrophs (Jansson and Persson 1982; El-Shinnawi *et al.* 1993). In nitrification, ammonium is oxidized to nitrate largely by autotrophic soil bacteria. In autotrophic nitrification, the conversion of ammonium takes place in two steps: the transformation of ammonium into nitrite (NO_2^-) by oxidizing bacteria such as *Nitrosomonas*, and the oxidation of nitrite into nitrate by *Nitrobacter* (Paul and Clark 1996; Watson *et al.* 1989). Oxidation of nitrite usually proceeds at a more rapid rate, so it is a relatively rare form of inorganic soil nitrogen (Stevenson and Cole 1999). Ammonium and nitrate make up the bulk of soil inorganic nitrogen, and are the principal forms available for bioremediation.

Ammonification:

$$R - NH_2 + H_2O \rightarrow NH_3 + R - OH + \mathit{energy} \tag{8.1}$$

Ammonia hydrolysis:

$$NH_3 + H_2O \rightarrow NH_4^+ + OH^- \tag{8.2}$$

Table 8.1 *Commonly used nitrogen fertilizers, nitrogen content, and chemical composition.*

Fertilizer	% Nitrogen	Composition
Ammonium nitrate	33	NH_4NO_3
Ammonium sulphate	20.5	$(NH_4)_2SO_4$
Calcium nitrate	16	$Ca(NO_3)_2$
Urea	45	$(NH_2)_2CO$
Anhydrous ammonia	82	NH_3
Diammonium phosphate	20	$(NH_4)_2HPO_4$
Ammonium polyphosphate	10–15	$(NH_4PO_3)_n$

First step of nitrification:

$$2NH_4^+ + 3O_2 \rightarrow 2NO_2^- + 2H_2O + 4H^+ \tag{8.3}$$

Second step of nitrification:

$$2NO_2^- + O_2 \rightarrow 2NO_3^- \tag{8.4}$$

Nitrate is an anion and, as such, is repelled by negatively charged soil colloids. Nitrate salts are highly soluble, so nitrate moves with soil water and can easily be leached through soil material. Furthermore, nitrate can be reduced through the process of denitrification (Eq. (8.5)). Denitrification, reduction of nitrate to nitrous oxide (N_2O) or dinitrogen (N_2) gas by heterotrophic anaerobic bacteria, can thus be a major mechanism for nitrogen loss from poorly aerated soil systems. Oxidation of organic substrates provides energy and carbon for denitrifying bacteria and nitrate acts as the terminal electron acceptor. Nitrate addition has been used to enhance anaerobic hydrocarbon degradation through this process, although the nitrate is used primarily as an electron acceptor rather than as a nutrient *per se* (Hutchins *et al.* 1991).

Denitrification:

$$2HNO_3 \underset{-2H_2O}{\overset{+4H}{\rightarrow}} 2HNO_2 \underset{-2H_2O}{\overset{+2H}{\rightarrow}} 2NO \underset{-H_2O}{\overset{+2H}{\rightarrow}} N_2O(gas) \underset{-H_2O}{\overset{+2H}{\rightarrow}} N_2(gas) \tag{8.5}$$

Where supplemental nitrogen is required for bioremediation, there are many fertilizer materials that can be used. Some common fertilizers are listed in Table 8.1. Ammonium nitrate, ammonium sulfate, and calcium nitrate are commonly used inorganic salts of nitrate or ammonium. All are highly soluble, and immediately provide nitrogen in a bioavailable form. There is some disagreement on the efficacy of various inorganic nitrogen salts for bioremediation.

Chang and Weaver (1997) noted a preference for ammonium versus nitrate by petroleum degrading soil bacteria. Brook *et al.* (2001) reported that ammonium sulfate was more effective than either ammonium nitrate or potassium nitrate. On the other hand, Rasiah *et al.* (1992) compared calcium nitrate, ammonium nitrate, sodium nitrate, ammonium chloride, and potassium nitrate, and found that adding nitrate salts to a soil contaminated with oil refinery waste increased biodegradation more than ammonium. They also reported that the counter-ion had an effect on degradation, with calcium nitrate being the most effective, and effectiveness decreasing as the counter-ion was changed in the following order: $Ca > Na > K > NH_4$. Wrenn *et al.* (1994) compared ammonium chloride and potassium nitrate in solution culture and found that ammonium chloride was much less effective, and attributed the difference between the two sources to acidity produced during nitrification of the ammonium nitrogen (see Eq. (8.4)). When pH was controlled, there was no difference between the two sources.

Urea is a simple organic nitrogen salt. It is highly soluble, but must be hydrolyzed by the urease enzyme to form ammonium before it can be utilized (Eq. (8.6)). Frankenberger (1988) noted that soil urease activity is inhibited by petroleum materials, limiting the effectiveness of urea as a nitrogen source in bioremediation. On the other hand, both Brook *et al.* (1997) and Lee and Silva (1994) found that hydrocarbon degradation rates were higher in urea amended soils than in soil fertilized with ammonium nitrate. Brook *et al.* (2001) reported that urea was superior to potassium nitrate at a C:N ratio of 40:1 and to ammonium nitrate or potassium nitrate at a C:N ratio of 20:1. If urea is used, attention should be paid to soil pH. During hydrolysis, soil pH can rise dramatically, and it can subsequently drop below the original soil pH as ammonium is nitrified (see Eq. (8.3)).

Urea hydrolysis:

$$CO(NH_2)_2 + 3(H_2O) \rightarrow 2NH_3 + CO_2 + 2H_2O \rightarrow 2NH_4^+ + CO_2 + 2OH^- \qquad (8.6)$$

In addition to inorganic nitrogen salts, there are a number of slow-release or controlled-release nutrient sources. These materials may be coated fertilizers (sulfur-coated urea and Osmocote® are examples), slowly soluble materials (such as metal ammonium phosphates), or materials that must be microbially mineralized to release nitrogen (organic fertilizers and urea formaldehydes, for example. Several studies have compared stimulation of bioremediation with controlled-release versus soluble nutrients. Cunningham (1993) found that MaxBac®, a fertilizer coated with a selectively permeable membrane, resulted in more rapid diesel fuel degradation than a conventional nutrient source. In contrast, a urea oligomer was less effective when added to a diesel fuel contaminated soil than either urea or ammonium sulfate (Brook *et al.* 2001). Croft *et al.* (1995) compared MaxBac with Inipol EAP22® (an oleophilic urea-based fertilizer), and reported that MaxBac fertilized soil was only marginally better than unfertilized soil, whereas soils fertilized with Inipol EAP22 showed a significantly better response. Oil-based fertilizers such as Inipol EAP22 are thought to be advantageous at coastal-contaminated sites where loss of soluble nitrogen can be extreme.

To maximize effectiveness, controlled-release nutrients should be released at a rate equivalent to microbial demand (see Chapter 4, Section 4.2.2.2, and discussed in Chapter 11, Section 11.3.2.2). It has been suggested that organic nitrogen sources that must be microbially decomposed to release nitrogen might best match microbial nitrogen demand during hydrocarbon degradation (Walworth *et al.* 2003). Walworth *et al.* (2003) demonstrated the effectiveness of a fish processing by-product as a controlled-release nutrient source. Lee and Silva (1994) also successfully used a urea-fish-meal mixture for degrading Prudhoe Bay crude oil.

Ammonia, a gas at atmospheric pressure and ambient temperatures, is an effective nitrogen source which can be injected into soil. Ammonia is highly water soluble. As it hydrolyzes to form the ammonium ion, hydrogen ions are consumed and soil pH can be raised to 9 or higher (Tisdale *et al.* 1993), which can adversely affect soil microorganisms in close proximity to the fertilizer (see Eq. (8.2)). At pH values of 8.0 and above, volatilization of ammonia gas can cause large losses of nitrogen if it is not incorporated into the soil. As in the case with urea, the final pH of soil fertilized with ammonia may be lower than that of unfertilized soil because of acidification caused by nitrification of ammonium to nitrate (see Eq. (8.4)).

8.2.3 Phosphorus in bioremediation

Phosphorus is generally recognized as the second most limiting nutrient in bioremediation, and many studies have demonstrated positive responses to phosphorus added to bioremediation systems. Responses to phosphorus and nitrogen are often additive, but phosphorus responses are usually smaller than those for nitrogen (Huntjens *et al.* 1986; Walworth and Reynolds 1995; Chang *et al.* 1996; Braddock *et al.* 1997). As with nitrogen, phosphorus requirements are often determined by calculating a carbon to phosphorus ratio, based on similar philosophy to that discussed above for nitrogen. Recommended or optimum C:P ratios range from 20:1 (Mills and Frankenberger 1994) to 100:0.2 (Huesemann 1994).

Mills and Frankenberger (1994) applied varying rates of potassium phosphate (K_2HPO_4), with C:P ranging from 12:1 to 117:1 (1000 to 100 mg P kg^{-1} soil), to a

Table 8.2 *Commonly used phosphorus fertilizers, phosphorus content, and chemical composition.*

Fertilizer	% Phosphorus	Composition
Ammonium polyphosphate	15–16	$(NH_4PO_3)_n$
Diammonium phosphate	23	$(NH_4)_2PO_4$
Potassium phosphate	18	K_2PO_4
Triple superphosphate	20	$Ca(H_2PO_4)_2$
Phosphoric acid	26	H_3PO_4

diesel fuel contaminated soil. Maximum respiration occurred in the soil with a C:P of 23:1 (500 mg P kg^{-1} soil), but respiration was reduced to less than that of the unfertilized control when the C:P was 12:1 (1000 mg P kg^{-1} soil). Reports of inhibition of petroleum biodegradation resulting from over-application of phosphorus fertilizers are rare, however. This is probably due to the extremely low solubility of most phosphorus salts. The mechanism controlling soil solution phosphorus in fertilized systems is generally the solubility of phosphorus compounds, rather than the amount of phosphorus present.

The concentration of phosphorus in soil solution is very low, generally about 0.05 mg l^{-1}, although this concentration is quite variable among soils. Organic phosphorus compounds account for 15 to 80% of total soil phosphorus (Tisdale *et al.* 1993). As with nitrogen, phosphorus in organic form must be mineralized before it can be utilized for bioremediation. The remainder of soil phosphorus is sorbed to mineral surfaces or contained in solid phase phosphorus compounds. In high pH soils (pH 6.5 or above), most inorganic phosphorus is complexed with calcium, whereas in low pH soils (pH 6.0 or below) phosphorus is complexed by aluminum and/or iron. Phosphorus fertilizers, upon addition to soil, begin to form these complexes and generally become less bioavailable over time.

There are few studies comparing phosphorus sources. Mills and Frankenberger (1994) found that diethylphosphate $(C_2H_7)_2HPO_4$ stimulated diesel fuel degradation significantly more than potassium phosphate (K_2HPO_4). Some common phosphorus sources are listed in Table 8.2.

If phosphorus is introduced via injection wells in *in situ* bioremediation systems, orthophosphate precipitation at or near the point of injection can occur, causing plugging of sediments. Aggarwal *et al.* (1991) studied several phosphorus formulations to determine the extent of plugging with each fertilizer. They found that use of trimetaphosphate and, to a lesser extent polyphosphates, can increase phosphorus solubility and reduce or delay phosphorus precipitation and sorption. Trimetaphosphate, polyphosphates, or phosphoric acid are also required if phosphorus is to be injected with water, as most other phosphorus fertilizers are too insoluble to be used in this manner.

8.2.4 Detrimental effects of excess nutrients

As previously indicated, many studies have demonstrated positive effects from supplying supplemental nitrogen; however, others report no response or negative effects. Although application of nutrients is not generally considered to pose a limitation to biodegradation (Morgan and Watkinson 1989), inhibition has been reported at high application rates of nitrogen fertilizer. Reports of specific inhibitory effects of excess nitrogen include an increased lag phase (Huntjens *et al.* 1986; Ferguson *et al.* 2003a) and preferential inhibition of aromatic degradation (Fayad and Overton 1995), although most data indicate overall inhibition of microbial respiration and hydrocarbon degradation. Genouw *et al.* (1994) found that addition of 4000 mg N kg^{-1} soil inhibited microbial degradation of an oil sludge. However, microbial inhibition has also been reported at lower application rates. Huntjens *et al.* (1986) noted that 400 mg N kg^{-1} added to a sandy soil inhibited oil degradation in contaminated soil. Addition of 100, 200, or 300 mg N kg^{-1} soil to sub-Arctic taiga soils contaminated with Prudhoe Bay crude oil stimulated biodegradation compared to unfertilized controls, although the greatest stimulation was seen at the lowest fertilizer level (100 mg N kg^{-1} soil) (Hunt *et al.* 1973). Similarly, Braddock *et al.* (1997) reported that addition of 100 mg N kg^{-1} soil resulted in more rapid respiration than did 200 or 300 mg N kg^{-1} soil (Figure 8.1). In fact, addition of 300 mg N kg^{-1} soil resulted in respiration rates equivalent to unfertilized soil. Recognition of the deleterious effects of excess nitrogen and understanding the mechanism of nitrogen imposed microbial inhibition are keys to proper nitrogen management.

Most nitrogen fertilizers are, as noted above, composed of nitrate and/or ammonium salts which are highly water soluble and these quickly dissolve into free water present in soil. This increases the salt concentration of the soil solution, and lowers the soil osmotic potential, which can inhibit microbial activity.

Soil water potential (energy) is a measure of the physical and chemical potential of the soil water. There are two major components of soil water potential in unsaturated soils. The matric potential is a result of the attraction between soil particles (primarily those in the clay size range, less than 0.002 mm) and water; the osmotic potential results from the interaction of water molecules and dissolved salts. Both of these potentials affect soil microbes and other soil organisms. Soil water matric potential is the more frequently measured component, measured via tensiometer or pressure plate and it is this potential that is used to construct the familiar soil moisture release or moisture retention

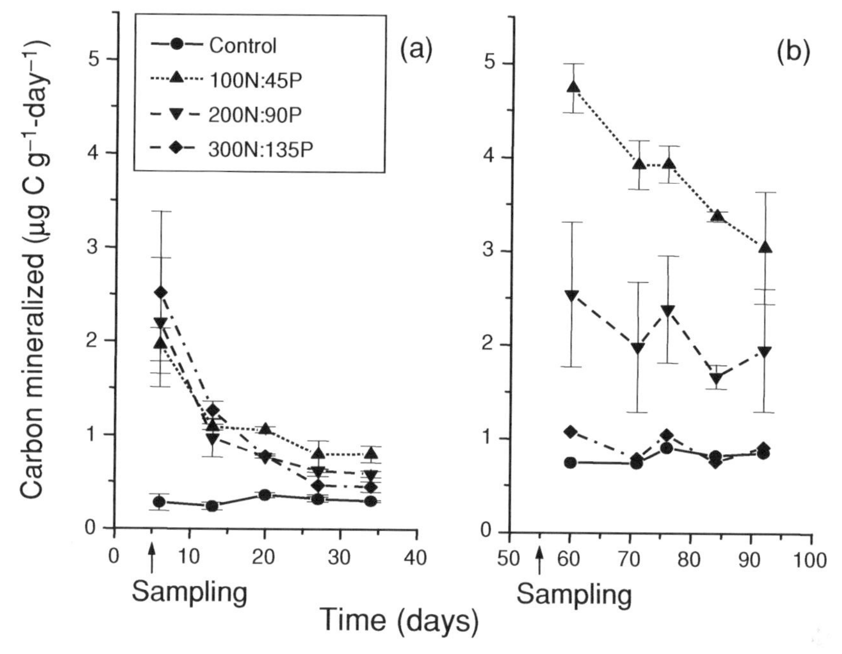

Figure 8.1. Soil respiration (expressed as C mineralized) in laboratory microcosms: (a) soil collected from field mesocosms immediately following treatment; (b) soil collected 6.5 weeks after fertilizer application. Values on both plots are expressed as the mean ± 1 standard error ($n = 3$) (from Braddock *et al.* 1997).

curves that relate soil moisture content and potential. The shape of these curves is, as one would expect, dependent on soil texture and particle specific surface area. Osmotic potential, on the other hand, is more challenging to measure and it is difficult to separate from matric potential. The combined matric and osmotic potential (total water potential) can be measured with a thermocouple psychrometer or similar devices that measure the relative humidity of soil vapor, but osmotic potential can not be easily measured individually. The matric potential can be measured separately and then be subtracted to determine osmotic potential. However, in moist soils, the matric potential is generally negligible (field capacity of a soil is the amount of water held at −0.01 to −0.03 MPa), so a measurement of the combined matric and osmotic potential can be used as an estimate of soil water osmotic potential in all but very dry soils.

Soil water potential is measured in units of pressure (bars, kPa, or MPa) and is negative in unsaturated soils: the drier the soil, or the saltier the soil solution, the more negative the soil water potential. As nitrogen fertilizer is added, total soil water potential declines as a result of increasing salinity of the soil water (Figure 8.2). The drier a soil, or the less water it contains, the smaller is the quantity of water for a given amount of fertilizer to dissolve in. The impact of

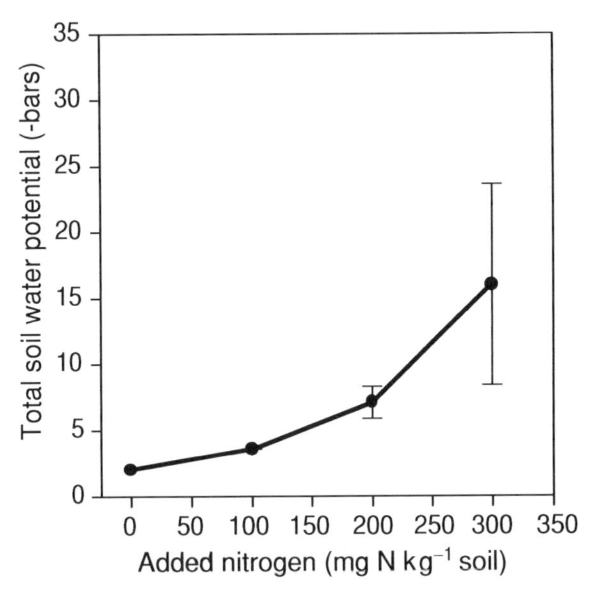

Figure 8.2. Total soil water potentials (measured via thermocouple psychrometry) in soil samples collected from field mesocosms 6.5 weeks following fertilizer application. Values are expressed as the mean $\pm$ 1 standard error ($n = 3$) (from Braddock *et al.* 1997).

nitrogen fertilizer on soil water potential is greater in dry than in moist soils. In Figure 8.3 it can be seen that the effect of ammonium nitrate on soil water potential is dependent on the level of soil moisture.

In lieu of measuring soil water osmotic potential, the contribution of nitrogen fertilizer to osmotic potential can be estimated (Walworth *et al.* 1997). By dividing the amount of nitrogen added (or the soil inorganic nitrogen concentration) by the soil moisture content (Eq. (8.7)), one can calculate an estimate of the nitrogen concentration in the soil solution, which has been termed N_{H_2O}:

Calculation of N_{H_2O}:

$$N_{H_2O} = \frac{\text{mg N}}{\text{kg H}_2\text{O}} = \frac{\text{mg N}}{\text{kg}\,soil} \div \frac{\text{kg H}_2\text{O}}{\text{kg}\,soil} \tag{8.7}$$

Nitrogen concentration is calculated as a function of soil water rather than a function of dry soil weight, which is the conventional notation. This approach has significant limitations, however. It does not take into account contributions of other soil salts to osmotic potential. In soils with saline contaminants, or in saline soils, non-nitrogenous salts can impose a limitation on biodegradation (Haines *et al.* 1994; Rhykerd *et al.* 1995). It is reasonable to ignore phosphorus

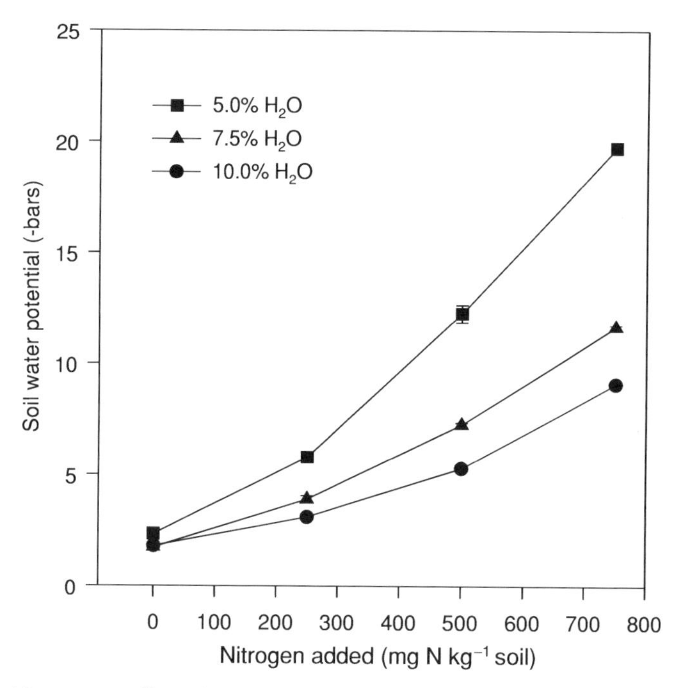

Figure 8.3. Effect of nitrogen on soil water potential (measured via thermocouple psychrometry) in an Alaskan loamy sand adjusted to three moisture levels. Values are expressed as the mean ± 1 standard error ($n = 3$) (from Walworth *et al.* 1997).

fertilizers in this calculation as they have lower solubilities than nitrogen fertilizers, and so contribute much less to osmotic potential (Tisdale *et al.* 1993).

Walworth *et al.* (1997) added ammonium nitrate to a sub-Arctic diesel-contaminated soil containing between 5 and 10% moisture by weight. The rate of microbial respiration was dependent on both the amount of nitrogen added to the soil, as well as the soil moisture level (Figure 8.4). If N_{H_2O} exceeded approximately 2500 mg N kg^{-1} H_2O, microbial activity was reduced over that in soil supplied with lower levels of nitrogen. In a study of the effects of nutrients on the hydrocarbon bioremediation potential of Arctic microbes, Braddock *et al.* (1997; 1999) showed that amendment with approximately 4000 mg N kg^{-1} soil H_2O stimulated carbon mineralization, whereas 8000 or 12 000 mg N kg^{-1} soil H_2O provided less or no stimulation. (In this *ex-situ* study, water content was not controlled so N_{H_2O} varied during the study.) Mohn and Stewart (2000) also found that nitrogen amendments to Arctic soils of small applications stimulated dodecane mineralization, but that larger amendments of about 8000 mg N kg^{-1} soil H_2O were inhibitory. Ferguson *et al.* (2003a) reported that fuel oil degradation in an Antarctic soil was stimulated by addition of 1570 mg N kg^{-1} soil H_2O, but not by addition of 28 000 mg N kg^{-1} H_2O. More recently, Walworth *et al.* (2007)

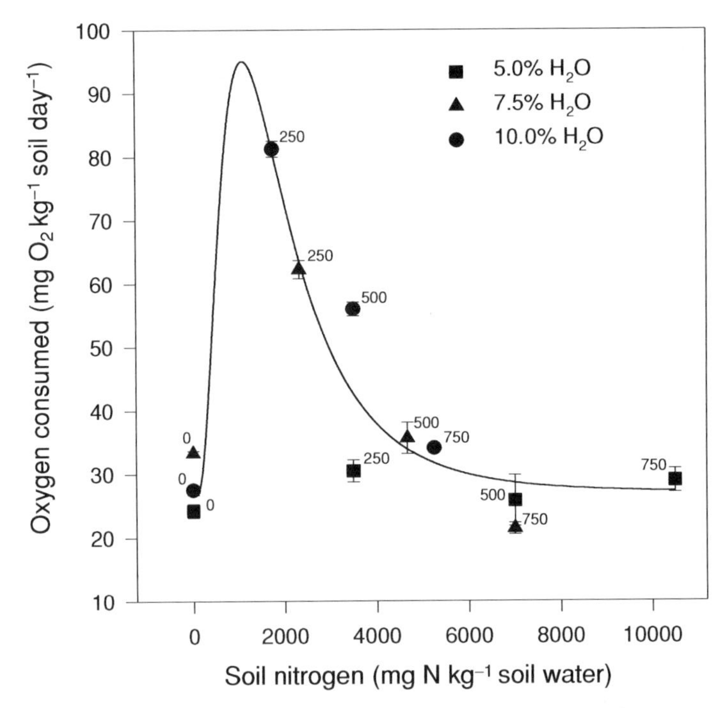

Figure 8.4. Nitrogen, expressed as a fraction of soil water (N_{H_2O}), versus oxygen consumption by petroleum degradation. The curve is a log-normal regression. Values are expressed as the mean $\pm$ 1 standard error ($n = 3$). Numbers next to data points represent nitrogen dose (mg N kg^{-1} dry soil) (from Walworth *et al.* 1997).

reported maximum oxygen consumption with application of 600 to 1200 mg N kg^{-1} soil H_2O and maximum petroleum degradation at 600 mg N kg^{-1} soil H_2O in a soil from sub-Antarctic Macquarie Island.

In terms of soil water potential, an osmotic potential drop of 0.50 MPa can reduce microbial petroleum degradation by roughly 50% (Braddock *et al.* 1997; Walworth *et al.* 1997). This is true regardless of whether the osmotic potential is decreased through application of a fertilizer salt such as NH_4NO_3 or a non-fertilizer salt such as NaCl (Walworth *et al.* 1997). In general, populations and activity of hydrocarbon degraders and heterotrophs can be reduced by osmotic stress (Braddock *et al.* 1997).

Maintaining N_{H_2O} below 2500 mg N kg^{-1} H_2O (or limiting osmotic drop to less than 0.50 MPa) limits the dose of inorganic nitrogen fertilizers that can be added to a soil. Use of sparingly soluble nitrogen sources could permit addition of higher nitrogen doses while minimizing osmotic stress. Data of Mohn and Stewart (2000) indicate that addition of 17 143 mg N kg^{-1} soil H_2O as non-water-soluble Inipol EAP22 to an Arctic tundra soil enhanced dodecane mineralization, whereas use of 25 714 mg N kg^{-1} soil H_2O as a combination of water-soluble urea and diammonium phosphate did not. In another Arctic soil, 5265 mg N kg^{-1} soil H_2O as Inipol EAP22 stimulated mineralization, and 7897 mg N kg^{-1} soil H_2O as urea and diammonium phosphate did not. These data suggest that Inipol EAP22 with limited water solubility did not inhibit mineralization, although more water-soluble forms of nitrogen fertilizers did. In addition to inhibiting contaminant biodegradation, excess nitrogen can adversely impact the environment. If inorganic nitrogen enters surface waters, stimulation of algal growth can result in eutrophication. In drinking water supplies, nitrate nitrogen poses a direct health concern and is therefore a regulated water contaminant in some countries.

High levels of phosphorus can also inhibit biodegradation. Mills and Frankenberger (1994) attributed inhibition caused by 1000 mg P kg^{-1} soil (C:P = 12:1) to osmotic effect of the fertilizer, however soil osmotic potential was not measured. The low solubility of most phosphorus fertilizers and the strong tendency for phosphates to sorb to soil minerals diminishes the likelihood of negative impacts of excess phosphorus.

8.3 Recommendations for fertilizer use

Nitrogen must be properly managed to provide optimum conditions for bioremediation. Published data point out clearly that an adequate nitrogen supply can increase populations of hydrocarbon degraders, decrease lag time, increase degrader activity, and maximize rates of biodegradation. However, overapplication can reduce or eliminate responses to added nitrogen. C:N ratios may provide a means for calculating the total amount of nitrogen needed to degrade a given quantity of hydrocarbon, but they are not useful for determining the proper amount of nitrogen to add to maximize degradation rate (see Chapter 4, Section 4.2.2.2). It is suggested here that water-soluble nitrogen additions be based on calculation of N_{H_2O} (see Eq. (8.7)), and that no more than 2000 mg N kg^{-1} soil H_2O be added, regardless of contaminant concentration.

If greater levels of nitrogen are required, it is recommended that the application be split into multiple smaller applications, and that inorganic soil nitrogen be monitored and concentrations be kept below 2000 mg N kg^{-1} soil H_2O. Alternatively, higher amounts of nitrogen can be added with controlled release nutrient sources.

For example, consider a sandy soil with a water holding capacity of 5% by weight and contaminated with 10 000 mg diesel fuel per kg soil. Using the frequently cited optimum C:N ratio of 10:1, the calculated nitrogen requirement would be 1000 mg N kg^{-1} soil. Using Eq. (8.7), one can calculate that this would result in a N_{H_2O} value of 20 000 mg N kg^{-1} H_2O, a level that would almost certainly be inhibitory. Limiting the nitrogen applied in a single dose to keep the N_{H_2O} level below 2000 mg N kg^{-1} soil would require that no more than 100 mg N kg^{-1} soil be applied at one time. Additional nitrogen could be added once inorganic nitrogen levels had dropped substantially.

If, on the other hand, the same level of contamination (10 000 mg diesel fuel per kg soil) was found in a silt loam soil with a water holding capacity of 30% by weight, application of 1000 mg N kg^{-1} soil suggested by the 10:1 C:N ratio would only raise the N_{H_2O} level to 3333 mg N kg^{-1} H_2O. This concentration is only slightly higher than the recommended 2000 mg N kg^{-1} H_2O maximum, and would not likely cause substantial inhibition. Basing nitrogen application on N_{H_2O} would suggest that the maximum nitrogen dose should not exceed 600 mg N kg^{-1} soil.

Inorganic nitrogen is highly mobile in soil systems. It is rapidly redistributed by water moving through soil. Therefore, placement of nitrogen fertilizer is not critical. Inorganic nitrogen salts are also well suited to application in irrigation water. On the other hand, inorganic nitrogen is easily lost from uncontained systems where leaching water can carry nitrogen out of the target zone.

Several factors should be considered in selection of a source of nitrogen for bioremediation. As indicated earlier, the form of nitrogen affects microbial response, but conflicting data from various studies make it difficult to recommend one nutrient source over others. The selected source should fit the application technique. If ammonia is used, it must be injected below the surface to minimize volatilization losses. Some attention should be paid to pH changes caused by fertilizer addition. Urea and ammonia can cause large pH increases when they hydrolyze in the soil (Equations (8.2) and (8.6)). Once hydrolyzed they, and all other ammonium sources, will slowly lower soil pH as the ammonium nitrogen gets nitrified (oxidized to nitrate) (see Eq. (8.3)). On the other hand, nitrate can be quickly lost through anaerobic organic carbon mineralization (denitrification) in poorly aerated soils.

Although phosphorus is often not as critical as nitrogen, it often provides some stimulation of biodegradation. Although there are few reports of inhibition from high rates of phosphorus application, Mills and Frankenberger (1994) noted that inhibition occurred with application of more than 500 mg P kg^{-1} soil. It is recommended that this be considered an upper application level.

Inorganic phosphorus salts do not move appreciably in soil. Therefore, phosphorus is very difficult to distribute within a mass of soil. Essentially, most phosphorus fertilizers must be placed where they are to be utilized; very little redistribution will occur after application. Only more soluble phosphorus sources, such as ammonium polyphosphates, can be applied with irrigation water, or will move with the flow of water.

8.4 Future research

Controlled release nutrients provide an opportunity for increased precision in nitrogen management for bioremediation. Greater quantities of nitrogen can be added with these materials, so site management could be decreased, which is highly desirable in remote Arctic and Antarctic sites. Potential surface and groundwater contamination can also be decreased. However, most controlled-release fertilizers have been developed for use in the horticultural and agricultural industries, and are not tailored for cold-climate applications. Little is known about their release rates under cold conditions. Effective use requires matching nitrogen release rate to microbial nitrogen demand, or their stability during freeze/thaw cycles. Therefore, release rates of appropriate materials in the Arctic environment need to be determined. Then, application rates can be matched to microbial nitrogen demand to optimize bioremediation rates. If nutrient release from currently available materials does not adequately match biodegradation demand, controlled-release materials may need to be tailored to cold-region conditions.

9

Landfarming

JAMES L. WALWORTH, C. MIKE REYNOLDS,
ALLISON RUTTER, AND IAN SNAPE

9.1 Introduction

Landfarming has been described as "a simple technique in which contaminated soil is excavated and spread over a prepared bed and periodically tilled until pollutants are degraded" (Vidali 2001) but, in practice, it can be either an *ex situ* or *in situ* technique. Landfarming generally uses a combination of volatilization and biodegradation to reduce hydrocarbon concentrations. For biodegradation to be effective, stimulating aerobic soil microorganisms is essential; this is commonly accomplished by adding nutrients and mixing the soil to increase aeration. Aerating the soil in this way also increases the loss of hydrocarbon contaminants to the atmosphere via volatilization. Volatilization of diesel and lighter hydrocarbons greatly assists the remediation process but it is less effective for heavier molecular weight hydrocarbons such as crude oil.

For *in situ* landfarming it is possible to treat only relatively shallow layers of soil where reasonable oxygenation can be maintained. In *ex situ* landfarming, excavated contaminated soil is spread as a thin layer in a treatment bed that is often lined with an impermeable layer to control leaching and runoff. *Ex situ* landfarming can be as simple as soil spread in a cleared area or it can be a major construction with contouring or drainage systems or both for removal of excess water. Plumbing can also be used for the application of water, either alone or in combination with nutrients or other amendments, to the landfarm surface. In more elaborate systems, mechanisms for the introduction of air can be provided, and there is a continuum in technological complexity with biopiles (Chapter 10).

Bioremediation of Petroleum Hydrocarbons in Cold Regions, ed. Dennis M. Filler, Ian Snape, and David L. Barnes. Published by Cambridge University Press.

The advantage of landfarming is its simplicity and relatively low cost. Although soil cleanup via landfarming may be slower than with other higher input systems, it is a commonly used cleanup method in temperate regions because of its low cost (see Chapter 1, Figure 1.1). Landfarming in cold regions entails some additional considerations. Freezing and thawing can drive water and contaminant movement in soil, adding to the difficulty of managing runoff and undertaking multi-season monitoring. In cold environments with high precipitation, soils dry slowly and oxygen diffusion is slower in wet soils. Although soil moisture management may be a critical consideration in cold-climate landfarming, its importance is only beginning to be appreciated. Of key importance, biodegradation rates and volatilization are known to be slower at low temperatures (Ferguson *et al.* 2003b; Snape *et al.* 2005). Soil can be heated but this is usually costly and often impractical. A goal of cold-climate landfarming is to alleviate other limitations (e.g. moisture, oxygen, nutrients) such that temperature becomes the limiting factor to bioremediation. By reducing those limitations, the soil-microbial-contaminant system can be poised to take full advantage of natural seasonal warm cycles. Nevertheless, cleanup times are likely to be longer in cold environments than in temperate climates. The relatively short remediation season requires careful management of various factors such as soil tillage, moisture, and nutrient levels to minimize treatment times and capitalize on the cost savings of simple passive treatment systems.

9.2 Review

9.2.1 Landfarming management factors

9.2.1.1 Excess moisture and insufficient soil aeration

Excess moisture can result in poor soil aeration and limit aerobic hydrocarbon degradation in landfarms (Fine *et al.* 1997; Huesemann and Truex 1996). This is of particular concern in cold regions with high rates of precipitation or organic soils with high water-holding capacity because evaporation and diffusion rates are lower.

Chatham (2003) noted some unsuccessful landfarming projects, which were mostly associated with poor soil aeration. At Drill Site 1 and 2 Relief Pits on the north slope of Alaska lack of demonstrable remediation was attributed, in part, to saturated soil conditions. Chatham (2003) also reported that soil contaminated with 2000 to 2500 mg diesel range organics (DRO) kg^{-1} at the Service City Pad in Alaska was not remediated due to poor aeration and soil temperature averaging 3.3 °C, although the pad was plumbed to permit injection of air. Throughout research in Alaska Reynolds (1993) and Reynolds *et al.* (1994, 1998) observed that longer hydrocarbon half-lives were associated with areas that tended to remain saturated for longer periods after rain, indicating that poor aeration may limit bioremediation.

9.2.1.2 *Insufficient soil water*

An adequate supply of water is critical for biological activity (Atlas 1981). Although microbial communities in soil are robust and can respond rapidly to favorable conditions, lack of biochemically available water decreases the rate of microbial processes. Landfarmed soils often can be irrigated on a practical scale, depending on local infrastructure. However, soils contaminated with large amounts of petroleum can be relatively hydrophobic and difficult to "wet." Lack of water can particularly limit hydrocarbon biodegradation in freely drained coarse-textured soils having low water-holding capacities. Ferguson *et al.* (2003a) conducted a microcosm study that included combinations of nutrients and water on soil from Old Casey Station in Antarctica contaminated with aged light diesel fuel. Although the greatest effects were observed in response to nitrogen (N) fertilization, increasing soil water content significantly raised rates of hydrocarbon degradation. Salts, including those that comprise N fertilizers, can reduce bioavailability of soil water and rates of soil hydrocarbon degradation (Chapter 4, Section 4.2.2.4 and Chapter 8, Section 8.2.2).

9.2.1.3 *Nutrients*

Nutrient requirements for bioremediation are well recognized and guidelines are discussed in Chapter 8. Nutrients can stimulate microbial activity and are relatively inexpensive to obtain. For remote sites, shipping and application costs can be significant but the cost of remobilization for successive nutrient applications could be significantly reduced through the use of advanced technologies such as controlled-release nutrients (CRNs). A benefit of landfarming is that fertilizers can be easily applied to the surface and incorporated in soil using common agricultural practices such as tilling.

Relatively few field-scale cold-region landfarm operations have evaluated biodegradation rates in relation to nutrient application rates. Sandvik *et al.* (1986) studied the response to varying rates of fertilizer (N-P-K: 25-3-6) in two *ex situ* landfarms with refinery oil sludge in Oslofjord, Norway. Oil sludge was incorporated to a depth of 25 cm, and subsequently tilled once per month. A significant, but modest, response to supplemental nutrients was observed (Table 9.1).

Applied in appropriate amounts N-fertilizer is beneficial; however, adding excessive fertilizer to soil or water can have detrimental effects, largely based on toxic response to osmotic potential (Walworth *et al.* 1997). Discussed in

Table 9.1 *Response to nutrient addition in two Norwegian landfarms (Sandvik et al. 1986)*

Nitrogen added (mg kg^{-1})		0	200	400	600
Soil TPH (mg kg^{-1})	Initial	After 33 months			
Site 1	23 000	22 000	21 000	18 000	17 000
Site 2	34 000	23 000	15 500	8500	6000

Chapter 8 for example, excess N-fertilizer in an Alaskan gravel pad was shown to reduce microbial populations and to inhibit hydrocarbon biodegradation (Braddock *et al.* 1997; 1999). The sensitivity of soils to over-fertilization is greatest in soils with low water-holding capacities where there is little water to dissolve fertilizers, whereas in wetter soils greater dilution reduces salt concentrations.

9.2.1.4 Soil temperature

Although low soil temperatures are often thought to be the main limiting factor for microbial degradation of soil petroleum contaminants in cold regions (Chapter 5), the factors discussed above – moisture, oxygen, and nutrient availability – are likely to be more limiting at many sites. Nevertheless, petroleum degradation rates do increase rapidly with rising soil temperature when other factors are optimized (Walworth *et al.* 2001; Ferguson *et al.* 2003b).

Passive soil warming of cold soil via the use of covers has been evaluated in a limited number of studies. At two sub-Antarctic sites on the Grande Terre (Kerguelen Archipelago) Delille *et al.* (2004b) covered soils with a black plastic sheet placed directly on the soil surface with a transparent plastic cover placed 10 cm above the soil. Plastic soil covers induced a small (2 °C) increase in the temperature of surface soil. There was a modest but significant increase in degradation of alkanes in covered soil, but degradation of polycyclic aromatic hydrocarbons (PAHs) was not affected. Kerry (1993) studied soil contaminated by Special Antarctic Blend (SAB) in the Vestfold Hills, Antarctica. Among the treatments studied was a plastic soil covering (opaque with 85% light transmission) that was perforated to allow aeration. Soil temperatures were up to 5 °C higher under plastic, but no stimulation of hydrocarbon loss was observed. It was concluded that, in addition to increasing soil temperatures the plastic covers inhibited evaporative losses, a trade-off that must be considered with the use of soil covers (see Section 9.2.1.7 this chapter, and Chapter 10, Section 10.3.2.2 for additional discussions).

9.2.1.5 *Soil microbial communities*

Considerable data are available on the effectiveness of microbial inoculants (bioaugmentation) in cold-climate landfarming. Bioaugmentation with non-indigenous soil microorganisms is prohibited by international treaty in Antarctica (see Chapter 1, Section 9.2.3.2), but has been a relatively common practice in Arctic and sub-Arctic landfarms. Nonetheless, the current consensus is that indigenous soil microbes capable of degrading contaminant hydrocarbons are usually present, their populations often enhanced by petroleum contaminants, fertilization, or both, and inoculation is unnecessary, expensive, or even counterproductive.

In a study conducted in north-central Alberta, soil contaminated with 6.7 l m^{-2} of crude oil was treated with bacteria selected from oil-contaminated soil with and without the addition of urea phosphate (~300 mg N kg^{-1})(Jobson *et al.* 1974). The authors noted that soil bacteria population levels were increased by addition of fertilizer, but not by inoculation with bacteria alone. After 307 days, including a winter season, the fraction of alkanes was reduced in plots treated with fertilizer, whether or not they were inoculated with petroleum-degrading bacteria. Similar results were obtained in a three-year landfarming experiment conducted at a site in the Canadian Northwest Territories (Westlake *et al.* 1977). At the Service City Pad in northern Alaska, plate counts of hydrocarbon-degrading bacteria after 56 days of landfarming operation were equivalent for un-inoculated plots and those that were inoculated with a consortium of hydrocarbon-degrading bacteria (1.2×10^7 versus 1.0×10^7, respectively) (Chatham 2003). In contrast, Ruberto *et al.* (2003) reported that inoculated B-2-2 *Acinebacter* persisted in a sub-Antarctic soil and resulted in increased total petroleum hydrocarbon (TPH) reduction over a 50-day period.

Additional studies indicate that hydrocarbon-degrading bacteria are generally present (Chapter 4) and that soil contamination and fertilization can increase their numbers and activity in Arctic (Piotrowski *et al.* 1992; Reimer *et al.* 2005), sub-Antarctic (Delille *et al.* 2003; Delille *et al.* 2004a; Ruberto *et al.* 2003) and Antarctic (Aislabie *et al.* 2001) soils (Table 9.2).

9.2.1.6 *Vegetation*

The rate and extent of bioremediation has been increased by the presence of plant communities in temperate region soils (Newman and Reynolds 2004), however there is little information about the role of plants in cold-region landfarming. Reynolds (2004) evaluated effects of vegetation on bioremediation in field sites at Barrow, Galena-Campion, and Annette Island, Alaska. Controls with no nutrients or vegetation were compared to N alone (2000 mg N kg^{-1} of soil H_2O, based on water content at −0.03 MPa) or in combination with vegetation. Soils were mixed at the beginning of the field studies, but not tilled

Table 9.2 *Microbial responses to hydrocarbon contamination and fertilization in Arctic, sub-Antarctic, and Antarctic soils*

Authors	Contaminant	Treatment	Response
Piotrowski *et al.* 1992	Diesel fuel	Surfactant, trace elements, vitamins, tillage	Active percentage of total bacteria increased with soil treatment.
Delille *et al.* 2003	Diesel fuel; crude oil	Inipol EAP-22	Oil-degrading bacteria numbers increased with fertilization in 2 of 2 soils, and with contamination in 1 of 2 soils.
Delille *et al.* 2004a	Diesel; Arabian light crude oil	Inipol EAP-22	Heterotrophic and hydrocarbon-degrading microorganisms increased with contamination and with fertilization.
Ruberto *et al.* 2003	Gas-oil	Nitrogen and phosphorus	Contamination increased hydrocarbon-degrading bacteria; N and P did not.
Reimer *et al.* 2005	Petroleum hydrocarbons	Soils contaminated with petroleum hydrocarbons had increased population densities of total heterotrophs and fuel degraders.	
Aislabie *et al.* 2001	Petroleum hydrocarbons	Numbers of hydrocarbon degraders, culturable bacteria, yeasts, and fungi increased in 2 of 3 contaminated soils. All categories of microbes were extremely low and response to petroleum contamination was minimal in an Antarctic dry valley soil.	

thereafter. Vegetation had a positive effect on petroleum depletion for both N alone or unfertilized control treatments. However, the effect of vegetation was subtle; it did not affect all petroleum fractions equally and was not detectable by standard monitoring techniques because of high soil spatial variability within sites and environmental variability between sites.

9.2.1.7 Soil mixing (tilling)

Soil mixing, generally by some form of tillage, is a common landfarming practice that can serve several purposes. Mixing aerates soil more efficiently than air injection, providing oxygen for aerobic biodegradation. It also redistributes water, nutrients, and contaminants, and reduces contaminant concentrations in hot spots that are otherwise difficult to degrade to low concentrations. Perhaps most significantly, mixing exposes contaminant hydrocarbons to the atmosphere and promotes volatilization. Ausma *et al.* (2002) measured volatile hydrocarbon fluxes from a landfarm soil contaminated with 7930 mg kg^{-1} of hydrocarbons in northern Ontario and found that emissions of C_9 to C_{14} alkanes increased five- to ten-fold for several days following soil tillage.

Volatilization is often overlooked. Several studies indicate that hydrocarbon losses via volatilization may be as, or more, important than losses via biodegradation, particularly in soils contaminated with low-molecular-weight compounds. During operation of a landfarm at the Sea Air Motive Pad in Deadhorse, Alaska, measured hydrocarbon loss was 14.0 mg kg^{-1} day $^{-1}$. Eleven percent (1.6 mg kg^{-1} day $^{-1}$) was attributed to biodegradation; the remaining 89% (12.4 mg kg^{-1} day $^{-1}$) was attributed to abiotic volatilization or leaching or both (Chatham 2003). Piotrowski *et al.* (1992) conducted a landfarming operation on diesel-fuel-contaminated soil on a coastal Alaska site north of the Arctic Circle. In a cell filled to a depth of 45 cm, treated with surfactant and micronutrients, and tilled weekly, they reported decreases from 11 491 to 6963 mg TPH kg^{-1} in the first season. Forty-four percent of hydrocarbons were lost in the first four weeks of operation; the authors estimated that volatilization could have accounted for most of this loss and as much as 30% of the original soil petroleum.

In a trial landfarm at Resolution Island in Nunavut, relative contributions of volatilization and biodegradation were evaluated using *n*-alkane to pristine and phytane ratios (Paudyn *et al.* 2005). Analysis of these ratios indicated significant biodegradation only in the plot that was fertilized. Reductions in hydrocarbon concentrations in a corresponding non-fertilized plot indicated similar potential rates of biodegradation and volatilization. The rate of contaminant loss was higher in the "tilled every day" plot versus the "tilled every four days" plot at Resolution Island where mean summer air temperature is 3 °C, the increase attributable to enhanced volatilization. In contrast, at a site in Ontario with mean temperature of 23–28 °C, Demque *et al.* (1997) determined that increasing the tillage rate from monthly to semi-monthly had no significant effect on volatilization rates.

9.2.2 *Case studies*

9.2.2.1 *Landfarming experiences in the Arctic*

Research studies, pilot-scale, field trial, and full-scale landfarming operations have been conducted in various Arctic locations. The following examples illustrate the range of sites and techniques applied in the Arctic.

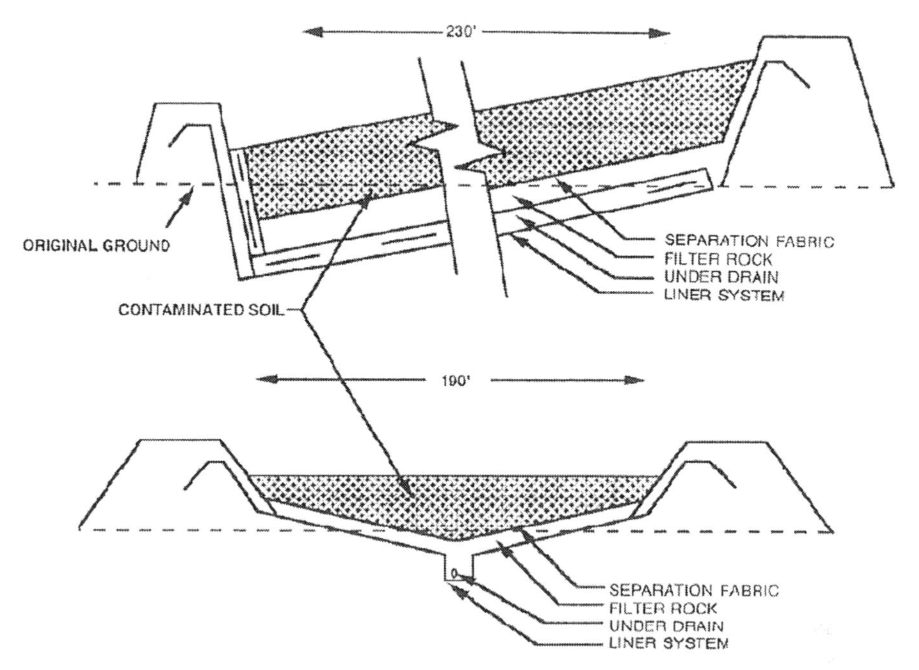

Figure 9.1. Fairbanks International Airport landfarm schematic (Reynolds *et al.* 1998).

Fairbanks International Airport, Alaska

In the early 1990s, a full-scale landfarm study was conducted with fuel-contaminated soil at the Fairbanks International Airport in Fairbanks, Alaska (Reynolds *et al.* 1998). The landfarm was constructed with a geotextile liner to prevent leaching, and a PVC pipe system to collect leachate for recirculation via surface irrigation and for fertilization (Figure 9.1). The native subsoil under the liner was graded and sloped so that the liner directed leachate into the collection system. Soil was tilled and approximately 4 mg N kg^{-1} (as ammonium nitrate) and 1 mg K kg^{-1} (as potassium sulfate) were added via the irrigation system weekly during the first season of operation. Water was added to keep the soil moisture content at 25–85% of field capacity. In the second season, fertilizer was applied monthly, at monthly rates of approximately 100 mg N kg^{-1} (ammonium nitrate), 34 mg P kg^{-1} (triple superphosphate), and 25 mg K kg^{-1} (potassium sulfate). In year two, excess water was a problem due to local weather conditions, so nutrients were applied as dry fertilizer.

A graded sampling regimen was used to account for spatial variability across the landfarm. Organic carbon levels were measured (Figure 9.2) and half-lives of organic carbon were calculated assuming first-order kinetics (Figure 9.3). Areas with long half-lives corresponded to soil that tended to remain saturated for

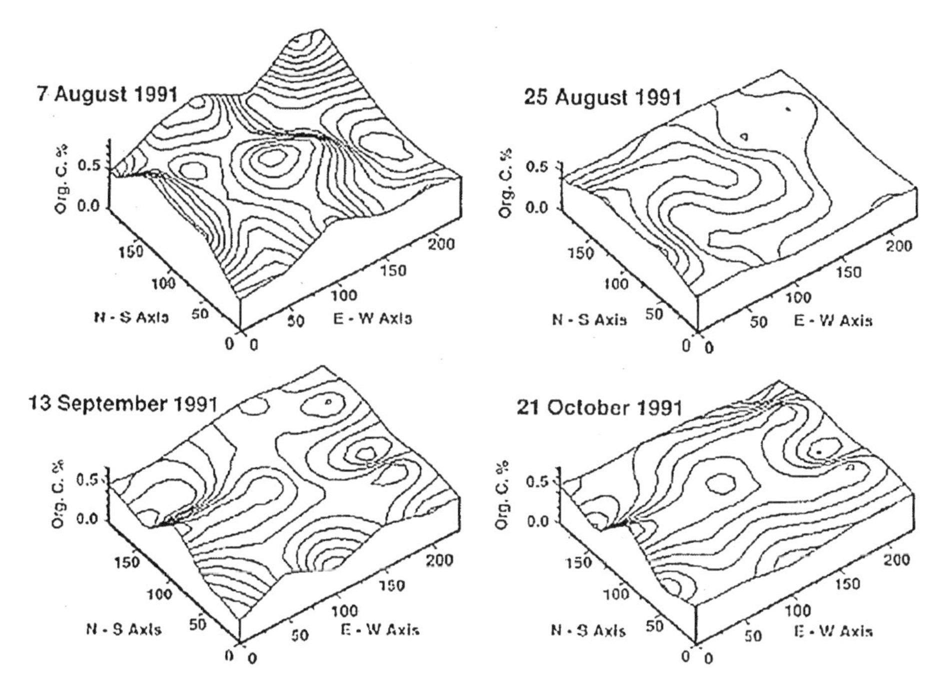

Figure 9.2. Soil organic carbon levels in the Fairbanks International Airport landfarm during season one (Reynolds *et al.* 1998).

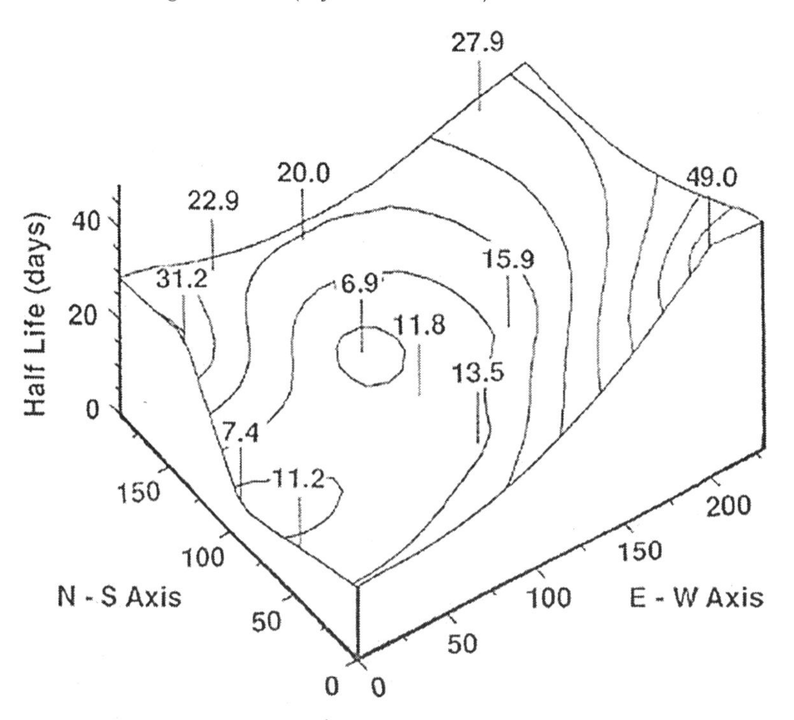

Figure 9.3. Calculated half-life of organic carbon at various locations in the Fairbanks International Airport landfarm (Reynolds *et al.* 1998).

longer periods after rain, suggesting poor aeration may have limited bioremediation. Results indicated average contaminant half-life integrated across the entire landfarm was nine days.

North Slope, Alaska

Chatham (2003) reported several recent, successful landfarms on gravel pads on the north slope of Alaska. DRO levels were reduced from 400–1680 mg kg^{-1} to 30–200 mg DRO kg^{-1} in 30 days in a landfarm on the Haliburton Pad with French drains and treatments including addition of fertilizer and enzymes, inoculation with hydrocarbon-degrading bacteria, and tillage. At the Service City Pad, DRO concentrations declined from 913–1053 mg kg^{-1} to 261–506 mg kg^{-1}, with the greatest decrease occurring in soil that was fertilized with approximately 86 mg N kg^{-1}, 19 mg P kg^{-1}, 36 mg K kg^{-1}, tilled every three days, and had soil moisture maintained at 30 to 40%.

Point Barrow, Alaska

Following a series of pilot studies (Braddock *et al.* 1997; 1999), a field-scale *in situ* landfarming operation was initiated in a gravel pad at Point Barrow, Alaska (McCarthy *et al.* 2004). A combination of ammonium phosphate (N-P-K:11-52-0) and urea (N-P-K:46-0-0) was applied such that application did not exceed 100 mg N kg^{-1} and 50 mg P kg^{-1} based on responses obtained in the pilot studies. The fertilizer was incorporated to approximately 1.5 m (the thawed depth) and the soil was tilled throughout the remediation period. Soil moisture was initially near field capacity (3 to 6% by mass) and frequent, light precipitation throughout the operation kept the soil moist. Initial diesel and gasoline range organics (GRO) ranged up to 1500 and 47 mg kg^{-1} respectively, and dropped to maximum levels of 530 and 13 mg kg^{-1}, with average levels of 380 and 10 mg kg^{-1}, respectively, in one-half of the landfarm after 31 days. In the other half, an additional fertilizer application was made after 34 days because apparent remediation rates were lower. Initial DRO and GRO concentrations in that half of the landfarm were 1400 and 140 mg kg^{-1}, respectively. After 55 days, maximum levels of DRO and GRO were below 684 and 25 mg kg^{-1}, and average levels were 430 and 15 mg kg^{-1}, respectively.

Pukatawagan, Manitoba (Canada)

Wingrove (1997) reported on an *ex situ* landfarm at Pukatawagan, Manitoba with up to 13 000 mg kg^{-1} of diesel fuel. The landfarm consisted of clay-lined cells that were sloped for drainage. Soil was tilled and nutrients and water content sampled weekly and adjusted as necessary (targets and added levels were not reported) except in an unfertilized control plot. After one month soil hydrocarbon levels were below 600 mg kg^{-1}, and after five months were below 230 mg kg^{-1}. Fertilizer application had no effect on remediation, suggesting that much of the loss was due to volatilization.

Resolution Island, Nunavut (Canada)

Paudyn *et al.* (2005) established a trial landfarm in the Canadian Arctic at Resolution Island, Nunavut in 2003, to assess treatment viability and the relative contributions of volatilization and bioremediation. The site has an average summer temperature of 3 °C. A landfarm was constructed on a leveled area with a slope of less than 5%, and filled with homogenized contaminated soil (initial concentration was 2670 mg DRO kg^{-1} soil) to a depth of 0.3 m. Four treatment regimes were established: a control plot (no tilling), daily tilling, tilling every four days, and tilling every four days with addition of fertilizer (urea and diammonium phosphate: 200 mg N kg^{-1}, 13.5 mg P kg^{-1}).

Soils were analyzed for DRO concentrations, and the ratios of the isoprenoids pristane and phytane and their linear counterparts were monitored. Hydrocarbon concentrations were reduced in all plots, including the control, after two seasons. The rate of remediation was highest in the fertilized plot, although in both the fertilized plot and the aerated daily plot hydrocarbon concentrations were < 500 mg kg^{-1} at the end of the second season. N-alkane:isoprenoid ratios in the fertilized plot were dramatically reduced upon addition of fertilizer but were unchanged in the three unfertilized plots, indicating that biodegradation was occurring only in the fertilized plot. Reductions in the hydrocarbon concentrations in "daily tilled" (over 80% loss) and "tilled every four days" (over 70% loss) plots clearly showed that volatilization was a significant mechanism for diesel fuel reduction. Additionally, loss in the fertilized plot (tilled every four days) was over 95%, versus approximately 45% in the control plot (no tilling and no fertilizer), suggesting that relative rates of bioremediation and volatilization were similar.

9.2.2.2 *Landfarming experiences in the Antarctic and sub-Antarctic*

There are few reports of field bioremediation trials of hydrocarbon-contaminated soils in the Antarctic and sub-Antarctic, and there have not yet been any full-scale landfarming operations. However, a number of small-scale studies have been conducted, dealing mainly with shallow remediation relative to Arctic sites.

Kerguelen Archipelago (France)

Delille *et al.* (2004a, 2004b) observed reduction in the levels of alkanes in both control soil and soil fertilized with Inipol EAP-22 at two sites on Grande

Terre, in the Kerguelen Archipelago. After one year, soils that were artificially contaminated with 2 l m^{-2} of diesel fuel or Arabian light crude oil lost nearly all alkanes and approximately 80 to 90% of PAHs. C_{18}: hopane and C_{18}: chrysene ratios decreased, consistent with biodegradation, and alkane loss was faster in fertilized than in unfertilized plots. Covering the soils with plastic to raise temperatures had a small effect on alkane degradation.

Macquarie Island (Australia)

On Macquarie Island, diesel fuel biodegradation was studied in a soil with high organic matter content (3.1% organic carbon) at a site overlain with tussock grass that was periodically saturated during high rainfall events (Rayner *et al.* 2007). The site contained soil with approximately 7000 mg DRO kg^{-1} soil. A field treatability evaluation with an air-sparge port proved unsuccessful because the shallow water table and thin soil cover led to channel development. Then a micro-bioventing system was tested for aerating the soil. Respiration rates were estimated by measuring oxygen depletion after air injection. Hydrocarbon degradation rates were between 3 and 25 mg $kg^{-1}day^{-1}$, with an average rate of ~10 mg $kg^{-1}day^{-1}$, of which approximately 1.3 mg kg^{-1} day^{-1} might reasonably be attributed to degradation of native soil organic matter.

Laboratory incubation experiments conducted on well-aerated soil from this site measured 20 and 30 mg DRO kg^{-1} day^{-1} loss during incubation in unfertilized and fertilized soil, respectively (Walworth *et al.* 2007). Diesel biodegradation was substantially increased by addition of small amounts of N-fertilizer (125 or 250 mg N kg^{-1} soil), but not by larger additions (375 to 625 mg N kg^{-1} soil). The lack of response to the higher rates of fertilization was attributed to depressed soil water osmotic potential.

King George Island

Ruberto *et al.* (2003) studied both abiotic and biotic losses in open vessels in a field experiment at Jubany Station on sub-Antarctic King George Island. The soil was artificially contaminated with approximately 15 000 mg kg^{-1} of gas-oil. There was 54 to 61% loss of TPH in the first 10 days, attributable to volatilization as this loss was observed even in an abiotic control. Over 54 days of study, 72% of the initial hydrocarbon was lost via abiotic processes. Biological losses of TPH were increased by inoculation of the soil with B-2–2 *Acinebacter*.

Casey Station, Antarctica

In 1998 Snape and co-workers (Snape *et al.* 2006a; Powell *et al.* 2006a; McIntyre *et al.* 2006) established a small-scale long-term landfarming trial at Old Casey Station in Antarctica. The trial was designed to evaluate the utility of

Table 9.3 *Trial details for a bioremediation field trial at Old Casey Station; FW is freshwater, TW is water from a contaminated waste disposal site, CRN is controlled release nutrients (LCRN = low application rate; HCRN = high application rate) and AER is aeration. Initial fuel concentrations in the soil were ~ 16 000–20 000 mg SAB kg^{-1}.*

Treatment	Nutrients (mg kg^{-1} MaxBac®)	Other actions	Residual TPH (mg kg^{-1})	
			top	Bottom
Control	0		5780 ± 3090	11 360 ± 1810
LCRN + FW	2050	Fresh water occasionally	2480 ± 560	6240 ± 430
LCRN + TW	2050	Contaminated water occasionally	3320 ± 1020	6870 ± 1240
HCRN	7180		2010 ± 1380	5500 ± 530
HCRN + AER	7180	Tilled opportunistically	1850 ± 390	3960 ± 410
LCRN	2050		1960 ± 710	4720 ± 700

using CRNs during near-surface passive bioremediation by evaluating degradation rates and processes for five treatments relative to a control.

Soil was collected from Old Casey Station from a site that had been contaminated in 1982 by ~36 000 l of SAB diesel fuel. Sixteen years later contaminant concentrations of 16 000 to 20 000 mg SAB kg^{-1} soil were common in the upper catchment (Snape *et al.* 2006a). Collected soil was sieved, fertilized with MaxBac 25.5-4-0.5, homogenized and placed into tins (24 × 25 × 15 cm) that had perforated bases. Soil samples were taken annually from the tins from a depth of 3–5 cm during the five years of the experiment, and from 5 cm depth intervals at the completion of the trial.

A brief summary of the results for fuel concentrations is shown in Table 9.3. From this experiment it was concluded that volatilization was a highly efficient means of removing SAB from the top few centimeters of soil, tilling promoted volatilization and aerobic biodegradation, anaerobic biodegradation was an important remediation process at lower depths, low nutrient availability and diffusion of volatile hydrocarbons was rate-limiting in summer, little volatilization or biodegradation occurred in winter, and the effective treatment period was 6–8 weeks in summer when the ground was thawed and snow-free.

Vestfold Hills, Antarctica

Kerry (1993) evaluated loss of hydrocarbons from a soil contaminated with 6 l m^{-2} of SAB fuel in soil at Vestfold Hills, Antarctica. Hydrocarbon levels at a depth of 0 to 3 cm were reduced from over 20 000 mg kg^{-1} to 1096 mg kg^{-1} in fertilized soil, and to 4269 mg kg^{-1} in unfertilized soil after one year. In fertilized soil coated with gum xanthic and covered with plastic, the residual hydrocarbons were 5588 mg kg^{-1}, indicating that losses through volatilization were probably considerable. The trial plots were open to infiltration and subsurface runoff, and it is not known how much of the artificial spill simply migrated away.

9.3 Recent advances in landfarming in seasonally frozen ground

9.3.1 Nutrient management

Our understanding of nutrient requirements has increased in recent years. Nitrogen demands, as well as the potentially adverse impact on bioremediation from over-fertilization, are relatively well known (see Chapter 8). However, nutrient management in a landfarm can be problematic, particularly at remote locations and in coarse-textured soils. The issue of concern is how to maintain optimum nutrient levels for a long time, during periods of variable water and temperature at remote sites. CRNs can offer an alternative to highly soluble agricultural-grade fertilizers and help overcome these limitations. Limited field experiments have evaluated the performance of commercially produced CRNs. Snape *et al.* (2006a) successfully tested MaxBac (Table 9.3), a granular material comprised of prills with a semipermeable coating to control nutrient release, in a field trial at Casey Station, Antarctica. Walworth *et al.* (2003) tested fish bone meal for bioremediation of contaminated Alaska soil. Nutrient (N and P) release rates, which were controlled by breakdown of organic N compounds and dissolution of phosphates, were temperature-dependent and mirrored nutrient demand of microbial biodegradation. Mumford *et al.* (2006) used ammonium-saturated zeolites as a CRN (discussed further in Chapter 11, Section 11.3.2.2). Calcium phosphate was added to provide calcium ions for cation exchange. It was found that solution calcium concentration controlled the rate of N release from the zeolites. Use of CRNs in cold-region landfarming, although largely unstudied, might offer a practical mechanism for accurately managing nutrients in cold, remote sites.

9.3.2 Role of volatilization

Several studies (Chatham, 2003; Paudyn *et al.* 2005; Piotrowski *et al.* 1992) have illustrated the relative importance of volatilization versus biodegradation. As lighter petroleum fractions are removed by volatilization, biological processes become more important, depending on site conditions and soil and contaminant properties. In some circumstances volatilization is the primary mechanism of hydrocarbon loss and, when site-specific regulations permit, can be promoted through landfarm tilling. Recognition of the importance of volatilization could lead to improved landfarm design and allow more accurate prediction of treatment duration.

9.3.3 Containment

Cold-region landfarming is a relatively slow process in contrast to other cleanup methods. Therefore, contaminant containment must be a key management consideration, although preventing runoff or leaching of petroleum-contaminated water from landfarms can be difficult. Migration has most often been controlled by constructing a berm around the edge of the landfarm and lining the landfarm with geotextile fabric (Reynolds *et al.* 1994). Permeable reactive barrier (PRB) technology can also be used to mitigate the spread of hydrocarbon contamination. PRBs are passive, low-technology systems for adsorbing contaminants from runoff water. Because they do not require power to operate, they may be ideal for remote sites. Snape *et al.* (2001) tested a variety of PRB materials at Casey Station, Antarctica, and found that fine-grained reactive materials are less suitable than coarse-grained materials because their permeability characteristics may change during freeze-thaw cycles. Their results indicate that granular activated carbon significantly reduced concentrations of petroleum hydrocarbons and heavy metals in runoff water. Gore *et al.* (2006) evaluated stability of PRBs filled with activated carbon, raw clinoptilolite, and a nutrient-amended clinoptilolite exposed to freeze-thaw cycles under varying moisture conditions to simulate their use in cold regions. The activated carbon was relatively stable, whereas the clinoptilolite exhibited a tendency to crack and form fragments, which could indicate that long-term permeability of PRBs filled with this material might be problematic. There is a need for basic and applied research regarding PRB materials and design for development of this technology for application in cold environments.

9.3.4 Aeration

Despite the fact that inadequate oxygen is one of the most important limitations to biodegradation and overall landfarm performance, we are just beginning to explore many facets of managing landfarm aeration. The importance of the level of oxygen supply is demonstrated by basic studies such as that of Sierra and Renault (1995). However, we have limited information about oxygen management in landfarms. Rayner *et al.* (2007) were successful using

Table 9.4 *Landfarming design considerations*

Water management	Slope the base of the landfarm. Provide a leachate collection system for moisture control and leachate treatment. Prevent off-site migration of runoff and leachate or treat migrating water. Minimize the depth of soil in the landfarm. Provide irrigation if supplemental water is needed.
Nutrient management (see Chapter 8)	Optimizing nitrogen is key to efficient bioremediation. Supplemental phosphorus may speed bioremediation. In remote areas or in very coarse-textured soils, controlled release nutrients should be considered.
Soil mixing (tilling)	Regular mixing improves soil aeration, helping to ensure an adequate supply of oxygen at depth. Mixing increases volatilization, particularly of short chain hydrocarbons. Mixing redistributes moisture, nutrients, and contaminants and prevents or reduces concentration "hotspots."
Aeration	Aeration enhances volatilization and biodegradation of petroleum hydrocarbons. Aeration becomes increasingly important with lower temperatures.

micro-bioventing points placed very close together for shallow aeration. The effect of alternate aerobic/anaerobic regimes has not been studied, and the importance of zones of anaerobic degradation (e.g. at the bottom of a landfarm cell) is poorly understood. However, recent studies suggest the importance of anaerobic degradation of hydrocarbons (Powell *et al.* 2006b).

9.4 Guidelines and recommendations

9.4.1 Landfarming design considerations

Based on practical and research experience, we believe the design parameters shown in Table 9.4 are most important for successful cold-region landfarming.

9.4.2 Appropriate monitoring

Monitoring treatment efficiency in landfarms can be hampered by highly heterogeneous contaminant distribution, poor distribution of water and oxygen, the presence of frozen soil, and remote location. Detecting changes and attributing them to managed variables in a complex soil system requires appropriate sampling strategies (Chapter 7), meaningful chemical analyses, and judicious data analysis. Spatial variability within a landfarm can interfere with reliable sampling. Although soil mixing can help to alleviate this problem, spatial variability will always exist. Therefore, strategic sampling with suitable replication is essential.

9.4.2.1 *Soil sampling*

Because soil sampling for bioremediation has been extensively described by Paetz and Wilke (2005), only a brief summary is provided here. Primary considerations in soil sampling are (1) discrete versus composite sampling, (2) spatial distribution of samples, and (3) sampling schedule. Selection of a sampling scheme depends upon the sampling objectives. Consideration should be given to (1) the trade-off between precision and resolution (spatial and temporal) versus cost, and (2) the intended use of the resulting data (see Chapter 7, Section 7.2.3 for additional discussions).

Composite sampling combines numerous sub-samples into a composite sample that seeks to represent the entire sampled area. Individual sampling points can be selected either randomly or systematically. An advantage of composite sampling is reduced analysis cost; however composite sampling can preclude mapping contaminant spatial distribution. Soil field screening with organic vapor analysis prior to sampling can detect hot spots, and supports sampling pattern design and mapping of contaminant spatial distribution. Field screening with composite sampling is routinely used in the Arctic to assess contaminant reductions over time. A final discrete sampling event is necessary to substantiate that cleanup goals have been met, and to support site closure.

Paetz and Wilke (2005) discussed numerous sampling patterns, including random non-systematic patterns, stratified random sampling, circular grids, sampling along a linear source, and regular grids. Sample depth is also critical. If the ground is frozen, there is a temptation to sample only the surface. This can be highly misleading as the top few centimeters of soil can have substantially lower hydrocarbon concentrations than the soil below. Also, many undisturbed cold-regions soils have gravel or pebble-rich surfaces that can decrease contaminant concentration through dilution. In frozen soil it is preferable to sample the soils slightly below the surface at a depth of 3–5 cm (Snape *et al.* 2006a), to excavate a small pit, or to take cores to the full depth of the landfarmed soil.

Random sampling can be designed with a random number generator. Individual samples are often composited, but can be analyzed individually. Known coordinates of the samples provide a limited measure of spatial variability, but interpolation and spatial mapping are difficult. Modifications include non-systematic or non-regular patterns, where samples are collected on a grid in shapes resembling "N," "S," "W," or "X." In stratified random sampling, the area is divided into regular areas or cells and each cell randomly sampled.

Sampling on a regular grid (usually rectangular) is useful for constructing contour maps of contaminant levels. Samples can be discrete or, in stratified random sampling, sub-samples collected from grid cells can be composited to represent the concentration of each cell. Geospatial interpolation techniques such as kriging are useful for constructing contaminant concentration distribution maps and estimating the total contaminant in a volume of soil from grid sample data (Reynolds 1993; Reynolds *et al.* 1994, 1998). Grid sampling also facilitates repeated temporal sampling because sample points are relatively easy to locate, however soil mixing may move soil with respect to geospatial sampling points. Reynolds (1993) used estimates of the mass of contaminant in a soil volume with time-series kriging to develop degradation rates that integrated samples representing the entire area of the site to the depth sampled. The spatial distribution of rates was useful for identifying site limitations. Variations of grid sampling in non-rectangular grid patterns include sampling on a grid of equilateral triangles, which reduces between-sample spacing; circular grids with grid points arranged on concentric circles around suspected contaminant sources to concentrate samples near the source; and sampling to concentrate sample points near a linear source, such as a leaking pipeline.

An obvious disadvantage of grid sampling is that numerous samples must be analyzed to construct a reliable concentration profile map of a landfarm. The associated analytical costs are proportionally higher than for more limited sampling. As with treatment time and cost (Chapter 1, Figure 1.1), there is a balance between resolution of information and cost of sampling.

9.4.2.2 *Soil analysis*

As with sample collection, selection of analytical techniques for contaminants in landfarm soil samples depends upon the purpose of the analyses, driven by stakeholder or regulatory requirements. For example, to demonstrate that requirements have been met, the regulator may dictate analyses for specific compounds or groups of compounds with a high level of quality assurance/quality control and appropriate methodology or particular analytical methods. For detailed studies, individual pathways of contaminant loss may be monitored to evaluate relative losses via biotic and abiotic mechanisms (Snape *et al.* 2006a; Prince *et al.* 2002; McIntyre *et al.* 2006) (Chapter 7, Section 7.3.1). These may include direct measurement of volatile losses (Ausma *et al.* 2001, 2002) and fractional hydrocarbon analyses (McIntyre *et al.* 2006).

Because rates in cold regions can be, on average, slower than for warmer regions, determining if a landfarm is "working" by measuring changes in residual contaminant level or compound ratios can be difficult. An alternative is to try to measure such associated processes as biological activity, specific microbial processes, or microbial populations (SERDP 2005) (discussed in Chapter 4). Demonstrating that desired processes "are active" is beginning to be an acceptable practice for process monitoring but not for final contaminant concentration goals. The trade-off is that monitoring biological processes is a relatively new concept and currently may require more frequent sampling until our understanding is more complete.

In addition to contaminant assays, soil samples from landfarms are also used to evaluate such soil parameters as soil moisture content, nutrient levels, soil salinity, and pH. Knowledge of these parameters is critical for optimum landfarm management. Ability to manage these soil properties is lost without a comprehensive soil sampling and analysis program.

9.4.2.3 *Data analysis strategies*

The practical goal of monitoring landfarms is to confirm that cleanup goals have been met, to show that processes leading toward cleanup goals are ongoing, or to identify areas of a landfarm that might need further or different management action. For confirmation that cleanup goals have been attained, the data analyses are typically described in the project work plan and are generally straightforward. For landfarm process monitoring, the data analysis may be more involved. One approach is to detect change in contaminant content or composition. To increase our ability to detect change in complex systems such as landfarms, options include optimizing sample collection design and density, increasing the resolution of the chemical analysis, or employing techniques to improve statistical precision (see Chapter 7, Section 7.3.4).

As an example of using all three approaches, Reynolds *et al.* (2004b) compared low-input remediation systems during a multi-year study at three locations in Alaska. Composite sampling was used to reduce the number of chemical analyses and costs, but chemical analyses were more rigorous than typical. They included analysis of fraction-specific hydrocarbons that were in turn normalized to a biomarker, such as hopane or decalin to reduce uncertainties due to concentration variance, and expressed as depletion values.

9.5 Future research

The greatest need for cold-region landfarming research is for well-documented field-scale projects, particularly in the southern hemisphere. These studies need to be designed with suitable statistical replication and appropriate monitoring. Field-scale spatial and temporal variability should be measured and interpreted. Both biotic and abiotic hydrocarbon loss processes should be measured to provide delineation between hydrocarbon loss mechanisms. Systematic studies are needed to optimize tilling to maximize rates of hydrocarbon loss at remote sites. In addition, field-scale projects with good temperature monitoring are required. Practical temperature limitations for landfarms are still unknown.

There is a specific need for study of the management of soil moisture and aeration in landfarming. The need for providing good drainage is evident, but responses to varying degrees of soil aeration have not been well documented, and the potential for *in situ* aeration systems has not been fully explored. Also, better management of runoff will open up greater possibilities for landfarming by reducing risks of off-site contaminant migration. Low input landfarming, which can be seen as a slightly higher level of management than natural attenuation, may be an attractive alternative to higher input options. For example, this management scheme might be combined with the use of PRBs to treat runoff water as it leaves the landfarm site.

Better management of nutrient delivery, as by CRN systems, is needed for cold-region landfarming. Reducing the need for active management will permit landfarming in less accessible areas and allow application of higher doses of fertilizer without causing inhibition of bioremediation processes or off-site migration of nutrients. However, CRNs developed for temperate regions should not be applied in cold regions without consideration of the potential effects of temperature on nutrient release rates. Material coatings and structures of fertilizer materials can fail when confronted with the frequent freeze-thaw cycles prevalent in cold regions (Gore *et al.* 2006). CRN materials need to be developed specifically for cold-climate conditions and tested in realistic field conditions (see Chapter 11, Section 11.3.3.2).

There is growing interest in cold-region landfarming because it is a relatively inexpensive and effective method for dealing with contaminated soils in remote, inaccessible areas. For treating petroleum hydrocarbons in these areas landfarming is an attractive, inexpensive alternative. Research priorities such as those outlined above need to be addressed to fill gaps in our knowledge, and research results must be applied to provide better integration between science and technology. As technical designs are improved, landfarming in cold regions will probably become a preferred remediation method.

10

Thermally enhanced bioremediation and integrated systems

DENNIS M. FILLER, DAVID L. BARNES, RONALD A. JOHNSON, AND IAN SNAPE

10.1 Introduction

It is well established that microbial activity is slower at low temperatures, and that there is a corresponding decrease in biodegradation rates (Paul and Clark 1996; Walworth *et al.* 1999; Scow 1982; Ferguson *et al.* 2003b; discussed in Chapter 4). As temperatures fall to near the freezing point of water, biomineralization of hydrocarbons practically ceases. Evaporation rates are also slower at low temperature, although diesel products and more volatile fuels continue to volatilize below 0 °C. For most cold regions, soil is typically unfrozen for only 6–8 weeks, affording a short *in situ* or passive *ex situ* treatment season. Even when the ground is thawed, temperatures are generally lower than optimal for hydrocarbon-degrading bacteria (Braddock *et al.* 2001; Rike *et al.* 2003).

At their simplest, thermally enhanced bioremediation schemes aim to increase microbial activity by increasing soil temperatures and extending the period when the ground is unfrozen. Modern integrated designs go much further – they typically incorporate some form of venting to promote volatilization, and deliver nutrients, oxygen, and water to hydrocarbon-degrading bacteria in attempts to optimize bioactivity. They are also designed to prevent off-site migration of contaminants and nutrient-enriched waters.

Relative to other remediation options, thermally enhanced bioremediation is a low-cost treatment option (see Chapter 1, Figure 1.1). It is typically much cheaper than bulk extraction and disposal or on-site combustion/desorption treatments, perhaps by a factor of five or more, but approximately two to four times more expensive than landfarming (Chapter 9). However, true cost

Bioremediation of Petroleum Hydrocarbons in Cold Regions, ed. Dennis M. Filler, Ian Snape, and David L. Barnes. Published by Cambridge University Press.

comparisons are very difficult. Factors such as remoteness, risks to human health and the environment, importance of preventing off-site migration, the cost of energy production, desired treatment time, and contaminant concentrations and prescribed target cleanup levels are site, regional, and country specific.

This chapter focuses on the fundamentals of thermally enhanced bioremediation and develops the concept of an integrated remediation system. By synthesizing observations from laboratory, bench-scale, and field pilot studies we attempt to explain some of the design concepts from first principles. The suitability of construction materials, implementation considerations, and aspects of monitoring and validation are illustrated by reference to the few well-documented case studies from Alaska and Canada. Consideration of thermally enhanced bioremediation for possible use in other cold regions is given in the context of ease of technology transfer, legislative requirements, and cost of energy production. We try to indicate some of the economic and environmental considerations that will aid cost–benefit analyses.

10.2 Review

Bioremediation relies on microorganisms in soil to degrade organic contaminants. Efficient treatment is predicated on aerobic conditions, availability of hydrocarbon degraders, suitable soil water and nutrient regime, and type and concentration of contaminants. Temperature is also of major importance as it directly influences the rate of microbial activity, and treatability studies using Arctic, Antarctic and alpine soils demonstrate that low temperatures significantly inhibit biodegradation rates (Aislabie *et al.* 1998; Braddock *et al.* 2001; Ferguson *et al.* 2003b; Rike *et al.* 2003; Walworth *et al.* 1999). Because of this strong temperature dependence, it is particularly difficult to predict the duration of *in situ* biological treatments, by landfarming for example, because biodegradation rates are intrinsically controlled by weather.

Where landfarming involves tilling the ground, or where air is actively passed through soil in the case of bioventing, remediation is achieved through a combination of volatilization and biodegradation. Although petroleum hydrocarbon (PHC) losses via such schemes are usually attributed to biological processes, substantial remediation actually occurs through abiotic volatilization (Chapter 7, Section 7.3.1), another process subject to weather conditions.

For many, the distinction between PHC remediation by biodegradation or volatilization, or a combination of both, is moot; economical remediation is the prime objective. However, the distinction is important for future attempts to optimize treatment methods, especially where integrated schemes specifically target both processes to achieve fast and thorough PHC removal.

10.2.1 Development of thermally enhanced bioremediation

Because summers are short, and the active layer above permafrost thaws for only a couple of months annually, artificial soil warming is necessary for successful bioremediation of PHC within a few treatment seasons. Thermally enhanced bioremediation (TEB) implies bioremediation enhanced by an engineered system(s) to increase volatilization and biostimulation with aeration, soil warming, and nutrient and water management (Chapter 8). TEB is highly efficient in reducing light to medium-range hydrocarbon fractions quickly, and can evaporate otherwise recalcitrant fractions. TEB design criteria includes thaw protection of permafrost, prevention of off-site migration of contaminants and/or nutrient-laden waters, and facilitation of convenient monitoring of remediation progress.

Case studies: Early assessments of bioventing and TEB in Canada and Alaska

Results from bioventing demonstration and optimization projects in Alaska and Canada between 1992 and 1996 indicated better biodegradation rates of PHC with warmer soil temperatures and extended periods of biological activity (Moore *et al.* 1995; Thomas *et al.* 1995). Moore *et al.* (1995) reported significant PHC reductions (5000 to 2000 mg kg^{-1}) of natural gas condensate in wet glacial-fluvial sands and gravels mixed with clayey-silt fill, from combined bioventing and soil-vapor-extraction treatment at a Gulf Strachan Gas Plant northwest of Calgary, Alberta, Canada. Thomas *et al.* (1995) evaluated bioventing potential at three spills from tanks at the Alyeska Valdez Marine Terminal facility in Valdez, Alaska. Sand and gravel soils in lined containment cells were contaminated with oily water to 7800 mg diesel kg^{-1} soil. Treatment monitoring indicated increased PHC degradation with warmer weather, and no appreciable decrease in biodegradation rates during early winter at this coastal site.

A bioventing pilot study conducted from 1991 to 1995 at Eielson Air Force Base, near Fairbanks, Alaska, was the first attempt to manipulate bioremediation with soil warming schemes at a cold site. This trial enhanced soil temperatures by using sawdust as insulation over a warm-water circulation system embedded in silt over sandy-gravel contaminated with JP-4 jet fuel to 5000 mg PHC kg^{-1} soil. Leeson *et al.* (1995) reported the period of annual effective treatment was increased by 25% increased biodegradation rates were attributed to soil warming, and initial soil diesel concentrations were reduced by 60% within three years.

Table 10.1 *Results of TEB field study at University of Alaska Fairbanks site (Filler 1997). Concentrations in mg kg^{-1}*

	GRPH		Total BTEX	
	Bioventing with active warming	Bioventing with passive warming	Bioventing with active warming	Bioventing with passive warming
Initial concentration	2400	1000	730	545
Final concentration after 22 months	<10	<160	<5	<30
Cleanup levels	200	200	75	75

Notes: GRPH – gasoline range petroleum hydrocarbons; BTEX – benzene, toluene, ethylbenzene and xylenes.

Table 10.2 *Itemized costs (1997 US$) of treatment options for 3825 m^3 of contaminated soil for the University of Alaska Fairbanks study (Filler 1997)*

Item	Conventional bioventing	Bioventing with passive warming	Bioventing with active warming
Construction (planning, earthwork, installations)	56 200	106 700	124 100
Monitoring and O&M	107 500	77 000	76 600
Reporting (misc. expenses inclusive)	39 000	36 800	26 900
Total	202 700	220 500	227 600
Unit cost ($/m^3)	53	57.6	59.5
Remediation time (years)	7[a]	4[a]	1.8

[a]Projections based on maximum contaminant levels and first year biodegradation rates. Costs under *conventional bioventing* and *bioventing with passive warming* columns based on actual costs and remediation time projections.

Following the heating scheme trialed by Leeson *et al.* (1995), Filler (1997) developed a thermal insulation system for engineered bioremediation at a fuel-contaminated site at the University of Alaska Fairbanks. Prototype insulation system designs were modeled under field conditions for traffic loading, freeze-thaw penetration, and two-dimensional heat transfer to assess the merits of application and treatment effectiveness. Test plots treated with insulation system designs and bioventing were monitored for thermal, biological, and chemical comparisons with a control area over two years. Statistics for this bioremediation project are summarized in Tables 10.1 and 10.2.

Case study: Integrated TEB at Prudhoe Bay, Alaska

The Deadhorse Lot 6 industrial site at Prudhoe Bay comprised 5050 m^3 of contaminated diesel soils. Initial contaminant concentrations were 250–11 000 mg diesel kg^{-1} soil and 37.5–1 000 mg gasoline kg^{-1} soil. After mixing during construction, the weighted average diesel concentration in the biopile was 2000 $\pm$ 200 mg kg^{-1} soil. The biopile comprised a pipe network for bioventing, fertilizer distribution, and heat dissipation for permafrost control, and was capped with a thermal insulation system that included a customized electric heat mat. Power was supplied by a 75 kVA transformer. Evolutionary operation was incorporated to optimize venting and heating, cyclical heating was afforded with thermostatic control and automation to reduce energy costs, fertilization rates were calculated based on mass balance relationships between soil, water, and contaminants, and the site was successfully remediated within three years (Filler and Carlson 2000; Filler *et al.* 2001). Treatment results and costs for this case study are summarized:

Table 10.3

	Concentration in mg kg^{-1}					Economics (2003 US$)	
Timeline	Diesel	Gasoline	RRO	BTEX	Benz.	Item	Cost
Baseline 1999	252–11 000 (avg. 2000 ± 200)	37–1000	–	0.6–45.6	<0.2		
August 2000 (ex situ portion)	38–397	<2–27	–	<1.1	<0.02	Assessment, design, and construction	220 700
Sept. 2001						Improvements	18 000
(*in situ* portion)[a]						O&M (2 yrs.)	96 600
13 sample location	<40	<27	<61	<1	<0.04	Reporting (2 yrs.)	36 500
1 sample location	92	217	<50	16.8	0.46	Final confirmation	30 000
1 sample location (duplicate results)	579 (711)	14 (62)	<50	1(<2)	<0.08	2-yr Total	401 800
Cleanup levels	500	100	2000	15	0.5	Remediation time	2.5 yrs.
						Unit cost (US$ m^{-3})	79.5

[a]Sampling performed after *ex situ* portion of biopile remediated and removed.
RRO – residual range organics.

TEB principles and lessons learned from the University of Alaska Fairbanks field study were improved for application at an industrial site in Prudhoe Bay, Alaska, in 1999. The plan combined bioventing, soil warming, and fertilization to treat the largest single bioremediation site in cold regions at that time. This case study is highlighted on the previous page.

Most recently, Rayner *et al.* (2007) installed a *micro-bioventing* system comprising small air-injection rods to treat old diesel fuel in soil at a contaminated site on sub-Antarctic Macquarie Island. Approximately 100 metric tons of moderately contaminated water-saturated peaty soil (average diesel concentration: 2800 mg kg^{-1}) was amended with nitrogen fertilizer and aerated with the injection rods. Initial results from the pilot study indicated that a target concentration of 200 mg kg^{-1} could be achieved within two years of seasonal operation of the micro-bioventing array at ambient soil temperature of about 5 °C.

10.3 Recent advances, guidelines, and recommendations

The following guidelines and recommendations offer an approach to developing integrated TEB systems for cold region applications. Emphasis is given to fundamental design considerations for key elements of a TEB system – heat management, ventilation, and water and nutrient delivery. Additional design considerations for permafrost protection and prevention of off-site migration are also described. Guidelines and recommendations for assessing microbiology are described in Chapter 4, chemical and biological validations in Chapters 6 and 7, multi-parametric monitoring in Chapter 7, and nutrient and water requirements in Chapter 8. This chapter focuses on aspects of engineering design.

10.3.1 Fundamental design considerations

The fundamental design principles of integrated TEB are summarized in Figure 10.1. PHC losses are usually realized through a combination of volatilization and biodegradation. Integrated engineered systems enhance both processes by simultaneously aerating soil and removing vapors, warming soil, and optimizing nutrient and water conditions for increased bioactivity. To optimize the four managed design parameters illustrated in Figure 10.1, it is important to understand the relationships between soil composition and mass transfer.

10.3.1.1 Soil composition and biopile homogenization

In the Arctic, sand-gravel roads and infrastructure routinely founded on built-up gravel pads are constructed over tundra. When impacted with petroleum contaminants, these soils can be easily treated with engineered bioremediation.

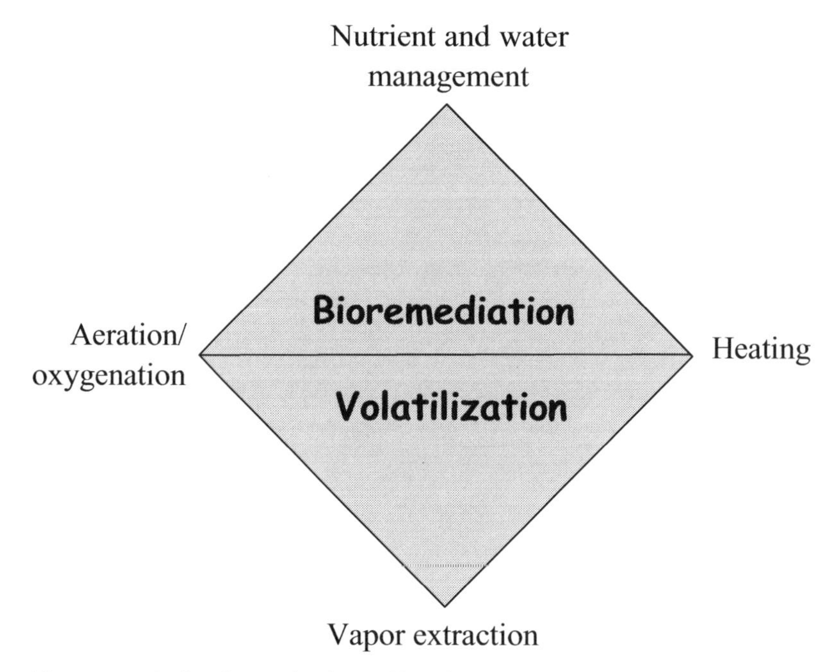

Figure 10.1. Design factors for thermally enhanced bioremediation.

More difficult to treat is impacted tundra sediments that exist through wind deposition or that underlie manmade pads and roads. These sediments often exhibit saturated conditions during the thaw season and are frozen in winter. In the high Arctic and Antarctica, the preponderance of petroleum contamination exists in nutrient-poor mineral soils associated with abandoned infrastructure, roads, and dumps within coastal margins.

Rocks and cobbles, gravels, sands, some silts, and unsaturated mixtures of these soil types are highly amenable to treatment, although for highly organic soils such as peat and tundra, treatment is dependent on degree of saturation. Highly saturated soils can be particularly difficult to oxygenate.

Variations in soil composition within a treatment zone can lead to considerable variability in treatment effectiveness for *in situ* techniques (Reynolds 1993). Thorough homogenization during biopile construction can overcome this to some extent. Soil homogenization should be attempted where one treatment system is constructed to remediate soils from co-located or different areas with significant contaminant and soil structure variability. For example, blending small amounts of peat and fine sand with coarse gravel would improve aeration and water-holding capacity, and make nutrient management easier. A strong argument for biopile homogenization is given when one considers thermal conductivity relative to heat diffusion, porosity relative to ventilation, and

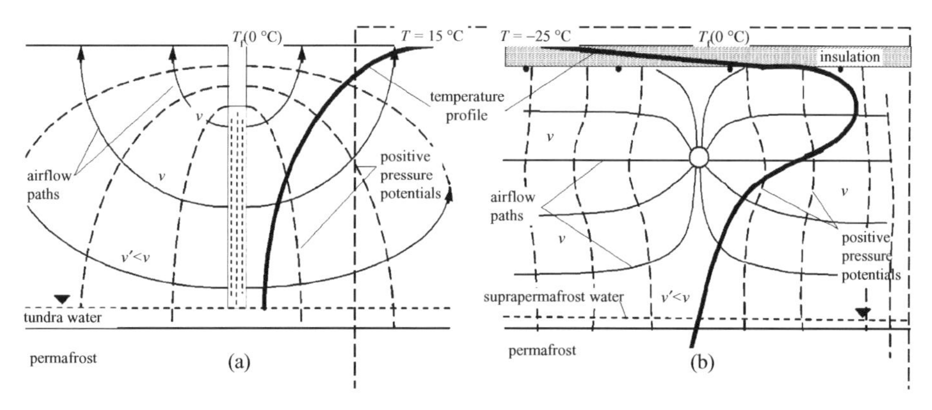

Figure 10.2. Subsurface airflow patterns and pressure distribution for a (a) vertical well *in situ* during summer operation, and (b) biopile horizontal pipe run in early winter.

water-holding capacity relative to water potentials and nutrient management (Chapter 8), with overall intent to optimize bioactivity and minimize treatment duration. Although soil mixing may not be possible at some sites that require *in situ* treatment, suitable TEB can be engineered nonetheless, if comprehensive assessment data is available.

10.3.1.2 Influence of soil type on airflow and bioventing

Contaminated gravels, sands, and silts with low to moderate (< 20%) moisture contents are amenable to bioventing as a consequence of favorable porosity and permeability. Clayey soil and silts with high moisture content are less permeable and are difficult to vent.

During bioventing, Johnson *et al.* (1990) found that under most conditions the governing flow equation for soil gas under a pressure gradient induced by a positive pressure at a well is the same as for groundwater flow. Consequently, the same analytical and numerical solutions to the groundwater flow equation are applicable to soil-gas flow in the vadose zone.

Seasonal airflow patterns and pressure potentials established at a vertical bioventing well, and for a horizontal well operating in a biopile at constant flow rates are illustrated in Figure 10.2. Summer and early winter temperature profiles are superimposed on the flow nets. Note that airflow velocity diminishes closer to the water surface in some soils, in response to increased saturation from capillary rise.

A field test for *in situ* bioventing requires installation of at least three observation wells positioned at various distances and locations away from an air-injection well. After inducing a constant flow rate at the injection well, pressures are monitored at the observation wells, thereby determining the effective radius of influence for the site. Assumptions of horizontal vapor flow, insignificant pressure changes within the treatment realm, and negligible changes in permeability caused by soil drying are valid for low-flow bioventing sites with shallow unsaturated zones, that are capped with insulation and confined by frozen ground or bedrock at depth. Applying Darcy's Law and the continuity equation to flow through porous soil, Johnson *et al.* (1990) developed the equation for steady-state air permeability (k) between injection and observation wells as

$$k = \frac{Q\mu_a \ln\left(\frac{R_W}{R_I}\right)}{m\pi P_{atm}\left[1 - \left(\frac{P_{W_{abs}}}{P_{atm}}\right)^2\right]} \quad \text{or} \quad k = \frac{2.3Q\mu_a}{4\pi \Delta P m}\log\left(\frac{t_2}{t_1}\right) \tag{10.1}$$

where Q is the venting rate, μ_a is the dynamic viscosity of air, m is the length of well screen, R_W is the well radius, R_I is the radius of influence, P_{atm} is atmospheric pressure, P_{wabs} is absolute vent well pressure ($P_{atm} + P_w$, the stabilized vent-well gauge pressure), ΔP is the hydrostatic law ($\Delta P = \rho_a g m$) where ρ_a is the density of air, and t_1 and t_2 are the test times at which corresponding gauge pressures P_1 and P_2 are measured.

Once the radius of influence is established, the number of injection wells and their placement can be determined. To account for variability in air permeability across a contaminated site, a factor of safety for design is afforded with overlap of well influence patterns. For remote sites, it may be impractical to conduct field tests for permeability and radius of influence. In such cases, intact soil samples can be collected and laboratory tested for permeability and soil characteristics (i.e. classification, porosity, and moisture content). Figure 10.3 is a general guide for bioventing configuration for use with granular soils exposed to low venting rates (470–2360 $cm^3\ s^{-1}$ or 1–5 cfm). However, whenever possible, site-specific field data should be collected for design of engineered bioremediation systems.

10.3.1.3 *Heat transfer in soil*

An understanding of the energy balance across the ground surface provides a basis for insulation system design and thermal modeling.

Heat transfer within unsaturated soil at a bioventing site is governed by conduction and convection, with contributions from surface radiation exchange (e.g. solar influx, long wave emissions from the earth, and evaporation). Solar warming is significant in summer, but decreases rapidly as the sun peaks lower on the horizon and the ground thermal gradient diminishes with the onset of winter. Heat addition through venting is realized when the injected air temperature is warmer than the ambient. Filler (1997) calculated the heat contribution from a single vent well at <1.3% that of the total contribution for a thermally enhanced system using an electrical heat mat operating in early winter at a large fuel-contaminated site in Fairbanks, Alaska. During winter, when vadose

(a)

Soil Type	Radius of Influence (R_I)
Sandy-gravel and gravels	7.5–12.5 m
Silty-sands	3–6 m
Silt	1.5–3 m

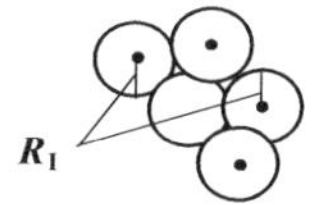

Plan view of *in situ* vertical well configuration.
(R_I a function of air injection rate)

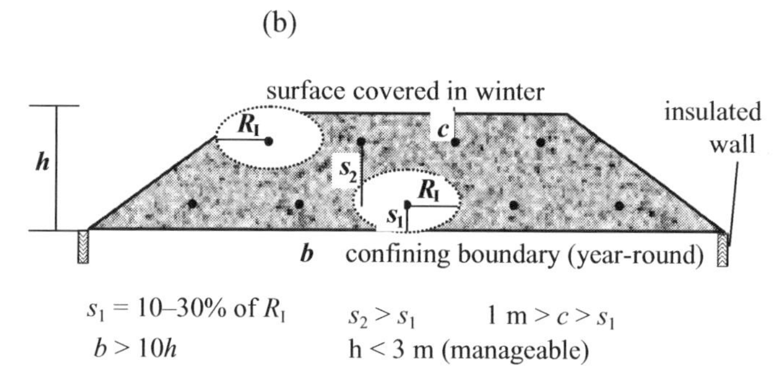

Biopile Notes:
Choose b and h such that pile side slopes afford easy access for O&M and monitoring.
PHC within clearance zone c will degrade faster from exposure to wind and sun during thaw season. R_I a function of biopile shape and confinement.

Figure 10.3. General guide for (a) bioventing wells and (b) pipes in cold regions.

zone soil temperatures are warmer than the injected air, venting will actually decrease soil temperatures. Heat loss through moisture evaporation is negligible when compared to the other losses.

As winter approaches, heat is lost from the treatment realm through conduction to the atmosphere as the top of the pile cools, and to cool tundra water or permafrost at the pile bottom. As the freeze-front progresses downward, ground heat is liberated as a function of decreasing temperature and moisture content, soil thermal conductivity and specific heat capacity, and the latent heat of water. The freeze-front will progress rapidly through dry sands and gravels, and more slowly through moist silt. The two-dimensional heat diffusion equation for homogeneous and isotropic soil under bioventing influence is:

$$\frac{\partial^2 T}{\partial z^2} + \frac{\partial^2 T}{\partial x^2} = \left(\frac{C}{K}\right)\frac{\partial T}{\partial t} + \frac{q_l + q_e - q_{hm}}{K} \qquad (10.2)$$

where C is volumetric specific heat, K is thermal conductivity, T is temperature, and q_l, q_e, and q_{hm} are latent heat, moisture evaporation, and heat mat contributions, respectively (Filler 1997). The heat mat (q_{hm}) contribution is omitted from Eq. (10.2) when only using insulation as a ground or biopile cover. For sands and gravels with low moisture contents, both the latent heat and evaporation contributions can be neglected, and the first-order differential equation becomes a function of soil thermal diffusivity (α or C/K) and surface and water table boundary conditions. Filler (1997) demonstrated that with active warming it can take a month to thaw 5 m of alluvium above the water table at an *in situ* sub-Arctic site, or a large gravel biopile in the Arctic. For soil realms undergoing treatment with and without the aid of mechanical warming, the respective simplified heat transfer equations are:

$$\frac{\partial^2 T}{\partial z^2} + \frac{\partial^2 T}{\partial x^2} = \left(\frac{C}{K}\right)\frac{\partial T}{\partial t} - \frac{q_{hm}}{K} \quad \text{and} \quad \frac{\partial^2 T}{\partial z^2} + \frac{\partial^2 T}{\partial x^2} = \left(\frac{C}{K}\right)\frac{\partial T}{\partial t} \tag{10.3}$$

10.3.1.4 Application of heat transfer models

When bioremediating remote sites that must remain active to traffic flow, or where a project is subject to fixed treatment duration, it is essential to model heat transfer within the treatment regime during the insulation system design phase. Once the thermal insulation system is installed at a remote site or is embedded below protective soil cover, it may be cost-prohibitive to modify or repair later.

Finite difference models can be used to approximate winter thermal regimes for large treatment areas where heat loss at boundaries is less important. For a well-designed heat mat under insulation, the interior region is governed by downward heat conduction; lateral heat losses at the boundaries can be significantly reduced with embedded insulated walls (Figure 10.3(b)).

Two-dimensional heat flow analysis is essential to bioremediation design where significant lateral heat losses are anticipated at the boundaries and/or permafrost degradation is a concern. The two-dimensional energy balance problem is well defined with finite element analysis. A strong finite element model should be able to handle multi-layered soil systems and account for latent heat of transformation effects (i.e. water-to-ice-to-water phase changes).

Solution to two-dimensional heat conduction models requires input data and defined boundary conditions for a discrete columnar element comprising the finite element mesh to be analyzed. For example, input data might include layer thickness, density and moisture content, latent heat, and frozen and unfrozen thermal conductivity and heat capacity for each individual insulation system and soil layer comprising the treatment regime. The respective surface and bottom boundary conditions will be functions of atmospheric temperature and the confining layer (i.e. permafrost, bedrock, water surface, or liner temperature).

Zarling and Braley (1988) developed the equation for sinusoidal soil surface temperature variation:

$$T_s(t) = T_{ms} - A_{os}\cos\left[\frac{2\pi(t - \varphi_s)}{365}\right]_{rad} \tag{10.4}$$

based on mean annual soil surface temperature (T_{ms}) and amplitude of soil surface temperature variation (A_{os}), functions of surface n-factors, periods of air-freezing and thawing seasons, and soil surface temperature phase lag (φ_s). In the absence of site-specific ambient air and surface temperatures, local climatic data for the most recent year, or published regional climate data can be used for determining reasonable approximations with Eq. (10.4). For a water surface boundary condition, the mean annual water temperature and its amplitude of variation will be used. For simplification, a constant temperature approximation may be used for the bottom boundary condition with little sacrifice to model accuracy. Side boundary conditions for an interior element will be zero heat-flux boundaries since they represent lines of symmetry with respect to the whole treatment regime. A summary of the heat conduction problem for the insulated and covered biopile depicted in Figure 10.3(b), recognizing the heat mat contribution, is summarized:

Surface boundary condition: $$T_s(t) = T_{ms} - A_{os}\cos\left[\frac{2\pi(t - \varphi_s)}{365}\right]_{rad} \tag{10.5}$$

Bottom boundary condition: $$T_{gw}(t) = T_{m(gw)} - A_{o(gw)}\cos\left[\frac{2\pi(t - \varphi_s)}{365}\right]_{rad} \tag{10.6}$$

Side boundary conditions: $$\frac{\partial T}{\partial x} = 0 \tag{10.7}$$

Heat mat boundary condition: $$q_{hm} = -xT + y \tag{10.8}$$

Note: The rate equation for a heat mat is typically linear, with the negative sign indicating a descending output trend with increased temperature, and the y-intercept identifying the maximum heat output.

10.3.1.5 *Load simulation*

In situ TEB schemes can be designed to accommodate traffic flow at an active property (Figure 10.4). Polystyrene or other board insulation, routinely used with road construction in Arctic regions, is the critical design element because of the potential for compression. A stress-strain analysis is performed based on the worst-case scenario of two of the heaviest vehicles parked on the site overnight.

Layer input data for a typical stress-strain analysis typically includes layer thickness, Young's Modulus, and Poisson's Ratio. The later two parameters can be obtained for insulation system materials from manufacturer's specifications. Field moduli of elasticity for soils can be determined from density and moisture

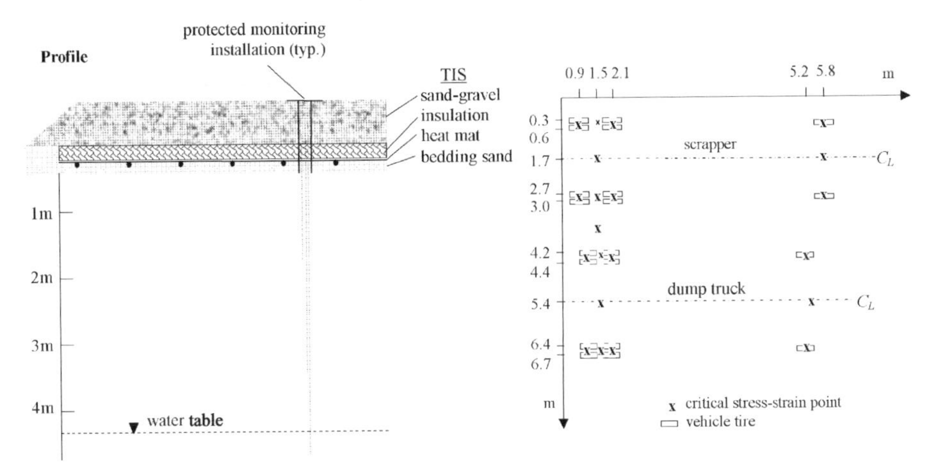

Figure 10.4. Example contaminated industrial site with thermal insulation system (ground application) and load simulation (after Filler 1997).

content data. Bowles (1988) presents empirical equations for estimating field moduli of elasticity based on drilling log blow counts (n); blow counts can be used to calculate or infer soil density. Poisson's Ratio for soils and n-density relationships can be found in classical soil and foundation texts. Since Poisson's Ratio for granular soils tends to decrease as soils freeze (Andersland and Ladanyi 1994), use of summer drilling log data will suffice for a stress-strain simulation.

ELSYM-5 (SRA Technologies) is a proven program for performing stress-strain analysis of multi-layered soil systems. Developed for the US Federal Highway System, and used routinely by the Alaska Department of Transportation for roadway/highway design, the software was designed specifically for soil-system applications. Required input data are the aforementioned soil characteristics and load applications. Critical stresses and displacement results, measured at the top and bottom of the insulation layer for the example depicted in Figure 10.4, will accurately gauge insulation compression (loss of R-value) from loading.

10.3.1.6 *Modeling freeze-thaw depths*

Type, thickness, and placement of insulation material over a contaminated site will govern freeze-thaw depths within a treatment regime. Filler (1997) demonstrated that for sub-Arctic sites subject to passive or mechanical warming, insulation thickness and placement governs subsurface warming in winter. Consequently, the design tables in Section 10.3.2.2 may be used without necessity to model freeze-thaw calculations for sub-Arctic applications. For bioremediation under more extreme conditions in the high Arctic and Antarctica, we strongly recommend performing freeze-thaw calculations for all thermal insulation system design. Software designed to perform freeze-thaw calculations will most likely be based on Stefan, Neumann, and/or Modified Berggren methodology. One such program, BERG-2, was developed to perform freeze-thaw analysis of engineered roadway systems (Braley 1984); Conner (1988) and Filler (1997) demonstrated its use for cold-region environmental applications.

10.3.1.7 Evolutionary operation

Evolutionary operation as it applies to bioremediation, is the process of optimizing systems to maximize treatment effectiveness and minimize operation and maintenance costs. For example, with respect to bioventing, the *tuning process* essentially increases the air-injection flow rate incrementally over time, while monitoring soil-gas quality and soil moisture contents. The idea is to optimize PHC volatilization without significantly diminishing biodegradation (Chapter 7, Section 7.3.1) through soil drying. An optimal system-wide flow rate would then become the basis for long-term bioventing operation.

Power requirements of heating components can also be optimized. Filler *et al.* (2001) demonstrated that significant energy-cost savings could be realized by cycling (or pulsing) power to a heat mat at a bioremediation site in the Arctic. With baseline data, evolutionary operation of any individual bioremediation system can be performed over several weeks during the treatment startup phase. Further seasonal optimizations can be achieved later during routine monitoring events.

10.3.2 Generic design of an integrated thermally enhanced bioremediation system

There is no limit to the number of bioventing wells and monitoring points for an *in situ* bioremediation site with vertical installations. The treatment regime will be defined by the extents of contamination, physical barriers, and operational requirements (e.g. sites that must accommodate traffic), and the number of air-injection wells will be defined by radius of influence. For biopiles (any combination of *in situ*/*ex situ* configuration), the air-injection pipe configuration is more a function of coverage relative to biopile shape than a particular radius of influence. Biopile size must be manageable with respect to treatability, weight, and permafrost control. If pipe runs are too long and with insufficient slope, friction losses coupled with low flow may diminish oxygenation and fertilization effects at the far ends. If a biopile is constructed too high, sheer weight of soil can subside the pile and damage bioventing and monitoring installations. To date, the largest known engineered biopile constructed on permafrost in the Arctic was approximately 55 m long, by 45 m wide, and 2.5 m tall (Filler *et al.* 2001).

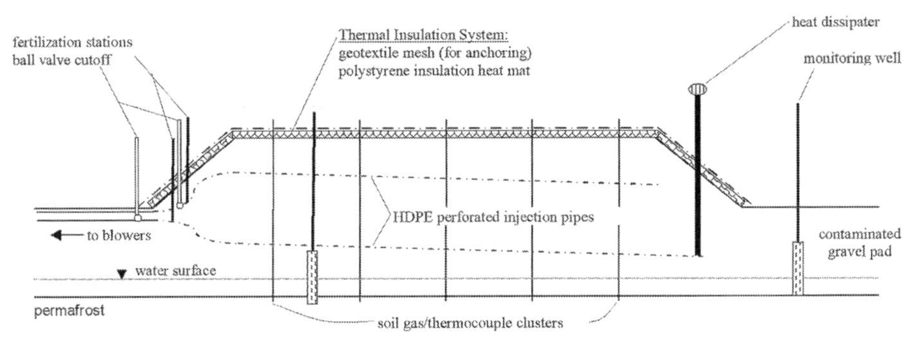

Figure 10.5. *In situ/ex situ* biopile profile: Arctic design.

The schematic in Figure 10.5 illustrates a biopile on permafrost configured with dual-purpose bioventing and fertilization pipes, soil gas and thermocouple cluster monitoring installations, monitoring wells for water quality, nutrient, and microbiological analyses, and an insulation system. Select pipes are retrofit with turbine ventilators at the far ends for heat extraction capability. Construction and operating recommendations for such a system are discussed in the following sections.

10.3.2.1 Bioventing and vapor extraction

Experience has shown that durable plastic pipes work well under load and extreme winter conditions. High-density polyethylene (HDPE) pipe should be used when flexibility is desired. ABS (acrylonitrile butadiene styrene) pipe is inflexible but will withstand heavier soil loads in larger biopiles. Corrugated drainage pipe (e.g. French drains) is not recommended for use with cold regions bioremediation because it will not hold shape with depth in a biopile and moisture can accumulate and freeze in corrugations.

Experience has shown that for small biopiles, ABS or PVC (polyvinyl chloride) pipe sufficiently accommodates 90-degree bends and is durable under soil loading and extreme winter temperatures. For larger biopiles, HDPE pipe is recommended to accommodate variable bending requirements associated with design and settlement. Connection fittings must be compatible with pipe runs to prevent leaks and avoid disconnect during construction and through settlement.

Pipe diameters for bioventing typically range from 2.5 to 10 cm, depending on blower size and project needs. If mechanized vapor extraction is afforded, small diameter (2.5 to 5 cm) pipe can be used; larger diameter pipes (to 10 cm) will accommodate commercially available, wind-driven devices often used with passive vapor extraction.

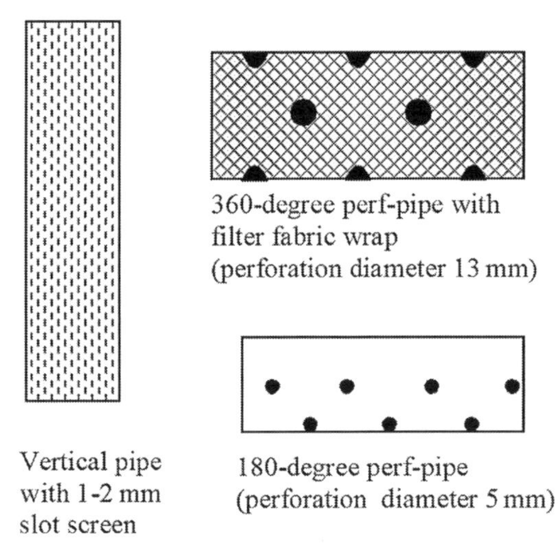

Figure 10.6. Perf-pipe configurations.

Type and size of perforations for air-injection pipe (perf-pipe) is a function of orientation and soil type. Figure 10.6 illustrates common perf-pipe configurations. For vertical installations, 1- to 2-mm slot-screen sections work well for oxygenating contaminated vadose zone. Not unlike their use with groundwater monitoring wells, screened pipe can be used in sands and gravels; a fine mesh filter fabric (or geofabric) wrap is recommended for use in finer soils. For horizontal installations, use of 180-degree perf-pipe or 360-degree perf-pipe with filter fabric is dependent on sediment intrusion considerations. With pipe that is not pre-fabricated, a uniform offset and staggered perforation pattern is recommended for optimal air distribution.

Regenerative blowers and rotary vane compressors are most commonly used for pressure applications in environmental remediation. Commercially available models include oilless and lubricated, single to four stage, miniature to large units. Some models are dual-function for bioventing and vapor extraction capabilities. Experiences with Japanese, German, and US models suggest that efficiencies and maintenance requirements are similar between brands. Self-lubricating, self-adjusting (oilless) models have endured continuous operation for up to three bioremediation treatment seasons without malfunction under sub-Arctic and Arctic conditions. Oilless models are highly recommended for most cold-climate applications because maintenance requirements can be as simple as a single filter/muffler change per season in low dust environments. Table 10.4 is a guide for blower selection when designing bioremediation systems for cold-region applications. In general, blower size and flow rate will be the selection criteria.

Table 10.4 *Blower specifications.*

	Size		Flow Rate		Pressure
Blower	(kW)	(hp)	(m^3 h^{-1})	(cfm)	(bar)
Low range (miniature)	0.25–0.75	1/3–1	to 10	to 20	to 0.65
Medium range (single unit to 1–3 vent pipes)	0.75–3.7	1–5	to 30	to 52	to 1.0
High performance (single unit to multiple vent pipes)	7–10	to 55	to 93.5	to 1.7	

10.3.2.2 Soil warming

In situ or biopile soil warming can be accomplished by passive means (i.e. with warm air injection, solar, and biological contributions), by active means such as electrical heating with heat mats, and radio frequency heating with vertically embedded probes, or a combination of heating schemes. Experiences in Alaska have demonstrated that radio frequency heating is not a cost-effective method of soil warming for bioremediation applications; radial heating is less efficient than a heating grid beneath an insulated cap. Soil warming with heat mats under insulation has proven to be a cost-effective means of enhancing treatment of PHC under sub-Arctic and Arctic conditions.

Thermal insulation systems

Heat mat efficiency is a function of heating element capacity and spacing. Prefabricated or custom-designed models can be used for environmental applications. For example, loop-configured, convection cell heat mats manufactured for use with foundations are typically designed for 208V AC power with output densities between 107 and 172 Wm^{-2}. However, prefabrication and close spacing of heating elements generally renders these models more expensive and excessive than field constructed mats. Components of custom-designed heat mats include heat trace or heat cables, foundation-grade wire mesh (heat transfer grid), and fasteners. Commercially available, self-regulating heat trace is manufactured as standard and corrosion-resistant models. Metal sheathed and mineral insulated, continuous output heating cables can also be fabricated to specific lengths. Heating elements are typically spaced apart and laid in serpentine fashion, and fastened on or beneath wire mesh to form the heat mat. When designing a heating scheme, choice of insulation and thickness will have direct bearing on heating efficiency and annual energy costs. Table 10.5

Table 10.5 *Insulation specifications (after Filler and Carlson 2000)*

Insulation	Average R-value[a] (m^2 °C W^{-1})	Insulation thickness[b] (cm)	Physical properties: Maximum compressive strength (kN m^{-2})	Maximum operating temperature (°C)	Moisture absorption (%)
Sawdust	0.35	Variable	<35	66	>30
R-Gard® (rigid board w/facer)	0.68	5, 7.6 & 10	97	74	4[c]
Expanded Polystyrene	0.83	5, 7.6 & 10	276	74	3–4[c]
Polyisocyanurate (foam)	1.22	Variable	172	74	15–20
Styrofoam	0.87	5, 7.6 & 10	414	74	<2[c]
Polyurethane Foam	0.99	5, 7.6 & 10	69	66	<30
Extruded Polystyrene	0.87	5, 7.6 & 10	345	82	<2[c]

Notes: Insulation listed from cheapest to most expensive (top to bottom) on per unit of measure basis.

[a]Per 2.5-cm of thickness.

[b]Commercially available sizes.

[c]Under optimum loading condition.

Table 10.6 *Heating elements, spacing, and unit costs (after Filler and Carlson 2000)*

Model W/m (W/lf)	Spacing (m)[a]	110–120V Use ($/m)	110–120V Use ($/$m^2$)[b]	208–240V Use ($/m)	208–240V Use ($/$m^2$)[b]
		Heat Trace (self-regulating)			
10 (3)	1	17	17	20	20
	2	17	9	20	10
16.5 (5)	1.5	24	16	27	18
	3	24	8	27	9
26 (8)	2.5	27	11	30	12
	3	27	9	30	10
33 (10)	3	29	10	33	11
	4	29	7	33	8
		Heat Cable (continuous output)			
33 (10)	3	na	na	15	5
50 (15)	4	na	na	17	4

Notes: Unit costs in 2003 US dollars, rounded to nearest whole dollar.

[a]Center-to-center spacing between circuits/lineations.

[b]Unit length cost divided by spacing.

summarizes thermal and physical properties for commercially available insulation that can be used for environmental applications. Table 10.6 compares corrosion-resistant heating elements (standard models with less protection are not recommended for geo-environmental application) on a cost per unit length and cost per unit area basis for various spacing scenarios.

Covers

A thermal insulation system cover may be a hand compacted, sand-gravel layer meeting a transportation specification at an active site, or a lightweight, plastic or polymeric geogrid with close-spaced mesh for anchoring insulation against wind at a remote site. Aircraft cable interwoven through anchor mesh and secured with eyebolts embedded in the ground, used tires, and sandbags have been used to supplement anchoring at Arctic sites.

The following general comments provide guidance for thermal insulation system design:

- Thicker insulation reduces heat loss to the atmosphere and enhances heat conduction through soil. The rate of heat loss decreases as insulation thickness increases:

$$q = \frac{A(T_{\mathrm{ht}} - T_{\mathrm{a,s}})}{xR} \qquad (10.9)$$

  where q is heat loss, A the insulated area, T_{ht} and $T_{\mathrm{a,s}}$ are the respective heat trace operating temperature and atmospheric or cover soil temperature, and x and R are insulation thickness and thermal resistance.
- Insulation finished with impermeable or semi-impermeable facer is preferred for geo-environmental applications. Surface facers minimize moisture intrusion and potential loss of thermal resistance.
- Use of higher-energy-rated heating elements increases spacing between circuits and insulation thickness, but reduces capital investment and annual operating costs. Spacing identified in Table 10.6 can be used as guidance for thermal insulation system design. 110–120V heat elements should never be combined with 208V or 240V elements for safety reasons.
- Heat cable is typically cheaper than pre-fabricated heat trace (or heat tape). Lower-cost heating cable is recommended for fast thawing and heat cycling applications.
- Operational temperatures of heating elements and board insulation are similar. Heat trace and cables can exceed the manufacturers listed maximum operating temperature while powering up during cycling. In some instances, heat elements piggy-backed on top of wire mesh, rather than to the underside, can cause contact melting

of insulation. Contact melting will be local; the depth of melting into insulation is directly related to the height at which a heat element rises above the mesh between any two fasteners. Poor fastening *en masse* can lead to diminished insulating capacity and create channels for heat loss.

- Heat cycling (or pulsing) based on set point temperatures requires remote thermostatic control of continuous output heat cable or self-regulating heat tape. Benefits of heat cycling are reduced power costs, capacity for differential soil warming, and permafrost control.

Permafrost control

Serious consideration must be given permafrost control with thermally enhanced bioremediation. Irrespective of environmental policy and/or permit requirements regarding environmental impact, unregulated heating of a biopile can degrade permafrost and result in settlement and damage to monitoring and pipe installations. Permafrost control within a bioremediation regime can be accomplished passively or with thermostatic control of heating. Horizontal aeration pipes can be retrofit with wind-activated turbine ventilators (Figure 10.5) to dissipate heat from a biopile. With *in situ* bioremediation and vertical air injection, screened dry wells can be installed throughout the vadose zone during the construction phase. Dry wells extend a few centimeters above the top of the thermal insulation system, are plugged with insulation and capped until thermal monitoring indicates need for heat dissipation. Dry wells can be uncapped and unplugged for passive heat dissipation, retrofit with risers and turbine ventilators, or mechanically pumped for more aggressive heat extraction.

An efficient way to prevent permafrost degradation is to regulate heating by thermostatic control. With thermal-monitoring clusters, the deepest thermocouples or thermistors are embedded at or near the permafrost boundary. A treatment regime is divided into zones, with one thermocouple dedicated to monitor the top of permafrost in each zone. All monitoring points are brought to the surface, joined and routed to zone controllers at the power station. Integrated controllers have functional capacity to establish set-point temperatures for the controlling thermistor(s), and programming to independently cycle heating elements off and on within the zones. Ability to control permafrost over extended treatment periods and energy cost savings realized through power cycling far outweigh the capital cost associated with a thermostatically controlled heating system.

10.3.2.3 Power supply

Thermally enhanced biopiles with PHC contamination have been successfully bioremediated within three years in the Arctic with use of electric power and diesel-electric generators (discussed further in Chapter 11, Section 11.2.1). Manipulation of the treatment regime can lengthen the annual period of effective treatment to six or seven months; sub-Arctic bioremediation sites can experience year-round treatment with properly designed insulation systems. Efficient thermal insulation system design is a function of heat element spacing, heat mat output, and insulation thickness. For a large integrated system in the Arctic, 20–30% of the overall capital investment is likely to be associated with active warming systems, and energy costs will be 15–35% of the annual operating and maintenance budget, depending on transport and fuel prices (Filler and Carlson 2000). For very cold remote sites such as those in the high Arctic and Antarctica, energy costs are likely to constitute a higher proportion of the operating and maintenance budget, driving the need for alternative energy sources for remote bioremediation.

10.3.2.4 *Prototype: thermally enhanced bioremediation system for Antarctica*

Full-scale TEB has not yet been undertaken in Antarctica. Using fuel spills at Casey Station, Australia has conducted a laboratory and field-based bioremediation feasibility study and concluded that increasing temperatures, by whatever means, will substantially increase remediation rates (Ferguson *et al.* 2003b). Of key importance, mesophilic organisms that will degrade hydrocarbons are naturally present in the soil. In laboratory microcosm trials, two species of mesophilic bacteria, *Pseudomonas spp.* and *Paenibacillus spp.* responded well to nutrient addition and aeration, indicating that natural consortia will degrade diesel at temperatures higher than the usual psychrotropic optimum (~ 25 °C). However, fuel costs for energy production are extremely high (~A$1/kWh) and bulk heating of soil may be cost-prohibitive. For this reason Australia is considering an integrated system that utilizes waste heat from water recycling, coupled with additional passive heating (Figure 10.7).

10.4 Conclusions and future research

Thermally enhanced bioremediation and integrated systems that involve vapor extraction, nutrient and water delivery, and monitoring are one of the low-cost remediation options currently available for cold regions. Where there are high demands on resource development, when timely land-transfer is required, or where the need for environmental protection is high, such schemes are probably the best overall compromise between cost and time. Treatment times are not as good as dig-and-haul or high temperature thermal treatments, but the costs and environmental risks are much lower. Conversely, lower cost treatment options such as air-sparging and bioventing without significant thermal

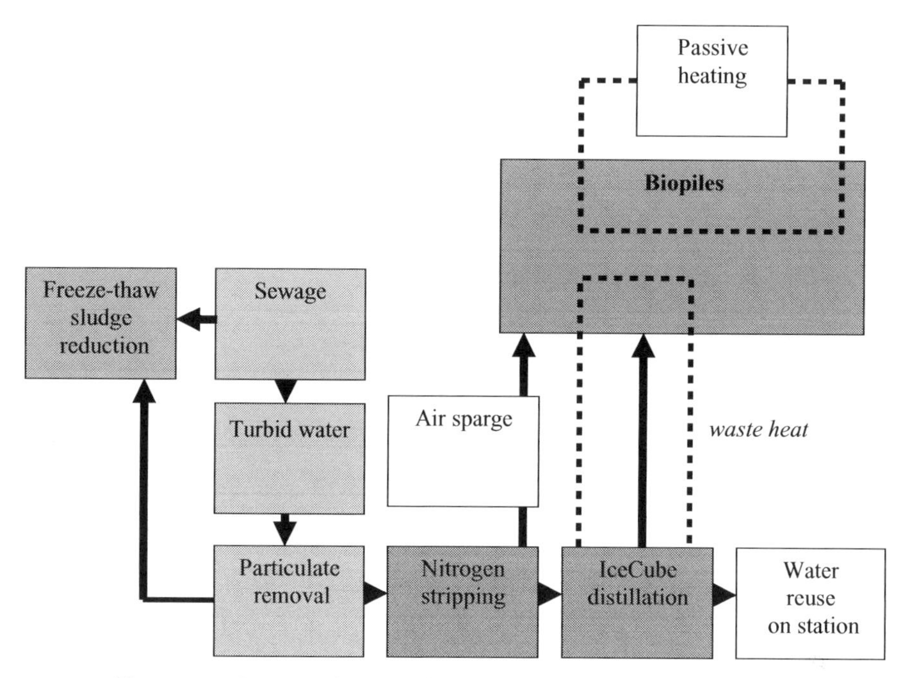

Figure 10.7. Conceptual model of integrated biopiles for Casey Station, Antarctica.

management, or landfarming (Chapter 9) and natural attenuation, will be less costly yet slower. Longer treatment times may still be acceptable for the management of certain sites, especially those that are remote, where infrastructure costs are high, and the risks to human health and the environment are low.

Significant recent advances have been made with respect to our understanding the importance of heating cold soils to enhance effective bioremediation outcomes, and how to achieve favorable heat transfer and temperature through clever use of thermal insulation systems and cost-effective power generation. On the horizon are conceptual designs for more efficient co-generated power stations, smaller, virtually pollution-free alternative energy supplies, and satellite-based remote sensing capabilities for potential use with environmental applications in polar regions. It is hoped that these developments will make thermally enhanced schemes more attractive for remote spills.

11

Emerging technologies

DALE VAN STEMPVOORT, KEVIN BIGGAR,
DENNIS M.FILLER, RONALD A. JOHNSON, IAN SNAPE,
KATE MUMFORD, WILLIAM SCHNABEL, AND
STEVE BAINBRIDGE

11.1 Introduction

In this book, current scientific knowledge and practical experiences with bioremediation of petroleum-contaminated soils in cold regions are reviewed and compiled. We now more fully understand the inter-relationships between cold temperatures, soil and water properties, and biological processes. This aids decision making about practical remediation treatment for petroleum-contaminated sites in cold regions. Landfarming and enhanced bioremediation schemes have emerged as viable soil treatment methods that offer a number of advantages over other methods. Nevertheless, work still needs to be done to optimize these methods, and with regards to evaluating phytoremediation and rhizosphere enhancement potentials for cold soils.

Two emerging technologies have been identified that could offer significant cost savings; low-cost heating and controlled-release nutrient systems are described briefly here (see also Chapter 8). In addition, natural attenuation has received little rigorous evaluation for use in cold soils. The main limitation for natural attenuation in cold regions is the low rate of degradation, coupled with off-site migration that can be relatively rapid in soils or gravel pads that have a poor adsorption capacity. Permeable reactive barriers are one groundwater treatment technology that could buy time for slower *in situ* techniques such as natural attenuation to take place. An outline of emerging permeable-reactive barrier technology is presented here, although full-scale trials are not yet complete. It is possible that such *in situ* techniques, when coupled with aeration,

Bioremediation of Petroleum Hydrocarbons in Cold Regions, ed. Dennis M. Filler, Ian Snape, and David L. Barnes. Published by Cambridge University Press.

sparging and biostimulation could offer methods for groundwater treatment in cold regions.

These emerging remediation techniques require additional work before consideration as viable treatment schemes for cold regions applications. They are discussed here in the context of recent developments, current status, and/or future needs. Groundwater treatment is also considered here in the context of emerging technology for cold regions.

11.2 Soil bioremediation

The effect of temperature on biodegradation rate is reviewed in Chapter 5 by Rike, Schiewer, and Filler. Low temperature is usually not the primary limitation to biodegradation, and as Walworth and others described in Chapter 9, the actions taken during landfarming are designed to overcome all other rate limitations so that temperature does become the limiting step. The issue then becomes a balance between the rate gains achieved through active heating versus the very large costs associated with energy production in cold regions, especially those that are remote. New advances in low-cost heating offer considerable potential for cost savings in cases where soil heating is needed.

11.2.1 Low-cost heating for engineered bioremediation

Energy technologies can be categorized as: (1) those needed to convert an energy source into electricity or heat, and (2) those used for energy storage. Possibilities under the first category for small scale applications include diesel-electric generators, wind turbines, hydro power, thermoelectric generators, photovoltaic (PV) devices, fuel cells, furnaces and boilers, and hybrid systems combining two or more of the above technologies. For the second, one can consider conventional fossil fuels, hydrogen, thermal storage, flywheels and batteries. Storage is a critical part of any technology where the energy resource is intermittent, such as wind and other solar energy. The first three devices need a generator to convert mechanical to electrical power while the next three convert heat, radiation, or a fuel directly into electricity. Furnaces and boilers convert the energy source into heat which can be used directly in a forced air or hydronic heating system. Of the four storage technologies mentioned, only the latter can produce electricity directly. Energy storage densities can vary from 38 kWh kg^{-1} for hydrogen down to 0.02 kWh kg^{-1} for lead acid batteries (Ristinen and Kraushaar 1999).

Currently, most electric power in rural Alaska is generated by diesel-electric generators (DEGs), which can produce electric power at efficiencies of over 30% if operated at suitable loads. A problem in the past in Alaska has been the lower efficiencies and prematurely short lifetimes of DEGs operated frequently at low loads (less than 25%). These systems can also be used in a cogeneration mode, producing both heat and electricity with an overall efficiency of over 60% (Johnson 1990). The cost of electricity produced in rural Alaska using this technology has ranged from US 10 cents to as much as US $1.00/kWh. If a DEG or other heat engine/generator is being used to provide on-site electricity, one could supply an amount of heat approximately equal to the electrical output by capturing heat from either the jacket water or the exhaust gas. The former has the advantage of not creating problems with respect to stack corrosion associated with impurities such as sulfur in the fuel.

Which combination of the above (or other) technologies is best for a particular engineered remediation application will depend on site-specific conditions. In some cases, it may be best to use a fuel (including hydrogen in the future) to produce heat to be used directly in the remediation process and then use a small generator to supply electricity. In other cases, it may be better to have a larger device producing electricity with heat being co-generated as a by-product. Freeze protection is always an issue in cold climates.

One form of renewable energy plentiful in Interior Alaska for 6–8 months per year is sunshine. In Fairbanks, a collector tilted at the latitude angle (64°) experiences insolation averaging 3.3 kWh m^{-2} day^{-1} over the 9 month period from February through October (NREL 2006). The values range from 2.4 in February to 5.3 in May. This is based on data plus modeling over a 30 year period and includes cloud cover effects. Hence, an average day's insolation spread over, say, 8 hours, with a collector efficiency of 70%, could supply heat at the rate of about 290 W per m^2 of collector area. With heating mats used in bioremediation consuming of the order of 100 W m^{-2}, one square meter of collector could replace the heat supplied by about 3 m^2 of heat mats. With thermally enhanced bioremediation, heating is cycled to keep energy costs down and to prevent permafrost degradation. Therefore, uneven solar warming during each day as well as over the 9 month period corresponding to the solar cycle, is not a negative attribute. At the same time that heat is being supplied by incident solar radiation, electricity could be supplied by PV panels. Alternatively, electricity can be supplied during periods of reduced or interrupted solar collection. At an efficiency of, say, 12%, and the above quoted average solar insolation of 3.3 kWh m^{-2} day^{-1}, each square meter of solar cell could produce about 400 Wh day^{-1} of electric power. This could be used to blow heated air through heat tubes buried in the ground and/or assist with bioventing requirements.

To illustrate the concept of solar air heating, we built a 0.65 m $\times$ 1.3 m energy collection box at the University of Alaska Fairbanks with the inlet air supplied by a 12-volt DC blower driven by a 45-watt PV panel. The back of the box facing the sun was painted black and the front was covered with acrylic. The collection box was mounted vertically and facing south. On February 18, 2003, with the ambient temperature varying from −27 to −7 °C, the heated air temperature ranged from −27 to 11 °C at flow rates from 6.8 to 44 m^3hr^{-1}. The maximum blower power was 21 W and heat was delivered at a maximum rate of about 300 W. The 30-year average solar insolation for a collector set at a 90-degree angle for February is 2.5 kWh m^{-2} (NREL 2006).

11.2.2 Micro-bioventing

Micro-bioventing implies small diameter air-injection wells strategically arrayed within a treatment realm to optimize subsurface aeration. Installed vertically or diagonally as a function of aeration efficiency, wells can be configured as a single array (small treatment area) or in multiple arrays (larger treatment areas), and arranged to target hot spots. Experience indicates that well material and perforations are important considerations. For example, consider the design of a micro-bioventing scheme to treat contaminated soil and underlying suprapermafrost water in a gravel pad on tundra. Design considerations for the micro-well include a construction material that is both durable and cheap, and a good conductor of heat, a perforation pattern that accommodates both unsaturated and saturated soil, and screening of the well through fine sediment that is encountered at the base of the gravel pad. A 2.5cm-diameter copper micro-well is designed with slot perforations (Chapter 10, Figure 10.6) for unsaturated soil, followed by a section with somewhat larger-diameter circular perforations for sparging the saturated zone, of which the portion that is embedded within the sediment layer is wrapped with a filter pack, filter-fabric, or other geotextile material to keep fines from clogging the well. A short length of heat tape might be installed in each well and powered only for short periods in spring and late summer to extend the duration of seasonal bioventing.

With primary consideration given cost-efficiency, future engineered bioremediation designs will couple low-cost heating devices (Section 11.2.1) with energy-efficient blowers, each configured to an array of micro-bioventing wells, and fertilization (Chapter 8) for optimal treatment.

11.2.3 Phytoremediation and rhizosphere enhancement

Phytoremediation is the destruction, removal, or immobilization of soil contaminants brought about by plants and associated organisms. In general terms, phytoremediation acts through one or more of the following types of processes:

- *Phytoextraction/sequestration/degradation*: This process refers to the uptake of soil-borne contaminants into the plant matrix, followed by sequestration or degradation via vegetative metabolic processes.
- *Phytovolatilization*: This process describes the translocation of contaminants from the soil to the surrounding atmosphere through the vegetative transpiration stream.
- *Immobilization/stabilization*: Immobilization refers to the enhanced binding of contaminants to the soil and/or exterior root matrix brought about by root exudates, associated microbial activity, or physical stabilization of the soil matrix.
- *Hydraulic control*: In this process, plant species with low water use efficiency (e.g. hybrid poplars) are employed to reduce infiltration of precipitation or influx of groundwater through a contaminated zone, thus reducing the migration of contaminants through groundwater transport.
- *Rhizodegradation (rhizosphere-enhanced phytoremediation)*: Rhizodegradation describes the transformation of contaminants in the soil proximal to the roots (rhizosphere) by organisms associated with vegetative species (e.g., bacteria, mycorrhizal fungi).

In cold and temperate regions, the bulk of the research to date with regard to phytoremediation of petroleum hydrocarbons has focused upon processes associated with rhizodegradation. Due to the low water solubility of petroleum hydrocarbons and their high concentrations at many petroleum spill sites, contamination often remains in the soil exterior to the vegetative matrix. Moreover, since there is increased microbial activity in vegetated soils when compared to soils devoid of vegetation, it follows that encouraging the proliferation of vegetation at petroleum spill sites could potentially reduce contaminant concentrations via rhizodegradation.

In a recent field study designed to test the efficacy of petroleum rhizodegradation in cold regions, Reynolds (2004) demonstrated that the cultivation of cold-tolerant grasses increased the degradation rates of petroleum compounds within the soils compared to non-vegetated control soils at three different sites in Alaska. Although the results of the study were clear, the author pointed out that rhizodegradation results at similar sites are not always readily quantifiable, nor is the implementation of a rhizodegradation strategy always appropriate. In general, rhizodegradation is most appropriate as a long-term, relatively low-cost remedy at sites where the risk to potential receptors is relatively low. The success of a rhizodegradation strategy is dependant upon the soil temperature, length of the growing season, depth of the root zone, water content of the soil, level of contamination, availability of nutrients, soil chemistry, and microbial population among other factors. In summary, although it is likely that the application of vegetation to petroleum-contaminated soils in cold regions exerts a positive impact on contaminant degradation in the soil near the root zone, the level of impact is highly dependent upon the factors described above.

The utilization of plants at remediation sites is ultimately dependent upon whether or not the plants will actually grow at the sites. Consequently, plant selection is a key factor in the success of a phytoremediation project. In cold regions, native species often prove more reliable compared to non-native species due to their ability to flourish in the extreme environment. Tolerance to the contaminants in question is another crucial factor governing plant selection. The University of Saskatchewan has developed a web-available database describing the demonstrated tolerance and efficacy of various vegetative species for use at petroleum sites (PhytoPet© 2007).

11.2.4 Site-specific cleanup criteria with institutional controls

The State of Alaska has adopted a risk-based approach for contaminated soil and groundwater that allows for site-specific criteria in determining the level of remediation effort required. First, the basis for this risk-based approach is established exposure-pathway (i.e., ingestion, inhalation, and migration to groundwater) cleanup levels. Then, these exposure-pathway cleanup levels are refined to reflect the vast differences in climatic conditions that are found in a state that is 1 717 854 sq km in area. For environmental regulation, Alaska is divided into three climatic zones: (1) the "Arctic zone" is that area of the state north of 68 degrees North latitude, and any site south of latitude 68 degrees that is underlain by continuous permafrost; (2) the "under 40 inches (or 102 cm) of precipitation per year zone"; and (3) the "over 40 inch zone." Therefore, established petroleum hydrocarbon cleanup levels for gasoline range organics (C_6 to C_{10} compounds), diesel range organics (C_{10} to C_{25} compounds), and residual range organics (C_{25} to C_{36} compounds) vary according to exposure pathway and climatic zone.

Alaska environmental regulation stipulates that a petroleum hydrocarbon may not remain in the environment at a concentration above its maximum allowable concentration, unless the responsible party for cleanup demonstrates that the petroleum hydrocarbons will not migrate from the site and will not pose a significant risk to human health, safety, and welfare, or to the environment.

Part of the determination of risk-based cleanup (RBC) levels involves initial assumptions about site conditions such as future land use, and in the case of groundwater contamination, use of the aquifer as a drinking water source. Thus, once RBC levels for soil and groundwater are reached at the contaminated site, no further remediation is required as long as the initial assumptions continue into the future. The status of *conditional closure with institutional controls* is then granted by the regulatory agency. Institutional controls can be: (a) a *deed notice* that advises potential future property owners of contaminated soil or groundwater left in place, (b) *deed restrictions*, which advise against the use of the aquifer for drinking water purposes, or (c) an *advisory* stipulating that contaminated soil will have to be treated if it is identified for excavation. A more complex institutional control, called an *Equitable Servitude and Easement*, establishes certain restrictions on the future use of the property and requires the responsible party for cleanup to take further remedial action if the initial land use assumptions change in the future. For example, an *Equitable Servitude and Easement* institutional control would accompany an industrial land use scenario that reverts to residential land use.

Further research on four-phase contaminant transport in soil, and considerations for ground aspects (e.g., tundra versus taiga, permafrost versus intermittent permafrost or no permafrost, frost-susceptible soil versus non-frost-susceptible soil, wet versus dry, and vegetation cover) may ultimately be included in the determinations of site-specific cleanup levels with institutional controls.

11.3 Groundwater treatment

11.3.1 Initial indicators about potential for groundwater treatment

Although applications of bioremediation of hydrocarbons at cold-climate sites have generally been limited to utilization of aerobic processes in soils, a growing number of studies have indicated that aerobic and anaerobic biodegradation of hydrocarbons are significant processes in groundwater at cold-climate sites (Herrington *et al.* 1997; Armstrong *et al.* 2002; Ulrich *et al.* 2006). Almost all of the information on the bioremediation of hydrocarbons in groundwater at cold-climate sites has been obtained in studies of intrinsic bioremediation (i.e., natural attenuation), the unassisted biodegradation of the hydrocarbons by cold-adapted microorganisms. Anaerobic biodegradation of hydrocarbons may involve sulfate reduction, nitrate reduction, reduction of ferric iron or manganese, or methanogenesis (also referred to as methane fermentation: Stumm and Morgan 1996). Laboratory tests have generally indicated that mineralization/ biodegradation rates tend to be slower under anaerobic conditions compared to aerobic conditions (Wilson *et al.* 1986; Barker *et al.* 1987; Cross *et al.* 2003; Salminen *et al.* 2004), though there have been exceptions, where aerobic and anaerobic rates were found to be similar (Billowits *et al.* 1999).

11.3.1.1 Rates of biodegradation of PHC in groundwater

The effect of temperature on bacterial degradation is traditionally accounted for by using the Q_{10} rule (Diaz-Ravina *et al.* 1994). The Q_{10} relationship is related to the Arrhenius equation (Chapter 5) and is the ratio of a first-order rate constant at a specific temperature to the rate constant at a temperature 10 °C lower. The Q_{10} value is calculated using the following equation (Metcalf and Eddy 1991):

$$Q_{10} = \left[\frac{k_2}{k_1}\right]^{10/\Delta T} \tag{11.1}$$

where k_1 and k_2 are the rate constants calculated at temperatures T_1 and T_2 and ΔT is the change in temperature from T_2 to T_1. A general rule of thumb is that reaction rate doubles for every 10 degree Kelvin rise in temperature; therefore the expected Q_{10} value is 2 (Metcalf and Eddy 1991). A Q_{10} value less than 2 means that the rates at low temperatures are higher than predicted by Equation (11.1). This provides an estimate of biodegradation rates that might be expected in groundwater at cold temperatures.

In previous reviews of the groundwater literature, the effect of temperature on rates of hydrocarbon degradation has largely been ignored. Aronson *et al.* (1997) reported that "temperature and redox environment did not appear to be correlated to the anaerobic biodegradation of benzene in aquifer environments." In their overview on rates of hydrocarbon degradation in groundwater, Suarez and Rifai (1999) stated the conventional Q_{10} assumption, that below 10 °C growth rates of microorganisms tend to double with every 10 °C increase in temperature (i.e., $Q_{10} = 2$), but provided no further discussion on temperature. The majority of reported rates are based on first-order kinetics:

$$C = C_0 e^{-kt} \tag{11.2}$$

where C_0 is the initial dissolved hydrocarbon concentration, C is the concentration after time interval t, and k is a first-order rate constant (units of time^{-1}). Others have simulated biodegradation using the zero-order equation (Suarez and Rifai 1999).

11.3.1.2 Role of temperature: laboratory tests under aerobic conditions

Bradley and Chapelle (1995) challenged the conventional "$Q_{10} = 2$" assumption. In microcosm tests, they found that biodegradation of toluene at 5 °C in samples from a jet-fuel-contaminated sand aquifer from Adak, Alaska occurred at a faster rate (factor of two) than biodegradation of toluene at 20 °C in petroleum-contaminated sediment from South Carolina. The groundwater temperature at the Alaska site was reported to be 4–6 °C. Bradley and Chapelle concluded that intrinsic remediation might be viable for cold sites, and that the biodegradation rates may be faster than would be normally expected because the microorganisms are adapted to the cold environment. In tests with hydrocarbon-contaminated soil and groundwater from Alaska, Braddock *et al.* (2001) found that the rate of microbial growth (heterotrophs and hydrocarbon degraders) was enhanced as the temperature increased from 4 to 10 °C, and was highest at either 10–15 °C or 25 °C. In nutrient amended tests on groundwater, Cross *et al.* (2003) inferred a 38% increase in the rate of dodecane biodegradation over the temperature range 10 to 28 °C. Van Stempvoort *et al.* (2004) found that the overall rates of losses of total hydrocarbons and the C_6–C_{10} fraction in groundwater increased by 80 and 50%, respectively, when the temperature was increased from 4 to 23 °C.

11.3.1.3 *Role of temperature: laboratory tests under anaerobic conditions*

Cross *et al.* (2003) found that with an increase in temperature from 10 to 20 °C the degradation rate of total extractable hydrocarbons (C_{11}–C_{30}) doubled. In O_2-limited batch tests, Van Stempvoort *et al.* (2004) found that an increase in temperature from 5 to 23 °C had no observable effect on the hydrocarbon loss/degradation rate.

11.3.1.4 *Role of other factors: laboratory tests*

Several investigators have reported results of laboratory experiments that indicated that hydrocarbon-biodegradation rates in groundwater increase with addition of various nutrient mixtures (Braddock and McCarthy 1996; Braddock *et al.* 2001; Billowits *et al.* 1999; Soloway *et al.* 2001; Cross *et al.* 2003). Others reported mixed results. For example, Van Stempvoort *et al.* (2004) found that addition of $(NH_4)_2HPO_4$ increased the rate of hydrocarbon degradation in batches amended with ferrous sulfate, but not in batches amended with ferric iron (FeO_2H). Whyte and co-workers (Whyte *et al.* 1998; Billowits *et al.* 1999; Soloway *et al.* 2001) found that nutrient addition increased hexadecane mineralization in aquifer sediment microcosms, but not in other tests with groundwater samples. Laboratory experiments have provided evidence that nitrate (Cross *et al.* 2003; Fan *et al.* 2006), sulfate (Cross *et al.* 2003; Van Stempvoort *et al.* 2004; Fan *et al.* 2006) or ferric iron amendments (Van Stempvoort *et al.* 2004) may serve as sources of electron acceptors to enhance degradation.

11.3.1.5 *Field evidence on biodegradation rates*

Salanitro (1993) inferred that rates of biodegradation of aromatic hydrocarbons (BTEX) in groundwater at relatively cold sites in Michigan and Ontario, and corresponding microcosm experiments at 10–12 °C, were similar to those at warmer sites (e.g., Florida and Texas, USA).

Herrington *et al.* (1997) summarized the results of investigations of the natural attenuation of aviation/jet fuel plumes in groundwater at five US Air Force Bases, one in northern Michigan, and four in Alaska. The Michigan site groundwater temperatures ranged from 9.8 to 14.9 °C, and the Alaska groundwater temperatures ranged from 3.4 to 11.7 °C. The hydrocarbon plumes were apparently receding at all sites, linked to aerobic and anaerobic biodegradation processes. The estimated overall first-order BTEX degradation rates at these cold climate sites ranged from 0.19 to 2.99% day^{-1}, including aerobic and anaerobic processes. Herrington *et al.* noted that these biodegradation rates were similar to those reported for warmer sites.

Armstrong *et al.* (2002) analyzed monitoring data for contaminant plumes in groundwater at 124 sites in western Canada, mainly "upstream" oil and gas sites in Alberta, where groundwater temperatures are typically within the range of 5–10 °C. The estimated first-order rates for BTEX attenuation at the sites ranged from 0.0002 to 0.017 day^{-1}, with the majority of plumes ranging from 0.001 to 0.005 day^{-1}. They noted that these ranges are similar to and overlap with ranges of BTEX degradation rates reported for plumes at warmer sites, as documented in surveys conducted in the United States (Suarez and Rifai 1999). Armstrong *et al.* suggested that possible causes for marginally lower biodegradation rates for the oil and gas sites in Western Canada compared to sites surveyed in the United States were the cooler temperatures and presence of co-contaminants at the Canadian sites.

Ulrich *et al.* (2006) compiled data from field studies (published and previously unpublished reports) on rates of biodegradation of BTEX components in groundwater under anaerobic conditions where temperature data were available (Figure 11.1). They reported 95% confidence intervals for the rates (a) as reported, and (b) normalized to 5 °C and 10 °C, based on the conventional assumption that $Q_{10} = 2$. By normalizing the rates to 5 °C and 10 °C, they found that the 95% confidence limits were reduced implying that temperature was a significant factor in the scatter in rates. Approach (b) thus provided rates of natural attenuation of hydrocarbons in groundwater in cold-climate field settings that could be used by regulators and remediation practitioners.

Some researchers (Chiang *et al.* 1989; Westervelt *et al.* 1997) have reported higher rates for BTEX degradation in cold groundwater than the normalized ranges inferred by Ulrich *et al.* (Figure 11.1). In some of these cases (Chiang *et al.* 1989), aerobic biodegradation was inferred, which may account for the faster rates. Other researchers have reported low or insignificant rates of hydrocarbon degradation in groundwater at cold-climate sites (Mitchell and Friedrich 2001; Richmond *et al.* 2001). However, given the large uncertainty in such rate calculations, and the potential impact of factors other than temperature (toxicity,

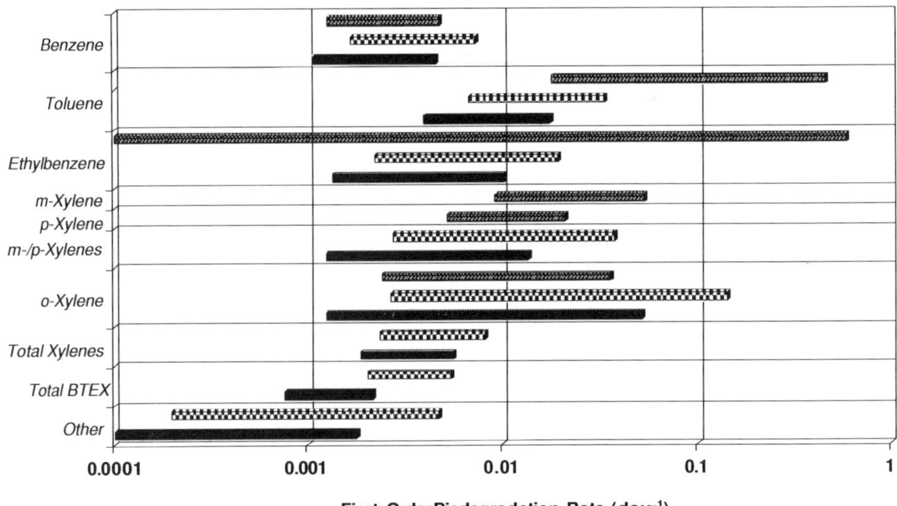

Figure 11.1. Rates of anaerobic biodegradation of hydrocarbons in groundwater, based on literature review of field data, where field temperatures were reported (modified after Ulrich *et al.* 2006).

nutrient limitation, etc.) it may be appropriate to undertake further study to confirm that rates reported to be either above or below the normalized ranges provided by Ulrich *et al.* are valid.

11.3.1.6 *Survey of published information on bioremediation of hydrocarbons in groundwater at cold climate sites*

Van Stempvoort and Biggar (2007) reviewed field investigations of bioremediation of hydrocarbon plumes in groundwater at 35 cold climate sites, including 24 in North America and 11 in Scandinavia and the Baltic region of Europe. This section provides a summary of the results of that review.

The majority of the contaminant plumes studied was in sand or sand/gravel aquifers. Others were in fractured rock (Carss *et al.* 1994; Van Stempvoort *et al.* 2006c; Eriksson *et al.* 2006), gravel fill over peat (Mitchell and Friedrich 2001), and silt and clay deposits (Van Stempvoort *et al.* 2002, 2007a,b). Geology did not appear to limit the occurrence of *in situ* bioremediation of hydrocarbons in groundwater at cold sites. There was evidence for significant intrinsic or enhanced *in situ* bioremediation of hydrocarbons in all types of geologic media investigated.

The mean annual air temperatures for the field investigation sites varied between −12 and 8 °C, whereas reported groundwater temperatures ranged from 0.3 to 15.8 °C. Typically, the reported groundwater temperatures were obtained in warm seasons. Most of these sites have either no permafrost or discontinuous permafrost, but three sites had extensive permafrost. The temperature of groundwater in the upper few meters below ground surface fluctuates seasonally, and some freezes during the winter. In contrast, the temperature of groundwater at depths of 10 m or more is relatively stable year-round. This vertical contrast in the thermal regime may affect the types of hydrocarbon-degrading microorganisms in groundwater. This topic appears to be a research gap, although a few laboratory studies have examined the effect of cyclic freezing on bioremediation of hydrocarbons in soils (Eriksson *et al.* 2001; Børresen *et al.* 2006).

For 21 of 35 cold climate field sites where the mean annual air temperatures were at or above 0 °C, all reported evidence for *in situ* bioremediation of hydrocarbons in groundwater. Of the remaining 14 sites, which had mean annual air temperatures below 0 °C (−12 to −0.3 °C), 12 reported significant *in situ* biodegradation of hydrocarbons in groundwater (intrinsic or enhanced). This information indicates that there may be potential to utilize *in situ* bioremediation as a technology to clean up petroleum-contaminated groundwater in cold regions, including sites with subzero mean annual temperatures and the presence of permafrost.

At 23 of the 35 field sites, the hydrocarbon plumes in groundwater were derived from fuel, including diesel, gasoline, jet/aviation fuel, or unidentified types. Sources at other sites were crude oil, refinery wastes, natural gas and/or gas condensates, landfill leachate, and others unknown or unidentified. Most studies reported concentrations of BTEX (benzene, toluene, ethylbenzene, and xylenes); some reported other aromatic species as total or total extractable hydrocarbons. *In situ* biodegradation of hydrocarbons was reported to be significant in all of the types of plumes that were investigated. However, a gap remains regarding relative biodegradability of hydrocarbon plumes derived from various types of petroleum sources in various hydrogeological settings under cold-climate conditions.

A majority (27 of 35) of the field studies summarized by Van Stempvoort and Biggar (2007) were investigations of intrinsic bioremediation. Twenty of these studies emphasized hydrogeochemistry, specifically investigation of the distribution of electron acceptors, such as oxygen, sulfate, nitrate, or of the products of electron-accepting processes, such as dissolved ferric iron and manganese, sulfide, or methane. Typically lower concentrations of electron acceptors were detected in the contaminant plumes, and/or higher concentrations of the resulting product species. Most of these studies reported enriched dissolved iron or depleted nitrate or sulfate in the plumes, evidence for the role of anaerobic bacteria in the degradation of hydrocarbons. Others reported evidence for aerobic hydrocarbon degraders or methanogens. The role of methanogens may have been under-reported because methane was not analyzed in most of the studies.

Some of the studies reviewed by Van Stempvoort and Biggar (2006) reported the detection of "probable/putative biodegradation metabolites in the contaminant plumes," such as partially oxidized petroleum hydrocarbons (Eganhouse *et al.* 1993; Eriksson *et al.* 2006), short-chain fatty acids (Van Stempvoort *et al.* 2006, 2007b), higher alkalinity (Klonowski *et al.* 2005; Van Stempvoort *et al.* 2007b), or higher alkalinity/hardness ratio (Eriksson *et al.* 2006), which is an apparent indicator of the mineralization of hydrocarbons to CO_2. Others have reported shifts in the isotopic composition of dissolved inorganic carbon associated with this mineralization process (Baedecker *et al.* 1993; Van Stempvoort *et al.* 2002, 2005). Van Stempvoort and others reported isotope data for S and O in sulfate that suggested that sulfate had been an important electron acceptor in hydrocarbon degradation in cold groundwater (Van Stempvoort *et al.* 2005; Van Stempvoort *et al.* 2006).

The review (Van Stempvoort and Biggar 2007) reported that some studies included microbial analyses, including enumeration techniques for total bacteria, and others for specific types of hydrocarbon-degraders, sulfate reducers, iron-reducers, nitrate-reducers and/or methanogens (Carss *et al.* 1994; Lai *et al.* 2001; Cross *et al.* 2003). Typically the microbial populations (total or specific groups) were more abundant or more active in hydrocarbon plumes compared to background groundwater. Others reported both microbial enumeration procedures and molecular analyses (Whyte *et al.* 1998; Billowits *et al.* 1999; Purkamo *et al.* 2004; Eriksson *et al.* 2006). As publications of molecular analyses of microorganisms in hydrocarbon plumes at cold-climate sites increase, it is becoming apparent that some strains or types are widespread geographically (Van Stempvoort and Biggar, 2007).

Six of the field studies reviewed by Van Stempvoort and Biggar (2007) reported the use of active bioremediation techniques to clean up hydrocarbon-contaminated groundwater at cold climate sites. These involved:

- *ex situ* aeration with biostimulation and either groundwater recirculation (Carss *et al.* 1994; Shields *et al.* 1997) or a bioreactor (Mitchell and Friedrich 2001);
- *in situ* biosparging with biostimulation (Soloway *et al.* 2001);

- *in situ* aeration with bacterial inoculation and addition of "biogenic" substances (Pawelczyk *et al.* 2003); or
- bioventing (Barnette *et al.* 2005).

The outcome of the biosparging approach was not reported. The other five active bioremediation applications were reported to be successful, based on disappearance or reduction of BTEX concentrations (Carss *et al.* 1994; Mitchell and Friedrich 2001; Barnette *et al.* 2005), a decline in TPH or oil concentrations (Carss *et al.* 1994; Mitchell and Friedrich 2001; Pawelczyk *et al.* 2003), oxygen loss (Carss *et al.* 1994), an increase in abundance of hydrocarbon-degrading bacteria (Carss *et al.* 1994), and/or shrinkage of the hydrocarbon plume (Shields *et al.* 1997).

Some researchers have suggested that *in situ* biostimulation with nutrients or electron acceptors might enhance the bioremediation of hydrocarbon-contaminated groundwater at cold climate sites, based either on field studies (Curtis and Lammey, 1998; Van Stempvoort *et al.* 2007a,b) or laboratory tests (Billowits *et al.* 1999; Cross *et al.* 2003).

A recent study by Rayner *et al.* (2007) examined the effect of micro-bioventing on the petroleum-hydrocarbon concentrations in shallow groundwater on sub-Antarctic Macquarie Island. They found that air injection led to a significant decrease in groundwater fuel concentrations.

11.3.2 Practical and developing water treatment methods

Permeable reactive barriers and control release nutrients are two methods being considered for use in Antarctica and Alaska. Permeable reactive barrier trials are underway at contaminated sites at Casey Station, Antarctica; a trial is being planned for a North Slope (Alaska) contaminated site. Use of two-phase partitioning bioreactors for the treatment of contaminated water in cold regions is also under consideration. Bioreactor work is left for future discussion as we are not aware of any planned or active bioreactor trials at this time.

11.3.2.1 Permeable reactive barriers

A permeable reactive barrier (PRB) is a continuous, *in situ* permeable treatment zone of reactive materials designed to intercept and remediate the contaminated portion of a groundwater plume. Generally they have low operating and maintenance costs/requirements, are ideal for deployment in remote locations and can be used for long-term operation. Plume treatment is ensured by the installation of a large PRB structure, or through the use of a passive flow system such as a *funnel and gate* (Starr and Cherry 1994) or a *trench and gate* (Bowles *et al.* 2000).

A variety of mechanisms may be utilized to remove contamination as the plume passes through the reactive media. These are either physical-chemical (contaminant entrapment, attenuation by adsorption and/or precipitation), or involve biological processes (biotic conversion to benign reaction products). There are a number of considerations when designing a PRB: (1) reactive material selection, (2) barrier physical parameters, and (3) barrier fate; each of these in turn can be strongly influenced by cold region specific processes, especially ground freezing (Snape *et al.* 2001).

Material selection

The freezing point of the contaminant for a range of water concentrations and the effect of low temperatures on reactivity can influence PRB performance (Woinarski *et al.* 2002, 2003, 2006). In general, the sorption of polar and ionic solutes is related to a thermodynamically favorable change in enthalpy, and is therefore temperature dependent. However, the sorption of non-polar organic solutes is thought to correspond to a thermodynamically favorable change in entropy, and is less temperature dependent. In addition, lower temperatures generally slow the reaction kinetics. Both equilibrium partitioning and slow kinetics combine to require larger barrier widths in cold environments to allow sufficient residence time within the barrier for the particular contaminant attenuation reaction(s) to be completed.

Hydraulic conductivity and particle size are critically important. The barrier must have a permeability that is equal to, or greater than, that of the surrounding subsurface, so that the contaminated plume passes through and is free draining to prevent icing and channel formation. The barrier must thaw before the surrounding subsurface at the start of the summer period or remain reasonably permeable whilst frozen, to afford treatment of the contaminated water during the early melt period (Gore *et al.* 2006a). If external heating is used to create an artificial thaw, insulation should be used to avoid heat loss to the surrounding soil or aquifer to ensure the barrier remains keyed into the permafrost.

The physical and chemical stability of the material is also important (Gore *et al.* 2006b). Freeze-thaw cycles may induce material break-up, and the subsequent formation of fine material may reduce barrier hydraulic conductivity through the clogging of pores, lower the mass of material in the barrier, and lead to ice lens formation through the migration of water to the freezing front.

Other important factors include environmental compatibility to avoid the release of further contaminants from the introduced material, availability and ability to transport the media to the field site, and cost.

Figure 11.2. Trial barrier with PRB, Casey Station, Antarctica.

Barrier physical parameters

The reaction zone size is determined by the flux of water and the rate of the reaction between the contaminant and the PRB media. The residence time needs to be sufficiently long to reduce contaminant concentrations to acceptable levels. Shape and orientation will be designed according to the size of the catchment, the distribution of contaminants, and the estimated flux of water. A trial barrier in Antarctica has wings that are up to 10 m long, and a series of trial treatments that are enclosed in 1 m deep cages that total a 5 m wide by 1.8 m long reaction zone (Figure 11.2).

Barrier fate

If the barrier is located in a sensitive environment, it should be designed to allow removal once the contaminant source zone has been remediated and the plume attenuated. PRBs need regular monitoring to ensure that reactive media are replaced before down-gradient contaminant breakthrough occurs. Presently, the utility of PRBs for petroleum remediation in cold regions is in the pilot testing stage; preliminary results look encouraging.

11.3.2.2 Controlled-release nutrients

The benefit of using controlled-release nutrients (CRNs) is realized every day in millions of gardens through the world. Surprisingly, relatively little effort has been made to apply the approach to bioremediation, especially in cold regions, where there could be greatest benefit. The main benefit is that CRNs can reduce the need for multiple nutrient additions in impoverished soils with low water holding capacities. This feature can reduce the total remediation cost, prevent high levels of nutrients being dispersed into sensitive terrestrial and marine environments, and can consistently provide optimal nutrient concentrations for biodegradation in soil pore water (Chapter 8).

Several CRN mechanisms have been trialed in cold-region bioremediation. Some systems such as Inipol EAP 22 (Elf Atochem) are better described as slow release. The mix comprises oleic acid (hydrophobic phase), tri(laureth-4)-phosphate (phosphorus source and surfactant), 2-butoxyethanol (surfactant and emulsion stabilizer), and urea (nitrogen source) (Shultz 1996). Inipol EAP 22 was successfully applied to the oiled shoreline of the Exxon Valdez oil spill, and Delille and co-workers (Delille *et al.* 2002; Delille *et al.* 2003) trialed the product in Antarctica and in the sub-Antarctic with positive results. However, concerns have been expressed regarding its suitability for use in landfarming, and some of the surfactant and emulsion stabilizer is known to be toxic (www.uga.edu/srel/oil.htm).

An alternative system was developed by Grace-Sierra, with a customized blend for bioremediation (optimum N:P:K, no trace metal contaminants). MaxBac is a semipermeable membrane surrounding water-soluble inorganic N, P, and K which is applied in pellet form. Application rates are determined by maintaining a high C:N:P ratio, rather than an optimal N-concentration in soil water (Chapter 8). Nevertheless, Snape and co-workers have used MaxBac as a slow-release nutrient source in a landfarming trial in Antarctica (Chapter 9). Results from the trial indicate that nutrients are still present after five years in the ground. However, Gore and others (Gore *et al.* 2004) expressed some concern about how controlled the nutrient release was in freezing-thawing conditions. They observed that many pellets burst during the first few freeze-thaw cycles and a large proportion of nutrient was released during this process.

Zeolites have been identified as another possible substrate for a CRN system for use in either landfarming or to enhance biodegradation in a permeable reactive barrier system. Through use of an ion exchange mechanism, zeolites can increase nitrogen contents and can also be used to prevent excess nutrient dispersal.

A pilot field trial at Casey Station in Antarctica is currently being used to evaluate the performance of two zeolite CRNs and MaxBac in a PRB. Both zeolites were dominantly clinoptilolite ($(Ca, Na_2, K_2)_3[Al_6Si_{30}O_{72}] . 24H_2O$) which was the preferred mineral form because it is robust. This allows transport to remote locations, and is resistant to weathering and impacts on particle integrity, bulk

Table 11.1 *Nitrogen released from treatment cells from front third of barrier (cumulative flow through the barrier was 6640 m^3 in year 1)*

Treatment	Distribution of flow (%)	Total ammonium released as N, 0.6 m from front of barrier (kg)
Cage 1 – MaxBac	30	5.6
Cage 2 – Zeopro	19	6.2
Cage 3 – Zeopro and GAC	18	5.4
Cage 4 – Sand	19	0.5
Cage 5 – Ammonium loaded zeolite	15	0.8

density, and saturated hydraulic conductivity, which have been identified as possible issues associated with diurnal freezing and thawing of terrestrial environments in polar regions. One of the zeolite substrates is Zeopro, a commercially available ammonium and potassium loaded zeolite with a synthetic apatite coating, derived from St. Cloud Mine, Winston, New Mexico. For comparison, an ammonium loaded Castle Mountain zeolite (NSW Australia) was also used. Sand was used in one cell as a control and as a measure of unamended nutrient concentrations.

The design is a sequential permeable (bio-) reactive barrier where the CRNs were placed in the first segment of each treatment sequence, followed by the petroleum hydrocarbon sorption material (granulated activated carbon). The final segment contains sodium modified zeolite to capture any excess ammonium that is not metabolized by indigenous microorganisms during the biodegradation process. The barrier is currently being monitored for temperature, pH, electrical conductivity, dissolved oxygen, ammonium, nitrate/nitrite, phosphorus, and total petroleum hydrocarbons. Barrier material samples are collected yearly. The barrier also contains oxygen distributors to ensure that oxygen diffusion does not become rate limiting to biodegradation, and heat trace to control the freezing state of the system. Preliminary results from the barrier trial are shown in Table 11.1, in which ammonium concentrations throughout selected cells of the barrier are depicted.

11.4 Conclusions

Through trial and error, environmental practitioners have recognized that cold temperatures and freeze-thaw effects, either directly or indirectly, limit environmental cleanup. Conventional bioremediation trials have been successful in reducing PHC contamination in tundra and in high-Arctic mineral soils.

Enhanced bioremediation schemes implemented in Alaska and sub-Antarctic Macquarie Island have proven more effective at removing higher concentrations of PHC in soil in comparatively shorter time: 2 to 3 years in Arctic Alaska (Chapter 10); 1 to 2 years (estimated) on Macquarie Island (Rayner *et al.* 2007).

Environmental scientists are working hard to understand the interrelationships between cold temperatures, soil and water properties, and biological processes to define and optimize practical treatments. Primary objectives are to better match environmental outcomes with remediation technique, at minimal cost, and to develop low-cost, low-maintenance methods to treat spills in ecologically sensitive or resource-limited places. Other goals include development of groundwater treatment methods for use in cold regions, and use of alternative energies with engineered remediation toward improved efficiency and lower treatment costs. Bioremediation technology has progressed such that it can now be implemented for treatment of vast petroleum spills in cold climates around the world. Indeed, it is likely to be the only affordable remediation option for very large spills in Arctic Russia, the numerous legacy spills in Antarctica, and in the remote high Arctic.

References

Aggarwal, P. K., Means, J. L., and Hinchee, R. E. 1991. Formulation of nutrient solutions for in situ bioremediation. In *In Situ Bioremediation*, Hinchee, R. E. and Olfenbuttel, R. F. (eds.), Columbus, OH, Battelle Press, 51–66.

Aguirre-Puente, J. and Gruson, J. 1983. Measurement of permeabilities of frozen soils. *Proc. 4th Int'l. Conf. on Permafrost*, 5–9.

Aichberger, H., Hasinger, M., Braun, R., and Loibner, A. P. 2005. Potential of preliminary test methods to predict biodegradation performance of petroleum hydrocarbons in soil. *Biodegradation* **16**: 115–25.

Aiken, G. R., McKnight, D. M., Wershaw, R. L., and MacCarthy, P. 1985. An introduction to humic substances in soil, sediment, and water. In *Humic Substances in Soil, Sediment, and Water: Geochemistry, Isolation, and Characterization*, G. R. Aiken, D. M. McKnight, R. L. Wershaw, and MacCarthy, P. (eds.), New York, NY, Wiley-Interscience, 1–12.

Aislabie, J. 1997. Hydrocarbon-degrading bacteria in oil-contaminated soils near Scott Base, Antarctica. In *Ecosystem Processes in Antarctica's Ice-Free Landscape*. Lyons, W. B., Howard-Williams, C., and Hawes, I. (eds.), Rotterdam, Balkema Publishers Ltd., 253–8.

Aislabie, J. M., Balks, M. R., Foght, J. M., and Waterhouse, E. J. 2004. Hydrocarbon spills on Antarctic soils: effects and management. *Environ. Sci. Technol.* **38(5)**: 1265–74.

Aislabie, J., Baraniecki, C., and Foght, J. M. 2002. Distribution and diversity of phenanthrene-degrading bacteria from soils of the Ross Sea region, Antarctica. *Proc. 3rd Int'l. Conf. on Contaminants in Freezing Ground*, Australian Antarctic Division, 103.

Aislabie, J., Foght, J., and Saul, D. 2000. Aromatic-hydrocarbon degrading bacteria isolated from soil near Scott Base, Antarctica. *Polar Biol.* **23**: 183–8.

Aislabie, J., Fraser, R., Duncan, S., and Farrell, R. L. 2001. Effects of soil spills on microbial heterotrophs in Antarctic soils. *Polar Biol.* **24**: 308–13.

Aislabie, J., McLeod, M., and Fraser, R. 1998. Potential of biodegradation of hydrocarbons in soil from the Ross Dependency, Antarctica. *Appl. Microbiol. and BioTechnol.* **49**: 210–14.

Aksenov, V. I., Klinova, G. I., and Scheikin, I. V. 1998. Material composition and strength characteristics of saline frozen soils. *The 7th Int'l. Permafrost Conf.*, 1–4.

Aldrich, H. P. and Paynter, H. M. 1966. *Depth of Frost Penetration in Non-uniform Soil.* U.S. Army Cold Regions Research and Engineering Laboratory Special Report 104.

Alexander, M. 1999. *Biodegradation and Bioremediation.* San Diego, CA, Academic Press.

Allen-King, R. M., Barker, J. F., Gillham, R. W., and Jensen, B. K. 1994. Substrate- and nutrient-limited toluene biotransformation in sandy soil. *Environ. Toxicology and Chem.* **13**: 693–705.

Amann, R. I., Ludwig, W., and Schleifer, K-H. 1995. Phylogenetic identification and in situ detection of individual microbial cells without cultivation. *Microbiol. Rev.* **59(1)**: 143–69.

AMAP. 1998. AMAP Assessment Report: *Arctic pollution issues, Arctic Monitoring and Assessment Programme (AMAP)*, Oslo, Norway.

AMAP. 2006. Prospectus for the preparation of the Arctic Council's assessment of oil and gas activities in the Arctic (January 2006 version). Arctic Monitoring and Assessment Programme (AMAP), Oslo, Norway. viewed 11 August 2006, www.amap.no/MiscTempFiles/OGAOutline-January2006.doc.

Andersland, O. B. and Ladanyi, B. 1994. *An Introduction to Frozen Ground Engineering.* London, Chapman & Hall.

Andersland, O. B. and Ladanyi, B. 2004. *Frozen Ground Engineering.* American Society of Civil Engineers & John Wiley & Sons.

Andersland, O. B., Wiggert, D. C., and Davies, S. H. 1996. Hydraulic conductivity of frozen soils. *J. Environ. Eng.*, March, 212–16.

Anderson, D. M. and Tice, A. R. 1972. Predicting unfrozen water contents in frozen soils from surface area measurements. In *Frost Action in Soils*, Washington, DC, National Academy of Sciences, 12–18.

Arenson, L. U. and Sego, D. C. 2004. Freezing processes for a coarse sand with varying salinities. *Proc. 12th Int'l. Conf. on Ground Freezing*, Smith, D. W., Sego, D. C. and Lendzion, C. A. (eds.).

Arey, J. S., Nelson, R. K., Xu, L., and Reddy, C. M. 2005. Using comprehensive two-dimensional gas chromatography retention indices to estimate environmental partitioning properties for a complete set of diesel fuel hydrocarbons. *Analytical Chem.* **77**: 7172–82.

Armstrong J. E., Biggar K., Staudt W., *et al.* 2002. *Assessment of Monitored Natural Attenuation at Upstream Oil & Gas Facilities in Alberta: Final Report.* Canadian Association of Petroleum Producers, Research Report 2001-0010. Komex International Ltd., Calgary, AB, Canada.

Aronson, D., Philip, H., and Howard, P. H. 1997. *Anaerobic Biodegradation of Organic Chemicals in Groundwater: A Summary of Field and Laboratory Studies.* Final report prepared for American Petroleum Institute, Chemical Manufacturer's Association, National Council of the Paper Industry for Air and Stream Improvement, Edison Electric Institute, American Forest and Paper Association. Environmental Science Center, Syracuse Research Corporation, North Syracuse, New York.

Athey, P., Reeder, D., Lukin, J., McKendrick, J., and Conn, J. S. 2001. *Tundra Treatment Guidelines*, Alaska Department of Environmental Conservation.

Atlas, R. M. 1979. Measurement of hydrocarbon biodegradation potentials and enumeration of hydrocarbon-utilizing microorganisms using carbon-14 hydrocarbon-spiked crude oil. In *Native Aquatic Bacteria: Enumeration, Activity, and Ecology*, Costerton, J. W. and Colwell, R. R. (eds), Philadelphia, American Society for Testing and Materials. ATSM STP 695, 196–204.

Atlas, R. M. 1981. Microbial degradation of petroleum hydrocarbons: an environmental perspective. *Microbiol. Rev.* **45(1)**: 180–209.

Ausma, S., Edwards, G. C., Fitzgerald-Hubble, C. R., *et al.* 2002. Volatile hydrocarbon emissions from a diesel fuel-contaminated soil bioremediation facility. *J. Air & Waste Mgmt. Assoc.* **52**: 769–80.

Ausma, S., Edwards, G. C., Wong, E. K., *et al.* 2001. A micrometeorological technique to monitor total hydrocarbon emissions from landfarms to the atmosphere. *J. Environ. Qual.* **30**: 776–85.

Baedecker, M. J., Cozzarelli, I. M., Eganhouse, R. P., Siegel, D. I., and Bennett, P. C. 1993. Crude oil in a shallow sand and gravel aquifer – III. Biogeochemical reactions and mass balance modeling in anoxic groundwater. *Appl. Geochem.* **8**: 569–58.

Baker, G. C. and Osterkamp, T. E. 1988. Salt redistribution during laboratory freezing of saline sand columns. *5th Int'l. Symposium on Ground Freezing*, 29–33.

Baker, J. H. 1974. The use of temperature-gradient incubator to investigate the temperature characteristics of some bacteria from Antarctic peat. *British Antarct. Surv. B.* **39**: 49–59.

Balks, M. R., Holmes, D. J., and Aislabie, J. 2002. The fate and effects of hydrocarbons in Antarctic soil: preliminary results of an experimental fuel spill. In *Transactions of the 17th World Congress of Soil science*, Kheoruenromne, I. (ed), Bangkok, Thailand International Union of Soil Sciences, 320–1 to 320–9.

Banks, P. D. and Brown, K. M. 2002. Hydrocarbon effects on fouling assemblages: the importance of taxonomic differences, seasonal, and tidal variation. *Mar. Environ. Res.* **53**: 311–26.

Baraniecki, C. A., Aislabie, J., and Foght, J. M. 2002. Characterisation of *Sphingomonas sp.* Ant 17, an aromatic hydrocarbon-degrading bacterium isolated from Antarctic soil. *Microbial Ecol.* **43**: 44–54.

Barker, J. F, Patrick, G. C., and Major, D. 1987. Natural attenuation of aromatic hydrocarbons in a shallow sand aquifer. *Ground Water Monitor. Rev.* **7(1)**: 64–7.

Barnes, D. L. and Adhikari, H. 2006. Suprapermafrost ground water dynamics in gravel pads located in the Arctic. In *Contaminants in Freezing Ground: Proc. 5th Int'l Conf.*, Rike, A. G. Øvstedal, J., and Vethe, O. (eds.), Oslo, Norway: Norsk Geologisk Forening, 13.

Barnes, D. L. and Filler, D. M. 2003. Spill evaluation of petroleum products in freezing ground. *Polar Rec.* **39**: 385–90.

Barnes, D. L. and Wolfe, S. M. In press. Influence of ice on the infiltration of petroleum into frozen coarse grain soil. *Petroleum Sci. & Technol.*

Barnes, D. L., Wolfe, S. M., and Filler, D. M. 2004. Equilibrium distribution of petroleum hydrocarbons in freezing ground. *Polar Rec.* **40**: 245–51.

Barnette, M., Ziervogel, H., Das, D., Clark, J., and Hayden, K. 2005. Bioventing at a heating oil spill site in Yellowknife, Northwest Territories. *Proc. '05 Assessment and Remediation of Contaminated Sites in Arctic and Cold Climates (ARCSACC) workshop*, Edmonton, Canada, 207–16.

Bathurst, R. J., Rowe, R. K., Zeeb, B. A., and Reimer, K. J. 2006. A geocomposite barrier for hydrocarbon containment in the Arctic. *Int. J. Geoeng. Case Histories* **1**: 18–34.

Batley, G. E., Burton, G. A., Chapman, P. M., and Forbes, V. E. 2002. Uncertainties in sediment quality weight-of-evidence (WOE) assessments. *Hum. Ecol. Risk Assess.* **8**: 1517–47.

Bazilescu, I. and Lyhus, B. 1996. Russia Oil Spill. TED Case Studies, No. 265. Washington, DC, American University, viewed 11 August 2006, www.american.edu/ted/KOMI. HTM.

Bej, A. K., Saul, D., and Aislabie, J. 2000. Cold-tolerant alkane-degrading *Rhodococcos* species from Antarctica. *Polar Biol.* **23(2)**: 100–5.

Bekins, B. A., Warren, E., and Godsy, E. M. 1998. A comparison of zero-order, first-order, and Monod biotransformation models. *Ground Water* **36**: 261–8.

Bellona, 2006. Three times more oil spills in Komi Republic. Bellona, Oslo, viewed 11 August 2006, www.bellona.org/news/Three_times_more_oil_spills_in_Komi_Republic.

Berchet, V., Thomas, T., Cavicchioli, R., Russell, N. J., and Gounot, A. 2000. Structural analysis of the elongation factor G protein from the low-temperature-adapted bacterium *Arthrobacter globiformis* SI55. *Extremophiles* **4**: 123–30.

Berlow, E. L. 1999. Strong effects of weak interactions in ecological communities. *Nature* **398**: 330–4.

Biggar, K. W., Haidar, S., Nahir, M., and Jarrett, P. M. 1998. Site investigation of fuel spill migration into permafrost. *J. Cold Regions Eng.* **12**(2): 84–104.

Biggar K. W. and Neufeld, J. C. R. August, 1996. Vertical migration of diesel into silty sand subject to cyclic freeze-thaw. *Proc. 8th Int'l. Conf. Cold Regions Eng.*, Fairbanks, Alaska, 116–27.

Biggar, K. W., Van Stempvoort, D., Iwakun, O., Bickerton, G., and Voralek, J. 2006. Fuel contamination characterization in permafrost fractured bedrock at the Colomac mine site, NWT. In *Contaminants in Freezing Ground: Proc. 5th Int'l. Conf.*, Rike, A. G., Øvstedal, J., and Vethe, Ø. (eds.), Oslo, Norway, Norsk Geologisk Forening, 17.

Billi, D., Friedmann, E. I., Hofer, K. G., Grilli-Caiola, M., and Ocampo-Friedman, R. 2000. Ionizing-radiation resistance in the desiccation-tolerant cyanobacterium *Chroococcidiopsis. Appl. and Environ. Microbiol.* **66**: 1489–92.

Billowits, M. E., Whyte, L.G, Ramsay, J. A., Greer, C., and Nahir, M. 1999. An evaluation of the bioremediation potential of near surface groundwater contaminated with petroleum hydrocarbons in the Yukon. *Proc. '99 Assessment and Remediation of Contaminated Sites in Arctic and Cold Climates (ARCSACC) workshop*, Edmonton, Canada, 91–100.

Bockheim, J. G. and Tarnocai, C. 1998. Nature, occurrence and origin of dry permafrost. *Permafrost, 7th Int'l. Conf.*, Lewkowicz, A. G. and Allard, M. (eds.), Yellowknife, Canada, June 23–27, 57–63.

Børresen, M. H., Barnes, D. L., and Rike, A. G. 2006. Repeated freeze-thaw cycles and their effects on mineralization of hexadecane and phenanthrene in cold climate soils. *Proc. 5th Int'l. Conf. on Contaminants in Freezing Ground*, NGF Abstracts and Proceedings of the Geological Society of Norway, No. 2, p. 23.

Børresen, M., Breedveld, G. D., and Rike, A. G. 2003a. Assessment of the biodegradation potential of hydrocarbons in contaminated soil from a permafrost site. *Cold Reg. Sci. Technol.* **37**: 137–49.

Børresen, M. and Rike, A. G. 2003b. Effect of nutrient content on biodegradation of hydrocarbons in arctic soil. *Proc. 3rd Assessment and Remediation of Contaminated Sites in Arctic and Cold Climates* (ARCSACC) *Conf.*, Nahir, M., Biggar, K., and Cotta, G. (eds.), Edmonton, Canada, May 4–6, 220–6.

Bowles, J. E. 1988. *Foundation Analysis and Design*. New York, McGraw-Hill.

Bowles, M. W., Bentley, L. R., *et al.* (2000). In situ groundwater remediation using the trench and gate system. *Ground Water* **38**: 172–81.

Braddock, J. F., Harduar, L. N. A., Lindstrom, J. E., and Filler, D. M. 2000. Efficacy of bioaugmentation vs. fertilization only for treatment of diesel contaminated soil at an Arctic site. *Proc. 23rd Arctic and Marine Oilspill (AMOP) Technol. Seminar*, Vancouver, Environment Canada, 991–1002.

Braddock, J. F., Lindstrom, J., Filler, D. M., and Walworth, J. 2001. Temperature and nutrient effects on bioremediation of petroleum hydrocarbons in cold soils and groundwater. *Proc. '01 Assessment and Remediation of Contaminated Sites in Arctic and Cold Climates (ARCSACC) Workshop*, Nahir, M., Biggar, K., and Cotta, G. (eds.), Edmonton, Canada, 161–7.

Braddock, J. F., Lindstrom, J. E., and Prince, R. C. 2003. Weathering of a subarctic oil spill over 25 years: the Caribou Poker Creeks Research Watershed experiment. *Cold Reg. Sci. Technol.* **36(1–3)**: 11–23.

Braddock, J. F. and McCarthy, K. A. 1996. Hydrologic and microbiological factors affecting persistence and migration of petroleum hydrocarbons spilled in a continuous-permafrost region. *Environ. Sci. Technol.* **30**: 2626–33.

Braddock, J. F., Ruth, M. L., Catterall, P. H., Walworth, J. L., and McCarthy, K. A. 1997. Enhancement and inhibition of microbial activity in hydrocarbon-contaminated Arctic soils: implications for nutrient-amended bioremediation. *Environ. Sci. Technol.* **31(7)**: 2078–84.

Braddock, J. F., Walworth, J. L., and McCarthy, K. A. 1999. Biodegradation of aliphatic vs. aromatic hydrocarbons in fertilized Arctic soils. *Bioremediation J.* **3(2)**: 105–16.

Bradley, P. M. and Chapelle, F. H. 1995. Rapid toluene mineralization by microorganisms at Adak, Alaska: Implications for intrinsic bioremediation in cold environments. *Environ. Sci. Technol.* **29**: 2778–81.

Braids, O. C. and Miller, R. H. 1975. Fats, waxes, and resins in soil. In *Soil Components: Volume 1, Organic Components*, Gieseking, J. E. (ed.), New York, NY, Springer-Verlag, 343–68.

Braley, W. A. 1984. *A Personal Computer Solution to the Modified Berggren Equation*. Institute of Water Resources/Engineering Experiment Station-University of Alaska Fairbanks, Report No. AK-RD-85-19.

Broeze. R. J., Solomon, C. J., and Pope, D. H. 1987. Effects of low temperature on in vivo and in vitro protein synthesis in *Escherichia coli* and *Pseudomonas fluorescens. J. Bacteriol.* **134**: 861–74.

Brook, T. R., Stiver, W. H., and Zytner, R. G. 1997. Effect of nitrogen sources on the biodegradation of diesel fuel in unsaturated soil. *1997 CSCE/ASCE Environ. Eng. Conf.*, Edmonton, Alberta, Canada, July 22–26.

Brook, T. R., Stiver, W. H., and Zytner, R. G. 2001. Biodegradation of diesel fuel in soil under various nitrogen addition regimes. *Soil and Sediment Contam.* **10**: 539–53.

Brown, E. J. and Braddock, J. F. 1990. Sheen Screen, a miniaturized most-probable-number method for enumeration of oil-degrading microorganisms. *Appl. and Environ. Microbiol.* **56**: 3895–6.

Brown, K. W., Donnelly, K. C., and Deuel, J. 1983. Effects of mineral nutrients, sludge application rate, and application frequency on biodegradation of two oily sludges. *Microbial Ecol.* **9**: 363–73.

Brunner, W. and Focht, D. D. 1984. Deterministic three-half-order kinetic model for microbial degradation of added substrates in soil. *Appl. and Environ. Microbiol.* **47**: 167–72.

Burt, T. P. and Williams, P. J. 1976. Hydraulic conductivity in frozen soils. *Earth Sur. Proc.* **1**: 349–60

Burton, G. A., Batley, G. E., Chapman, P. M., *et al.*, 2002a. A weight-of-evidence framework for assessing sediment (or other) contamination: Improving certainty in the decision-making process. *Hum. Ecol. Risk Assess.* **8**: 1675–96.

Burton, G. A., Chapman, P. M., and Smith, E. P. 2002b. Weight-of-evidence approaches for assessing ecosystem impairment. *Hum. Ecol. Risk Assess.* **8**: 1657–73.

Bury, S. J. and Miller, C. A. 1993. Effect of micellar solubilization on biodegradation rates of hydrocarbons. *Environ. Sci. Technol.* **27**: 104–10.

Campbell, D. I., MacCulloch, R. J. L., and Campbell, I. B. 1998. Thermal regimes of some soils in the McMurdo Sound region, Antarctica. In *Ecosystem Processes in Antarctic Ice-free Landscapes*, Lyons, W. B., Howard-Williams, C., and Hawes, I. (eds), Rotterdam, Balkema, 45–56.

Carss, J. G., Agar, J. G., and Surbey, G. E. 1994. In situ bioremediation in Arctic Canada. *Proc. '93 Bioreclamation Symposium*. Boca Raton Florida, Lewis Publishers, **2(2)**: 323–8.

Cavicchioli, R., Thomas, T., and Curmi, P. M. G. 2000. Cold stress response in *Archaea. Extremophiles* **4**: 321–31.

CCME 1996. A Protocol for the Derivation of Environmental and Human Health Soil Quality Guidelines (PN 1332). Canadian Council of Ministers of the Environment.

CCME 2001. *Canada-wide Standard for Petroleum Hydrocarbons (PHC) in Soil: User Guidance.* Report 10-6162. Winnipeg: Canadian Council for Ministers for the Environment.

Chablain, P. A., Philippe, G., Groboillot, A., Truffaut, N., and Guespin-Michel, J. F. 1997. Isolation of a soil psychrotrophic toluene-degrading *Pseudomonas* strain: influence of temperature on the growth characteristics on different substrates. *Res. in Microbiol.* **148**: 153–61.

Chamberlain, E. J. 1983. Frost heave of saline soils. *4th Int'l. Conf. on Permafrost*, 121–6.

Chang, Z. Z., Weaver, R. W., and Rhykerd, R. L. 1996. Oil bioremediation in a high and a low phosphorus soil. *J. Soil Contam.* **5(3)**: 215–24.

Chang, Z. Z., and Weaver, R. W. 1997. Nitrification and utilization of ammonium and nitrate during oil bioremediation at different soil water potential. *J. Soil Contam.* **6(2)**: 149–60.

Chapman, P. M. 1986. Sediment quality criteria from the sediment quality triad: an example. *Environ. Toxicol. Chem.* **5**: 957–64.

Chapman, P. M., Ho, K. T., Munns, J., *et al.* 2002a. Issues in sediment toxicity and ecological risk assessment. *Mar. Pollut. Bull.* **44**: 271–8.

Chapman, P. M., McDonald, B. G., and Lawrence, G. S. 2002b. Weight-of-evidence issues and frameworks for sediment quality (and other) assessments. *Hum. Ecol. Risk Assess.* **8**: 1489–515.

Charbeneau, R., Johns, R., Lake, L., and McAdams, M. 1999. *Free-Product Recovery of Petroleum Hydrocarbon Liquids.* American Petroleum Institute Publication No. 4682.

Chatham, J. R. 2003. Landfarming on the Alaskan North slope – historical development and recent applications. *10th Annual Int'l. Petroleum Environ. Conf.*, Houston, TX, November 11–14, 2003. http://ipec.utulsa.edu/Conf2003/Papers/chatham_35.pdf.

Chattopadhyay, M. A. and Jagannadham, M. V. 2001. Maintenance of membrane fluidity in Antarctic bacteria. *Polar Biol.* **24**: 386–8.

Chiang, C. Y., Salanitro, J. P., Chai, E. Y., Colthart, J. D., and Klein, C. L. 1989. Aerobic biodegradation of benzene, toluene, and xylene in a sandy aquifer – Data analysis and computer modeling. *Ground Water* **27(6)**: 823–34.

Christensen, K. E. and Shenk, C. G. 2006. Observations of fuel transport from two fuel release events on multi-year sea ice with relevance to site assessment and closure. *Contaminants in Freezing Ground: Proc. 5th Int'l Conf.*, Rike, A. G. Øvstedal, J., and Vethe, Ø. (eds.), Oslo, Norway, Norsk Geologisk Forening, 25.

Chuvilin, E. M. and Miklyaeva, E. S. 2003. An experimental investigation of the influence of salinity and cryogenic structure on the dispersion of oil and oil products in frozen soils. *Cold Regions Sci. Technol.* **37**: 89–95.

Chuvilin, E. M., Naletova, N. S., Miklyaeva, E. C., Kozlova, E. V., and Istanes, A. 2001. Factors affecting the spreadibility and transportation of oil in regions of frozen ground. *Polar Rec.* **37(202)**: 229–38.

Clarke, P. J. and Ward, T. J. 1994. The response of southern hemisphere saltmarsh plants and gastropods to experimental contamination by petroleum hydrocarbons. *J. Exp. Mar. Biol. Ecol.* **175**: 43–57.

Collins, C. M., Racine, C. H., and Walsh, M. E. 1994. The physical, chemical and biological effects of crude oil spills after 15 years on a black spruce forest, Interior Alaska. *Arctic* **47(2)**: 164–75.

Colwell, R. R. and Walker, J. D. 1977. Ecological aspects of microbial degradation of petroleum in the marine environment. CRC Crit. Rev. *Microbiol.* **5**: 423–45.

COMNAP. 2006. Antarctic Facilities in Operation. *Council of Managers of National Antarctic Programs*. viewed 28 August 2006, (www.comnap.aq/operations/facilities/)

Conner, J. S. 1988. Case study of soil venting. *Pollution Eng.* **7**: 74–8.

Cookson, J. T. 1995. *Bioremediation Engineering: Design and Application*, New York, McGraw-Hill.

Council of the European Communities. 2004. Council Directive 75/439/EEC of 16 June 1975 on the disposal of waste oils. viewed 29 August 2006, http://eurlex.europa.eu/LexUriServ/LexUriServ.do?uri=CELEX:31975L0439:EN:HTML

Croft, B. C., Swannell, R. P. J., Grant, A. L., and Lee, K. 1995. The effect of bioremediation agents on oil biodegradation in medium-fine sand. In *Appl. Bioremediation of Petroleum Hydrocarbons*, Hinchee, R. E., (ed.), Columbus, OH, Battelle Press.

Cross, K., Biggar, K., Semple, K., *et al.* 2003. Intrinsic bioremediation of invert diesel fuel contaminating groundwater in a bedrock formation. *Proc. '03 Assessment and Remediation of Contaminated Sites in Arctic and Cold Climates (ARCSACC) Workshop*, Nahir, M., Biggar, K., and Cotta, G. (eds.), Edmonton, Canada, 227–42.

Cunningham, J. 1993. Increased biodegradation rates of diesel fuel in soil using controlled-release nutrients. In *Principles and Practices for Diesel Contaminated Soils, Volume II*, Kostecki, P. T., Calabrese, E. J., and Barkan, C. P. L. (eds.), Amherst, MA, Association for the Environmental Health of Soils, 29–42.

Curtis, F. and Lammey, J. 1998. Intrinsic remediation of a diesel fuel plume in Goose Bay, Labrador, Canada. *Environ. Poll.* **103(2–3)**: 203–10.

Dean, J. 1998. *Extraction Methods for Environmental Analysis*. New York, John Wiley and Sons.

DEC. 2005. 18 Alaska Administrative Code 75: Oil and Other Hazardous Substances Pollution Control. Alaska Department of Environmental Conservation.

Delille, D., Coulon, F., and Pelletier, E. 2004a. Effects of temperature warming during a bioremediation study of natural and nutrient-amended hydrocarbon-contaminated sub-Antarctic soils. *Cold Reg. Sci. Technol.* **40**: 61–7.

Delille, D., Coulon, F., and Pelletier, E. 2004b. Biostimulation of natural microbial assemblages in oil-amended vegetated and desert sub-Antarctic soils. *Microbial Ecol.* **47(4)**: 407–15.

Delille, D., Delille, B., and Pelletier, E. 2002. Effectiveness of bioremediation of crude oil contaminated subantarctic intertidal sediment: The microbial response. *Microbial Ecol.* **44**: 118–26.

Delille, D. and Pelletier, E. 2002. Natural attenuation of diesel-oil contamination in a subantarctic soil (Crozet Island). *Polar Biol.* **25**: 682–7.

Delille, D., Pelletier, E., Coulon, F., Feller, G., and Delille, B. 2006. Tools for bioremediation of sub-Antarctic soils exposed to petroleum hydrocarbons. *Newsletter for the Canadian Antarctic Research Network*, **21**: 11–16.

Delille, D., Pelletier, E., Delille, B., and Coulon, F. 2003. Effect of nutrient enrichments on the bacterial assemblage of Antarctic soils contaminated by diesel or crude oil. *Polar Rec.* **39(211)**: 309–18.

Deming, J. W. 2002. Psychrophiles and polar regions. *Current Opinions in Microbiol.* **5**: 301–9.

Demque, D. E., Biggar, K. W., and Heroux, J. A. 1997. Land treatment of diesel contaminated soil. *Can. GeoTechnol. J.* **34**: 421–31.

Denef, V. J., Park, J., Rodrigues, J. L. M., *et al.* 2003. Validation of a more sensitive method for using spotted oligonucleotide DNA microarrays for functional genomics on bacterial communities. *Environ. Microbiol.* **5**: 933–43.

Det Norske Veritas. 2003. *Russia Pipeline Oil Spill Study.* ESMAP Technical Paper 034-03. Joint UNDP/World Bank Energy Sector Management Assistance Programme, Norway, p. 82+app., viewed 11 August 2006, http://wbln0018.worldbank.org/esmap/site.nsf/files/034-03+Russia+Pipeline+Oil+Spill+Study+Report.pdf/$FILE/034-03+Russia+Pipeline+Oil+Spill+Study+Report.pdf

DIAND. 2002. *Contaminated Sites Program Management Framework, October 2002.* Department of Indian and Northern Affairs Development.

Diaz-Ravina, M., Frostegard, A., and Baath, E. 1994. Thymidine, leucine and acetate incorporation into soil bacterial assemblages at different temperatures. *FEMS Microbiol. Ecol.* **14**: 221–32.

Dibble, J. T. and Bartha, R. 1979. Effect of environmental parameters on the biodegradation of oil sludge. *Appl. and Environ. Microbiol.* **37**: 729–39.

Dirksen, C. and Miller, R. D. 1966. Closed-system freezing of unsaturated soil. *Soil Sci. Soc. of America* **30**: 168–73.

DND-NTI. 1998. *Agreement between Nunavut Tunngavik Incorporated and Her Majesty in the Right of Canada as represented by the Minister of National Defence for the Clean-up and Restoration of Distant Early Warning Sites within the Nunavut Settlement area. September 1998.*

Dörfler, U., Haala, R., Matthies, M., and Scheunert, I. 1996. Mineralization kinetics of chemicals in soils in relation to environmental conditions. *Ecotoxicol. and Environ. Safety* **34**: 216–22.

Dott, W., Feidieker, D., Kampfer, P., Schleibinger, H., and Strechel, S. 1989. Comparison of autochthonous bacteria and commercially available cultures with respect to their effectiveness in fuel oil degradation. *J. Indust. Microbiol.* **4**: 365–74.

Dragun, J. 1988. *The Soil Chemistry of Hazardous Materials.* Silver Spring, MD, Hazardous Materials Control Research Institute.

Duffie, J. A. and Beckman, W. A. 1991. *Solar Engineering of Thermal Processes, 2nd edn.* Wiley Interscience.

Dunne, J. A., Williams, R. J., and Martinez, N. D. 2002a. Food-web structure and network theory: the role of connectance and size. *Proc. Nat. Acad. Sci. USA* **99**: 12917–22.

Dunne, J. A., Williams, R. J., and Martinez, N. D. 2002b. Network topology and biodiversity loss in food webs: robustness increases with connectance. *Ecol. Lett.* **5**: 558–67.

Durant, N. D., Jonkers, C. A. A., and Bouwer, E. J. 1997. Spatial variability in the naphthalene mineralization response to oxygen, nitrate, and orthophosphate amendments in MGP aquifer sediments. *Biodegradation* **8**: 77–86.

Eckford, R., Cook, F. D., Saul, D., Aislabie, J., and Foght, J. 2002. Free-living nitrogen-fixing bacteria from Antarctic soils. *Appl. and Enviro. Microbiol.* **68**: 5181–5.

Edwards, D. A., Andriot, M. D., Amoruso, M. A., *et al.* 1997. *Development of fraction specific reference doses (RfDs) and reference concentrations (RfCs) for total petroleum hydrocarbons (TPH)*. Total Petroleum Hydrocarbon Criteria Working Group series; Volume 4, Amherst, MA, Amherst Scientific Publishers.

EEA-IMS. 2005. Progress in management of contaminated sites (CSI 015) – May 2005 Assessment. *European Environment Agency – Indicator Management Service*. viewed 18 August 2006, http://ims.eionet.europa.eu/IMS/ISpecs/ISpecification 20041007131746/IAssessment1116497286336/view_content.

Eganhouse, R. P., Baedecker, M. J., Cozzarelli, I. M., Aiken, G. R., Thorn, K. A., and Dorsey, T. F. 1993. Crude oil in a shallow sand and gravel aquifer – II. Organic geochemistry. *Appl. Geochem.* **8(4)**: 551–67.

EIA. 2000. Antarctica: Fact Sheet. Energy Information Administration, United States Department of Energy, viewed 13 November 2006, www.eia.doe.gov/emeu/cabs/antarctica.html.

Elliot, D. H. 1988. Antarctica – is there any oil and natural gas. *Oceanus* **31**: 32–8.

El-Shinnawi, M. M., Bayoumi, N. A., Aboel-naga, S. A., and Mohammed, S. S. 1993. Changes of nitrogen forms in different arid soils during incubation at varying moisture contents. *Egyptian J. Soil Sci.* **33(4)**: 435–61.

EMPCA. 1994. *Environmental Management and Pollution Control Act 1994*. Tasmania, Australia.

Environment and Food Agency Iceland. 2002. Contaminated Soil in Iceland. Environment and Food Agency, Iceland, viewed 3 October 2006, http://english.ust.is/infobase/pollution-prevention/WasteManagementinIceland/ContaminatedsoilInIceland.

Eriksson, M., Dalhammar, G., and Mohn, W. W. 2002. Bacterial growth and biofilm production on pyrene. *FEMS Microbiol. Ecol.* **40**: 21–7.

Eriksson, S., Hallbeck, L., Ankner, T., Abrahamsson, K., and Sjöling, Å. 2006. Indicators of petroleum hydrocarbon biodegradation in anaerobic granitic groundwater. *Geomicrobiol. J.* **23(1)**: 45–58.

Eriksson, M., Ka, J.-O., and Mohn, W. W. 2001. Effects of low temperature and freeze-thaw cycles on hydrocarbon biodegradation in Arctic tundra soil. *Appl. and Environ. Microbiol.* **67(11)**: 5107–12.

Eriksson, M., Sodersten, E., Yu, Z., Dalhammer, G., and Mohn, W. W. 2003. Degradation of polycyclic aromatic hydrocarbons at low temperature under aerobic and nitrate-reducing conditions in enrichment cultures from Northern soils. *Appl. and Environ. Microbiol.* **69**: 275–84.

Eschenbach, A., Wienberg, R., and Mahro, B. 1998. Fate and stability of nonextractable residues of [14C]PAH in contaminated soils under environmental stress conditions. *Environ. Sci. Technol.* **32**: 2585–90.

ESG. 1993. *The Environmental Impact of the DEW Line on the Canadian Arctic*. Environmental Sciences Group, Royal Military College, Kingston, Ontario, Canada.

Fan, X., Guigard, S., Foght, J. Semple, K., and Biggar, K. W. 2006. A mesocosm study of enhanced anaerobic biodegradation of petroleum hydrocarbons in groundwater from a flare pit site. *Proc. 59th Canadian Geotechnical Conf.*, Vancouver, Canada, Paper No. 346.

Farouki, O. 1981. *Thermal Properties of Soils.* U.S. Army Cold Regions Research and Engineering Laboratory Monograph 81–1.

Farr, A. M., Houghtalen, R. J., and McWhorter, D. B. 1990. Volume estimation of light nonaqueous phase liquids in porous media. *Ground Water* **28(1)**: 48–56.

Fayad, N. M. and Overton, E. B. 1995. A unique biodegradation pattern of the oil spilled during the 1991 gulf war. *Mar. Pollut. Bull.* **30(4)**: 239–46.

Ferguson, C. C. and Kasamas, H. 1999. *Risk Assessment for Contaminated Sites in Europe.* Policy Framework. Nottingham, LQM Press.

Ferguson, S. H., Franzmann, P. D., Revill, A. T., Snape, I., and Rayner, J. L. 2003a. The effects of nitrogen and water on mineralisation of diesel-contaminated terrestrial Antarctic sediments. *Cold Reg. Sci. Technol.* **37**: 197–212.

Ferguson, S. H., Franzmann, P. D., Snape, I., *et al.* 2003b. Effects of temperature on mineralisation of petroleum in contaminated Antarctic terrestrial sediments. *Chemosphere* **52(6)**: 975–87.

Filler, D. F. 1997. Thermally enhanced bioventing of petroleum hydrocarbons in cold regions. Doctoral Thesis, Dept. of Civil and Environ. Eng., University of Alaska Fairbanks.

Filler, D. M. and Barnes, D. L. 2003. Technical procedures for recovery and evaluation of chemical spills on tundra. *Cold Reg. Sci. Technol.* **37**: 121–35.

Filler, D. M. and Carlson, R. F. 2000. Thermal insulation systems for bioremediation in cold regions. *J. Cold Regions Eng.* **14(3)**: 119–29.

Filler, D. A., Lindstrom, J. E., Braddock, J. F., Johnson, R. A., and Nickalaski, R. 2001. Integral biopile components for successful bioremediation in the Arctic. *Cold Reg. Sci. Technol.* **32(2–3)**: 143–56.

Filler, D. M., Reynolds, C. M., Snape, I., *et al.* 2006. Advances in engineered remediation for use in the Arctic and Antarctica. *Polar Rec.* **42**: 111–20.

Fine, P., Graber, E. R., and Yaron, B. 1997. Soil interactions with petroleum hydrocarbons: abiotic processes. *Soil Technol.* **10**: 133–53.

Foght, J. and Aislabie, J. 2005. Enumeration of soil microorganisms. In *Soil Biology, Volume 5. Manual for Soil Analysis*, Margesin, R. and Schinner, F. (eds.), Springer-Verlag, 261–80.

Fourie, W., Barnes, D. L., and Shur, Y. 2007. The formation of ice from the infiltration of water in frozen coarse grain soils. *Cold Reg. Sci. Technol.* (in press).

Frankenberger, W. T. 1988. Use of urea as a nitrogen-fertilizer in bioreclamation of petroleum-hydrocarbons in soil. *Bulletin of Environ. Contam. and Toxicol.* **40(1)**: 66–8.

Franzmann, P. D., Zappia, L. R., Power, T. R., Davis, G. B., and Patterson, B. M. 1999. Microbial mineralisation of benzene and characterisation of microbial biomass in soil above hydrocarbon contaminated groundwater. *FEMS Microbiol. Ecol.* **30**, 67–76.

Garland, D. S. 1999. Quantifying biogenic interference in petroleum contamination tests of organic soil using pyrolysis-GC/FID, MS Thesis, University of Alaska Fairbanks.

Garland, J. L. and Mills, A. L. 1991. Classification and characterization of heterotrophic microbial communities on the basis of patterns of community-level sole-carbon-source utilization. *Appl. Environ. Microbiol.* **57**: 2351–9.

Garland, D. S., White, D. M., and Woolard, C. R. 1999. Contaminant analysis in tundra by pyrolysis-GC/FID. *Proc. 10th Int'l. Cold Regions Eng.: Putting Research into Practice*, American Society of Civil Engineers, Reston, VA, 352–62.

Gavril'ev, R. I. 2004. Thermal properties of soils and surface covers. In *Thermal Analysis, Construction, and Monitoring Methods for Frozen Ground*, Esch, D. C. (ed.), American Society of Civil Engineers, 277–94.

Genouw, G., Naeyer, F. D., Meenan, P. V., *et al.* 1994. Degradation of oil sludge by landfarming: a case study at the Ghent Harbour. *Biodegradation* **5**: 37–46.

Geosphere, & CH2MHILL. 2006. *Three- and Four-Phase Partitioning of Petroleum Hydrocarbons and Human Health Risk Calculations Technical Background Report.* Report prepared for the SOCWG.

Gibb, A., Chu, A., Wong, R. C. K., and Goodman, R. H. 2001. Bioremediation kinetics of crude oil at 5 °C. *J. Environ. Eng.* Sept. 818–24.

Gill, R. A. and Robotham, P. 1989. Composition, sources, and source identification of petroleum hydrocarbons and their residues. In *The Fate and Effects of Oil in Freshwater*, Green, J. and Trett, M. (eds.), New York, Elsevier Appl. Science, 11–40.

GNWT. 1998. *Guideline for Contaminated Site Remediation in the NWT*. Government of the Northwest Territories.

Goering, D. J. and Kumar, P. 1994. *Roadway Stabilization Using Air Convection Embankments*, Transportation Research Center, Report No. INE/TRC 94.01.

Goldstein, J., Pollit, N. S., and Inouye, M. 1990. Major cold shock protein of *Escherichia coli. Proc. Nat'l. Academy of Sci. USA* **87**: 283–7.

Goldsworthy, P. M., Canning, E. A., and Riddle, M. J. 2003. Soil and water contamination in the Larsemann Hills, East Antarctica. *Polar Rec.* **39**: 319–37.

Gore, D. B., Heiden, E. S., Snape, I., Nash, G., and Stevens, G. W. 2006a. Grain size of activated carbon, and untreated and modified granular clinoptilolite under freeze-thaw: applications to permeable reactive barriers. *Polar Rec.* **42(2)**: 121–6.

Gore, D. B., Heiden, E. S., Stevens, G. W., and Snape, I. 2004. Grain size of selected permeable reactive barrier materials under freeze-thaw (+/– water and diesel). *Proc. 4th Int'l. Conf. on Contaminants in Freezing Ground*, Filler, D. M. and Barnes, D. L. (eds.), Fairbanks, Alaska, 30 May – 3 June, 18.

Gore, D. B., Revill, A. T., and Guille, D. 1999. Petroleum hydrocarbons ten years after spillage at a helipad in Bunger Hills, East Antarctica. *Antarctic Sci.* **11**: 427–9.

Gore, D. B., Snape, I., Rayner, J. L., Dixon, E., and Stevens, G. W. 2006b. In *Hydraulics of permeable reactive barrier materials under freezing conditions, Proc. 5rd Int'l. Conf. on Contaminants in Freezing Ground Contaminants in Freezing Ground*, Rike, A. G., Øvstedal, J., and Vethe, Ø. (eds.), Norsk Geologisk Forening, Oslo, Norway, p. 37.

Gounot, A. M. and Russell, N. J. 1999. Physiology of cold-adapted microorganisms. In *Cold-Adapted Organisms*, Margesin, R., and Schinner, F. (eds.), Ecology, physiology, enzymology and molecular biology, Berlin, Springer-Verlag, 33–55.

Graham, D. W., Smith, V. H., Cleland, D. L., and Law, K. P. 1999. Effects of nitrogen and phosphorus supply on hexadecane biodegradation in soil systems. *Water, Air, Soil Poll.* **111**: 1–18.

Grechishchev, S.E., Chistotinov, L. V., and Shur, Y. L. 1980. *Cryogenic physics-geological processes and their forecast.* Moscow, "Nedra".

Grechishchev, S. E., Pavlov, A. V., and Ponomarev, V. V. 1998. Phase equilibrium and kinetics of saline soil water freezing, *7th Int'l. Permafrost Conf.*, 351–7.

Grigg, B. C., Assaf, N. A., and Turco, R. F., 1997. Removal of atrazine contamination in soil and liquid systems using bioaugmentation. *Pestic. Sci.* **50**: 211–20.

Gustafson, J. B., Tell, J. G., and Orem, D. 1997. Selection of representative TPH fractions based on fate and transport considerations. In *Total Petroleum Hydrocarbon Criteria Working Group series; Volume 3*, Amherst, MA, Amherst Scientific Publishers.

Haines, J. R., Kadkhokayan, M., Mocsny, D. J., *et al.* 1994. Effect of salinity, oil type, and incubation temperature on oil degradation. In *Applied BioTechnology for Site Remediation*, Hinchee, R. (ed.), Boca Raton, FL, Lewis Publishers, 75–83.

Hallet, B. 1978. Solute redistribution in freezing ground. *3rd Int'l. Conf. on Permafrost*, 86–91.

Hayward, S. A. L., Worland, M. R., Convey, P., and Bale, S. 2003. Temperature preferences of the mite, *Alaskozetes antarcticus*, and the collembolan, *Cryptopygus antarcticus* from the maritime Antarctic. *Physiol. Entomol.* **28**: 114–21.

Head, I. M., Saunders, J. R., and Pickup, R. W. 1998. Microbial evolution, diversity, and ecology: A decade of ribosomal RNA analysis of uncultivated microorganisms. *Microbial Ecol.* **35**: 1–21.

Helweg, A., Fomsgaard, I. S., Reffstrup, T. K., and Sorensen, H. 1998. Degradation of mecoprop and isoproturon in soil influence on initial concentration. *Int'l. J. Environ. Analytical Chem.* **70**: 133–48.

Herrington, R. T., Benson, L., Downey, D., and Hansen, J. 1997. Validation of fuel hydrocarbon attenuation in low-temperature groundwater environments. *Proc. In Situ and On-Site Bioremediation Symposium*, Columbus, OH, Battelle Press, **4(1)**: 303–8.

Hinzman, L. D., Bettez, N. D., Bolton, W. R., *et al.* 2005. Evidence and implications of recent climate change in Northern Alaska and other arctic regions, *Climate Change* **72**: 251–98.

Horiguchi, K. and Miller, R. D. 1980. Experimental studies with frozen soil in an 'Ice Sandwich' permeater. *Cold Reg. Sci. Technol.* **3**: 177–83.

Hoyle, B.L, Scow, K. M., Fogg, G. E., and Darby, J. L. 1995. Effect of carbon: nitrogen ratio on kinetics of phenol biodegradation by *Acinetobactor Johnsonii* in saturated sand. *Biodegradation* **6**: 283–93.

Huesemann, M. H. 1994. Guidelines for land-treating petroleum hydrocarbon-contaminated soils. *J. Soil Contam.* **3**: 299–318.

Huesemann, M. H. and Truex, M. J. 1996. The role of oxygen diffusion in passive bioremediation of petroleum contaminated soils. *J. Haz. Materials* **15**: 93–113.

Hunt, P. G., Rickard, W. E., Deneke, F. J., Koutz, F. R., and Murrman, R. P. 1973. Terrestrial oil spills in Alaska: environmental effects and recovery. *Proc. Joint Conf. on Prevention and Control of Oil Spills*, American Petroleum Institute and United States Coast Guard, Washington D.C., March 13–15.

Huntjens, J. L. M., Potter, H. D., and Barendrecht, J. 1986. The degradation of oil in soil. In *Contaminated Soil*, Assink, J. W. and Brink, W. J. V. D. (eds.), Dordrecht, Netherlands, Marinus Nijhoff, 121–4.

Hutchins, S. R., Sewell, G. W., Kovacs, D. A., and Smith, G. A. 1991. Biodegradation of aromatic-hydrocarbons by aquifer microorganisms under denitrifying conditions. *Environ. Sci. and Technol.* **25(1)**: 68–76.

Jackson, R. D. 1965. Water vapor diffusion in relatively dry soil: . Temperature and pressure effects on sorption diffusion coefficients. *Soil Sci. Soc. of America Proc.* **30**: 144–8.

Jansson, S. L. and Persson, J. 1982. Mineralization and immobilization of soil nitrogen. In *Nitrogen in Agricultural Soils*, Stevenson, F. J. (ed.), Madison, WS, American Society of Agronomy, 229–52.

Jobson, A., McLaughlin, M., Cook, F. D., and Westlake, W. S. 1974. Effects of amendments on the microbial utilization of oil applied to soil. *Appl. Microbiol.* **27**(1): 166–71.

Johansen, O. 1975. Thermal conductivity of soils. Ph.D. Dissertation, Norwegian Technical Univ., Trondheim; also, U.S. Army Cold Regions Research and Engineering Laboratory Transl. 637, July 1977.

Johnsen, A. R., Bendixen, K., and Karlson, U. 2002. Detection of microbial growth on polycyclic aromatic hydrocarbons in microtitre plates using the respiration indicator WST-1. *Appl. and Environ. Microbiol.* **68**: 2683–9.

Johnson, L. A., Sparrow, E. B., Jenkins, T. F., *et al.* 1980. *The Fate and Effect of Crude Oil Spilled on Subarctic Permafrost Terrain in Interior Alaska*, U.S. Environmental Protection Agency, Corvallis Environmental Research Laboratory, Office of Research and Development, EPA-600/3-80-040.

Johnson, P. C., Kemblowski, M. W., and Colthart, J. D. 1990. Quantitative analysis for the cleanup of hydrocarbon contaminated soils by in-situ soil venting. *Ground Water* **28(3)**: 413–29.

Johnson, R. A. 1990. Cogeneration and diesel electric power production. *The Cogeneration J.* 5: 44–60.

Jordán, F. 2001. Strong threads and weak chains? – a graph theoretical estimation of the power of indirect effects. *Community Ecol.* **2**: 17–20.

Kade, A., Walker, D. A., and Raynolds, M. K. 2005. Plant communities and soils in cryoturbated tundra along a bioclimate gradient in the Low Arctic, Alaska. *Phytocoenologia* **35**: 761–820.

Kandror, O., DeLeon, A., and Goldberg, A. L. 2002. Trehalose synthesis is induced upon exposure of *Echerichia coli* to cold and is essential for viability at low temperatures. *Proc. Natl. Acad. Sci.* **99(15)**: 9727–32.

Kane, D. L., Gieck, R. E., and Hinzman, L. D. 1990. Evapotranspiration from a small Alaskan arctic watershed. *Nord. Hydrol.* **21**: 253–72.

Kane, D. L. and Slaughter, C. W. 1974. Recharge of a central Alaska lake by subpermafrost groundwater. *Proc. 2nd In'l. Conf. on Permafrost*, Yakutsk, USSR, North American Contribution, 458–72.

Kennicutt, M. C. 2003. *Spatial and Temporal Scales of Human Disturbance: McMurdo Station, Antarctica. Final Report.* Geochemical and Environmental Research Group and Department of Geography, College of Geosciences, Texas A&M University and Marine Science Institute, The University of Texas at Austin.

Kerry, E. 1990. Microorganisms colonizing plants and soil subjected to different degrees of human activity, including petroleum contamination in the Vestfold Hills and MacRobertson Land Antarctica. *Polar Biol.* **10**: 423–30.

Kerry, E. 1993. Bioremediation of experimental petroleum spills on mineral soils in the Vestfold Hills, Antarctica. *Polar Biol.* **13**: 163–70.

Kersten, M. S. 1949. Thermal properties of soils. *University of Minnesota Engineering Experiment Station Bulletin*, no 28.

Khimenkov, A. N. and Brushkov, A. V. 2003. *Oceanic cryo-lithogenesis*, Moscow, "Nauka" (In Russian).

Kireeva, A. 2006. Murmansk Region oil-spill cleanup plan: just empty words on paper? Bellona Oslo, viewed 11 August 2006, http://bellona.no/bellona.org/english_import_area/energy/42017.

Klein, A. G., Kennicutt, M. C., Montana, P. A., *et al.* 2006. A long-term environmental monitoring program at McMurdo Station, Antarctica. In *2nd SCAR Open Science Conference 'Antarctica in the Earth System'*, Hobart.

Klonowski, M. R., Breedveld, G. D., and Aagaard, P. 2005. Natural gradient experiment on transport of jet fuel derived hydrocarbons in an unconfined sandy aquifer. *Environ. Geol.* **48(8)**: 1040–57.

Kolenc, R. J., Innis, W. E., Glick, B. R., Robinson, C. W., and Mayfield, C. I. 1988. Transfer and expression of mesophilic plasmid-mediated degradative capacity in a psychrotrophic bacterium. *Appl. and Environ. Microbiol.* **54**: 638–41.

Konrad, J. M. and McCammon, A. W. 1990. Solute partitioning in freezing soils. *Can. GeoTechnical J.* **25**: 108–18.

Konrad, J.-M. and Seto, J. C. T. 1991. Freezing of a clayey silt contaminated with an organic solvent. *J. Contam. Hydrol.* **8**: 335–55.

Kudriavtsev, V. A. (ed.). 1978. *General Permafrost Science (Geocryology)*, Moscow, Moscow State University (in Russian).

Kumar, G. S., Jagannadham, M. V., and Ray, M. K. 2002. Low-temperature-induced changes in composition and fluidity of lipopolysaccharides in the Antarctic psychrotrophic bacterium *Pseudomonas syringae*. *J. Bacteriology* **184**: 6746–9.

Lai, V., Biggar, K., Mullick, A., *et al.* 2001. Natural attenuation of 1,1,1 TCA and BTEX from a landfill in northern Alberta. *Proc. '01 Assessment and Remediation of Contaminated Sites in Arctic and Cold Climates (ARCSACC) Workshop*, Nahir, M., Biggar, K., and Cotta, G. (eds.), Edmonton, Canada, 215–25.

Laurie, A. D. and Lloyd-Jones, G. 2000. Quantification of phnAc and nahAc in contaminated New Zealand soils by competitive PCR. *Appl. and Environ. Microbiol.* **66**: 1814–17.

Leahy, J. G. and Colwell, R. R. 1990. Microbial degradation of hydrocarbons in the environment. *Microbiol. Rev.* **54(3)**: 305–15.

Lee, R. F. and Silva, M. 1994. Polycyclic aromatic hydrocarbon removal rates in oiled sediments treated with urea, urea-fish protein, or ammonium nitrate. In *Appl. Biotechnol. for Site Remediation*, Hinchee, R. E. (ed.), London, Lewis, 320–5.

Leeson, A., Hinchee, R. E., Kittel, J. A., and Foote, E. A. 1995. *Environics TOC Task 3 Bioventing Feasibility Study*, Eielson AFB site. Final report to Environmental Quality Directorate of the Armstrong Laboratory, Tyndall Air Force Base, Florida.

Lehner, C. A. 1995. Evaluation of controlled freezing to remove trapped residual NAPL. Unpublished MSc Thesis, Department of Civil and Environmental Engineering, Michigan State University.

Leszkiewicz, C. G. 2001. The effect of freeze-thaw temperature fluctuations on microbial metabolism of petroleum hydrocarbon contaminated Antarctic soil. Civil Engineering, University of New Hampshire, New Hampshire.

Lewis, D. L., Kollig, H. P., and Hodson, R. E. 1986. Nutrient limitation and adaptation of microbial populations to chemical transformations. *Appl. and Environ. Microbiol.* **51(3)**: 598–603.

Liebeg, E. W. and Cutright, T. J. 1999. The investigation of enhanced bioremediation through the addition of macro and micro nutrients in a PAH contaminated soil. *Int'l. Biodeterioration and Biodegradation* **44**: 55–64.

Lindstrom, J. E., Prince, R. C., Clark, J. C., *et al.* 1991. Microbial populations and hydrocarbon biodegradation potentials in fertilized shoreline sediments affected by the T/V Exxon Valdez Oil Spill. *Appl. and Environ. Microbiol.* **57**: 2514–22.

Line, M. A. 1988. Microbial flora of some soils of Mawson Base and the Vestfold Hills, Antarctica. *Polar Biol.* **8**: 421–7.

Linell, K. A. 1973. Long term effects of vegetation cover on permafrost stability in an area of discontinuous permafrost. *Proc. 2nd Int'l. Conf. on Permafrost, Yakutsk, USSR, North American Contribution*, 688–93.

Long, E. R. and Chapman, P. M. 1985. A sediment quality triad: Measures of sediment contamination, toxicity and infaunal community composition in Puget Sound. *Mar. Pollut. Bull.* **16**: 405–15.

Lunardini, V. J. 1978. Theory of n-factors and correlation of data. *Proc. 3rd Int'l. Conf. on Permafrost*, Edmonton, Alberta. Ottawa: National Research Council of Can., **1**: 41–6.

Mackay, D., Charles, M. E., and Phillips, C. R. 1974a. *The Physical Aspects of Crude Oil Spills on Northern Terrain*. Northern Pipelines, Task Force on Northern Oil Development, Environmental – Social Committee, Report No. 74-25.

Mackay, D., Charles, M. E., and Phillips, C. R. 1974b. *The Physical Aspects of Crude Oil Spills on Northern Terrain (Second Report)*. Northern Pipelines, Task Force on Northern Oil Development, Environmental – Social Committee, Report No. 73-42.

Mackay, D., Charles, M. E., and Phillips, C. R. 1975. *The Physical Aspects of Crude Oil Spills on Northern Terrain (Final Report)*. Arctic Land Use Research Program,

Northern Natural Resources and Environmental Branch, Department of Indian Affairs and Northern Development, INA Publication No. QS 8060-00-EE-A1.

Mahar, L. J., Wilson, R. M., and Vinson, T. S. 1983. Physical and numerical modeling of uniaxial freezing in a saline gravel. *4th Int'l. Conf. on Permafrost*, 773–8.

Manefield, M., Whiteley, A. S., Griffiths, R. I., and Bailey, M. J. 2002. RNA stable isotope probing, a novel means of linking microbial community function to phylogeny. *Appl. and Environ. Microbiol.* **68**: 5367–73.

Manilal, V. B. and Alexander, M. 1995. Factors affecting the microbial degradation of phenanthrene in soil. *Appl. Microbiol. and BioTechnol.* **35**: 401–5.

Margesin, R. 2000. Potential of cold-adapted microorganisms for bioremediation of oil-polluted Alpine soils. *Int. Biodet. Biodegrad.* **46**: 3–10.

Margesin, R., Labbe, D., Schinner, F., Greer, C. W., and Whyte, L. G. 2003. Characterization of hydrocarbon-degrading microbial population in contaminated and pristine alpine soils. *Appl. and Environ. Microbiol.* **69**: 3085–92.

Margesin, R. and Schinner, F. 1997a. Bioremediation of diesel-oil-contaminated alpine soils at low temperatures. *Appl. Microbiol. and BioTechnol.* **47**: 462–8.

Margesin, R. and Schinner, F. 1997b. Effect of temperature and oil degradation by a psychrotrophic yeast in liquid culture and in soil. *FEMS Microbiol. Ecol.* **24**: 243–9.

Margesin, R. and Schinner, F. 1997c. Efficiency of indigenous and inoculated cold-adapted soil microorganisms for biodegradation of diesel oil in Alpine soils. *Appl. and Environ. Microbiol.* **63**: 2660–4.

Margesin, R. and Schinner, F. 1997d. Laboratory bioremediation experiments with soil from a diesel-oil contaminated site – significant role of cold-adapted microorganisms and fertilizers. *J. Chem. Technol. Biotechnol.* **70**: 92–8.

Margesin, R. and Schinner, F. 1998. Oil biodegradation potential in alpine soils. *Arctic Alpine Res.* **30**: 262–5.

Margesin, R. and Schinner, F. 2001. Bioremediation (natural attenuation and biostimulation) of diesel-oil-contaminated soil in an alpine glacier skiing area. *Appl. and Environ. Microbiol.* **67**: 3127–33.

Mariner, P. E., Jin, M., and Jackson, R. E. 1997. An algorithm for the estimation of NAPL saturation and composition from typical soil chemical analysis. *Ground Water Monitor. Remed.* **17**: 122–9.

Master, E. R. and Mohn, W. W. 1998. Psychrotolerant bacteria isolated from Arctic soil that degrade polychlorinated biphenyls at low temperatures. *Appl. and Environ. Microbiol.* **64**: 4823–9.

McCarthy, K., Walker, L., and Vigoren, L. 2004. Subsurface fate of spilled petroleum hydrocarbons in continuous permafrost. *Cold Reg. Sci. Technol.* **38(1)**: 43–54.

McCarthy, K., Walker, L., Vigoren, L., and Bartel, J. 2004. Remediation of spilled petroleum hydrocarbons by *in situ* landfarming at an arctic site. *Cold Reg. Sci. Technol.* **40**: 31–9.

McCauley, C. A., White, D. M., Lilly, M. R., and Nyman, D. M. 2002. A comparison of hydraulic conductivities, permeabilities and infiltration rates in frozen and unfrozen soils, *Cold Reg. Sci. Technol.* **34**: 117–25.

McFarland, M. J. and Sims, R. C. 1991. Thermodynamic framework for evaluating PAH degradation in the subsurface. *Ground Water* **29(6)**: 885–96.

McIntyre, C., Harvey, P. M., Ferguson, S. H., *et al.* 2007. Determining the extent of biodegradation of fuels using the diastereomers of the acyclic isoprenoids. *Environ. Sci. Technol.* **41**: 2452–8.

McNamara, N. P., Black, H. I. J., Beresford, N. A., and Parekh, N. R. 2003. Effects of acute gamma irradiation on chemical, physical and biological properties of soils. *Appl. Soil Ecol.* **24**: 117–32.

Mercer, J. W. and Cohen, R. M. 1990. A review of immiscible fluids in the subsurface: properties, models, characterization and remediation. *J. Contam. Hydrol.* **6**: 107–63.

Mesarch, W. W., Nakatsu, C. H. and Nies, L. 2000. Development of catechol 2,3-dioxygenase-specific primers for monitoring bioremediation by competitive quantitative PCR. *Appl. and Environ. Microbiol.* **66**: 678–83.

Metcalf & Eddy, Inc. 1991. *Wastewater Engineering: Treatment, Disposal, Reuse*, 3rd edn. Boston, Massachusetts, McGraw Hill, Inc.

Metzger, L. O. Y., Munier-Lamy, C., Belgy, M. J., *et al.* 1999. A laboratory study of the mineralization and binding of ^{14}C labelled herbicide rimsulfuron in a rendzina soil. *Chemosphere* **39**: 1889–901.

Meyles, C. A. and Schmidt, B. 2005. Report on Soil Protection and Remediation of Contaminated Sites in Iceland: A Preliminary Study. *Environ. and Food Agency of Iceland*, viewed 18 August 2006, http://english.ust.is/media/skyrslur2005/Report_about_Soil_ Protection_and_Remediation_of_Contaminated_Sites_in_Iceland.doc.

MFE. 2003. *Contaminated Land Management, Guidelines No. 2.* Ministry for the Environment, Wellington, New Zealand.

Michel, V., Lehoux, I., Depret, G., *et al.* 1997. The cold shock response of the psychrotrophic bacterium *Pseudomonas fragi* involves four low-molecular-mass nucleic acid-binding proteins. *J. Bacteriol.* **179(23)**: 7331–42.

Mills, S. A. and Frankenberger, W. T. 1994. Evaluation of phosphorus sources promoting bioremediation of diesel fuel in soil. *Bulletin of Environ. Contam. and Toxicol.* **53**: 280–4.

Mitchell, I. and Friedrich, G. 2001. Multi-phase vacuum extraction at Bar-1 Komakuk Beach, Yukon Territory. *Proc. '01 Assessment and Remediation of Contaminated Sites in Arctic and Cold Climates (ARCSACC) Workshop*, Nahir, M., Biggar, K., and Cotta, G. (eds.), Edmonton, Canada, 84–97.

Mohn, W. W., Radziminski, C. Z., Fortin, M. C., and Reimer, K. J. 2001a. On site bioremediation of hydrocarbon-contaminated arctic tundra soils in inoculated biopiles. *Appl. Microbiol. and BioTechnol.* **57(1–2)**: 242–7.

Mohn, W. W., Reimer, K. J., Dalhammer, G., *et al.* 2001b. Bioremediation of Arctic soils contaminated by petroleum hydrocarbons. In *Assessment and Remediation of Contaminated Sites in Arctic and Cold Climates*, Nahir, M., Biggar, K., and Cotta, G. (eds.), Edmonton, Canada, 169–78.

Mohn, W. W. and Stewart, G. R. 2000. Limiting factors for hydrocarbon biodegradation at low temperature in arctic soils. *Soil Biol. and Biochem.* **32(8–9)**: 1161–72.

Moles, A., Rice, S. D., and Norcross, B. L. 1994. Non-avoidance of hydrocarbon laden sediments by juvenile flatfishes. *Neth. J. Sea Res.* **32**: 361–7.

Moller, S., Korber, D. R., Wolfaardt, G. M., Molin, S., and Caldwell, D. E. 1997. Impact of nutrient composition on a degradative biofilm community. *Appl. and Environ. Microbiol.* **63**: 2432–8.

Moore, B. J., Armstrong, J. E., Baker, J., and Hardisty, P. E. 1995. Effects of flow rate and temperature during bioventing in cold climates. In *In Situ Aeration: Air Sparging, Bioventing, and Related Remediation Processes,* Hinchee, R. E. *et al.* (eds.), Bioremediation Series, Book 3(2), 3rd Int'l. In situ and On Site Bioreclamation Symposium, Columbus, OH, Batelle Press, 307–14.

Morgan, P. and Watkinson, R. J. 1989. Hydrocarbon degradation in soils and methods for soil biotreatment. *Critical Rev. BioTechnol.* **8**: 305–33.

Morgan, P. and Watkinson, R. J. 1990. Assessment of the potential for *in situ* biotreatment of hydrocarbon-contaminated soils. *Water Sci. and Technol.* **22(6)**: 63–8.

Morita, R. T. 1975. Psychrophilic bacteria. *Bacteriol. Rev.* **29**: 144–67.

Mumford, K., Snape, I., Stevens, G., Rayner, J. L., and Walworth, J. L. 2006. Use of zeolite as a controlled release fertilization system for petroleum hydrocarbon remediation at low temperatures. *5th Int'l. Conf. on Contaminants in Freezing Ground,* May 21–25, 2006, Oslo, Norway.

Nakano, Y, Tice, A., and Oliphant, J. 1984. Transport of water in frozen soil: III. Experiments on the effects of ice content. *Adv. Water Res.* **7(1)**: 28–34.

Nelson, F. E., Shiklimanov, N. I., Hinkel, K. M., and Christiansen, H. H. 2004. Introduction: The Cirumpolar Active Layer Monitoring (CALM) workshop and the CALM II program, *Polar Geog.* **28**: 253–66.

NEPC. 2005. *Review of the National Environmental Protection (Assessment of Site Contamination) Measure: Issues Paper.* Adelaide, National Environmental Protection Council Service Corporation.

Newman, L. and Reynolds, C. M. 2004. Phytoremediation of organics. *Current Opinion in BioTechnol.* **15**: 225–30.

Niemeyer, T. and Schiewer, S. 2003. Effect of temperature and nutrient supply on the bioremediation rate of diesel contaminated soil from two Alaskan sites. *Proc. 3rd Assessment and Remediation of Contaminated Sites in Arctic and Cold Climates Conference (ARCSACC),* Edmonton, May 4–6, 212–19.

NREL, National Renewable Energy Lab, Solar Radiation Resource Information, http://rredc.nrel.gov/solar/ (accessed July 2006).

Oechel, W. C. and Billings, W. D. 1992. Anticipated effects of global change on carbon balance of arctic plants and ecosystems. In *Arctic Physiological Processes in a Changing Climate,* Chapin III, F. S., Jeffries, R. L., Shaver, G. R., Reynolds, J. F., and Svobada, J. (eds.), San Diego, CA, Academic Press, 139–68.

Oechel, W. and Vourlitis, G. 1995. Effect of global change on carbon storage in cold soils. In *Soils and Global Change,* Lal, R., Kimble, J., Levine, E., and Stewart, B. (eds.), New York, Lewis Publishers, 117–30.

Olovin, B. A. 1993. *Permeability of Perennially Frozen Soils. Novosibirsk, "Nauka"* (in Russian).

Osterkamp, T. E. and Romanovsky, V. E. 1999. Evidence for warming and thawing of discontinuous permafrost in Alaska, *Permafrost Periglac.* **10(1)**: 17–37.

Paetz, A. and Wilke, B.-M. 2005. Soil sampling and storage. In *Manual for Soil Analysis – Monitoring and Assessing Soil Bioremediation*, Margesin, R. and Schinner, F. (eds), Berlin, Springer-Verlag, 1–45.

Panicker, G., Aislabie, J., Saul, D., and Bej, A. K. 2002. Cold tolerance of *Pseudomonas* sp. 30–3 isolated from oil-contaminated soil, Antarctica. *Polar Biol.* **25(1)**: 5–11.

Pankow, J. F. and Cherry, J. A. 1996. *Dense Chlorinated Solvents and Other DNAPLs in Groundwater*, Portland, Oregon, Waterloo Press.

Paudyn, K., Poland, J. S., Rutter, A, and Rowe, R.K. 2005. Remediation of hydrocarbon contaminated soils in the Can. arctic with landfarms. *Proc. 4th Assessment and Remediation of Contaminated Sites in Arctic and Cold Climates Conference (ARCSACC)*, Edmonton, Alberta, 233–9.

Paudyn, K., Rutter, A., Rowe, R. K., and Poland, J. S. 2006. Remediation of hydrocarbon contaminated soils in the Canadian Arctic with landfarms. In *Contaminants in Freezing Ground: Proc. 5th Int'l Conf.*, Rike, A. G., Øvstedal, J., and Vethe, Ø. (eds.), Oslo, Norway, Norsk Geologisk Forening.

Paul, E. A. and Clark, F. E. 1996. *Soil Microbiology and Biochemistry*. San Diego, California, Academic Press.

Pawelczyk, A., Kazimierz Grabas, K., Barbara Kolwzan, B., and Steininger, M. 2003. Remediation of grounds at the former soviet military airfields contaminated by petroleum products. *Proc. 2nd European Bioremediation Conf.*, Chania, Crete, Greece, June 30–July 4, 2003, 184–7.

Pelletire, F., Prévost, D., Laliberté, G., and van Bochove, E. 1999. Seasonal response of denitrifiers to temperature in a Quebec cropped soil. *Can. J. Soil Sci.* **79**: 551–6.

Pelz, O., Chatzinotas, A., Andersen, N., *et al.* 2001. Use of isotopic and molecular techniques to link toluene degradation in denitrifying aquifer microcosms to specific microbial populations. *Archives of Microbiol.* **175**: 270–81.

PhytoPet© 2007. Phytoremediation of petroleum hydrocarbons web database developed by the University of Saskatchewan, www.phytopet.usask.ca/index.html.

Piotrowski, M. R., Aaserude, R. G., and Schmidt, F. J. 1992. Bioremediation of diesel contaminated soil and tundra in an Arctic environment. In: *Contaminated Soils: Diesel Fuel Contamination*, Kostecki, P. T. and Calabrese, E. J. (eds), Chelsea, MI, Lewis Publishers, 115–42.

Poland, J. S., Mitchell, S., and Rutter, A. 2001. Remediation of former military bases in the Canadian Arctic. *Cold Regions Sci. Technol.* **32**: 93–105.

Poland, J. S., Riddle, M. J., and Zeeb, B. A. 2003. Contaminants in the Arctic and the Antarctic: a comparison of sources, impacts, and remediation options. *Polar Rec.* **39**: 369–84.

Poland, J. S., Rutter, A., Rowe, K., McWatters, R., and Kalinovich, I. 2004. Design and application of a funnel and gate barrier system for PCB containment and remediation in the Canadian Arctic. *Contaminants in Freezing Ground: Proceedings of the 4th International Conference*, Filler, D. M. and Barnes, D. L. (eds.), Fairbanks, Alaska.

Pombo, S. A., Pelz, O., Schroth, M. H., and Zeyer, J. 2002. Field-scale ^{13}C-labeling of phospholipid fatty acids (PLFA) and dissolved inorganic carbon: tracing acetate assimilation and mineralization in a petroleum hydrocarbon-contaminated aquifer. *FEMS Microbiol. Ecol.* **41**: 259–67.

Potter, T. and Simmons, K. E. 1998a. *Composition of Petroleum Mixtures*. Amherst, MA, Amherst Scientific Publishers.

Potter, T. and Simmons, K. E. 1998b. *Analysis of Petroleum Hydrocarbons in Environmental Media*. Amherst, MA, Amherst Scientific Publishers.

Potts, M. 1994. Desiccation tolerance of prokaryotes. *Microbiol. Rev.* **58**: 755–805.

Pouliot, Y., Pokiak, C., Moreau, N., Thomassin-Lacroix, E., and Faucher, C. 2003. Soil remediation of a former tank farm site in western arctic Canada. *Proc. 3rd Assessment and Remediation of Contaminated Sites in Arctic and Cold Climates Conference (ARCSACC)*, Edmonton, May 4–6, 262–7.

Poulsen, M. M. and Kueper, B. H. 1992. A field experiment to study the behavior of tetrachloroethylene in unsaturated porous media. *Environ. Sci. Technol.* **26**(5): 889–95.

Powell, S. M., Bowman, J. P., and Snape, I. 2004. Degradation of nonane by bacteria from Antarctic marine sediment. *Polar Biol.* **27**: 573–8.

Powell, S. M., Bowman, J. P., Snape, I., and Stark, J. S. 2003. Microbial community variation in pristine and polluted nearshore Antarctic sediments. *FEMS Microbiol. Ecol.* **45**: 135–45.

Powell, S. M., Ferguson, S. H., Bowman, J. P., and Snape, I. 2006a. Using real-time PCR to assess changes in the hydrocarbon-degrading microbial community in Antarctic soil during bioremediation. *Microbial Ecol.* **52**: 523–32.

Powell, S. M., Ferguson, S. H., Snape, I., and Siciliano, S. D. 2006b. Fertilization stimulates anaerobic fuel degradation of Antarctic soils by denitrifying microorganisms. *Environ. Sci. Technol.* **40(6)**: 2011–17.

Powell, S. M., Harvey, P. M., Stark, J. S., Snape, I., and Riddle, M. J. 2007. Biodegradation of petroleum products in experimental plots in Antarctic marine sediments is location dependent. *Mar. Pollut. Bull.* **54**: 434–40.

Powell, S. M., Snape, I., Bowman, J. P., *et al.* 2005. A comparison of the short term effect of diesel fuel and lubricant oils on Antarctic benthic microbial communities. *J. Exp. Mar. Biol. Ecol.* **322**: 53–65.

Prince, R. C., Owens, E. H., and Sergy, G. A. 2002. Weathering of an Arctic oil spill over 20 years: the BIOS experiment revisited. *Mar. Pollut. Bull.* **44(11)**: 1236–42.

Pritchard, P. H. and Costa, C. F. 1991. EPA's Alaska oil spill bioremediation project. *Environ. Sci. Technol.* **25**: 372–9.

Pruthi, V. and Cameotra, S. S. 1997. Production and properties of a biosurfactant synthesized by *Arthrobacter protophormiae* – an Antarctic strain. *World J. Microbiol. and BioTechnol.* **13**: 137–9.

Purkamo, L., Salminen, J., and Jørgensen, K. 2004. Diversity of bacteria and archaea in petroleum hydrocarbon contaminated subsurface samples from Southern

Finland. Paper (in English) in publication by Finland's environmental administration: Maaperänsuojelu, Geologian tutkimuskeskuksen ja, Suomen ympäristökeskuksen, tutkimusseminaari 5.11.2004 (J. Seppälä, H. Idman, eds.), Ympäristönsuojelu Suomenympäristö 726, 40–6.

Rasiah, V., Voroney, R. P., and Kachanoski, R. G. 1991. Effect of N-amendment on C-mineralisation of an oily waste. *Water, Air Soil Poll.* **59(3–4)**: 249–59.

Rasiah, V., Voroney, R. P., and Kachanoski, R. G. 1992. Biodegradation of an oily waste as influenced by nitrogen forms and sources. *Water, Air Soil Poll.* **65(1–2)**: 143–51.

Ratkowsky, D. A., Lowry, R. K., McMeekin, T. A., Stokes, A. N., and Chandler, R. E. 1983. Model for bacterial culture growth rate though out the entire biokinetic temperature range. *J. Bacteriology* **154**: 1222–6.

Ratkowsky, D. A., Olley, J., McMeekin, T. A., and Ball, A. 1982. Relationship between temperature and growth rate of bacterial cultures. *J. Bacteriology* **149**: 1–5.

Rayner, J. L., Snape, I., Walworth, J. L., Harvey, P. M., and Ferguson, S. H. 2007. Petroleum-hydrocarbon contamination and remediation by microbioventing at sub-Antarctic Macquarie Island. *Cold Reg. Sci. Technol.* **48**: 139–53.

Reardon, K. F., Mosteller, D. C., and Rogers, J. D. B. 2000. Biodegradation kinetics of benzene, toluene, and phenol as single and mixed substrates for *Pseudomonas putida* F1. *BioTechnol. and Bioeng.* **69**: 385–400.

Redfield, A. C., Ketchum, B. H., and Richards, F. A. 1963. The influence of organisms on the composition of seawater. In *The Sea*, Hill, M. N. (ed.), New York, Wiley, 26–77.

Reimer, K. J., Colden, M., Francis, P., *et al.* 2003. Cold climate bioremediation – a comparison of various approaches. *Proc. 3rd Assessment and Remediation of Contaminated Sites in Arctic and Cold Climates Conference (ARCSACC)*, Edmonton, May 4–6, 290–300.

Reimer, K. J., Zeeb, B. A., Koch, I., *et al.* 2005. A critical review of bioremediation. *Proceedings of 4th Assessment and Remediation of Contaminated Sites in Arctic and Cold Climates Conference (ARCSACC)*, Edmonton, May, 195–232.

Reinuk, I. T. 1959. Condensation in the active layer of permafrost. Magadan, VNIIzoloto. (in Russian).

Revill, A. T., Snape, I., Jucieer, A., and Guille, D. 2007. Constraints on transport and weathering of petroleum hydrocarbons at Casey Station, Antarctica. *Cold Reg. Sci. Technol.* **48**: 154–67.

Reynolds, C. M. 1993. Field measured bioremediation rates in a cold region landfarm: spatial variability relationships. In *Hydrocarbon Contaminated Soils*, Kostecki, P. T. and Calabrese, E. J. (eds.), Chelsea, MI, Lewis Publishers, 487–99.

Reynolds, C. M. 2004a. Cyclic temperature effects on soil microbial activity and possible impacts on remediating contaminated soil. *Proc. 4th Int'l. Contaminants in Freezing Ground Conf.*, Filler, D. M. and Barnes, D. L. (eds.), Fairbanks, Alaska.

Reynolds, C. M. 2004b. *Technology Demonstration Final Report – Field Demonstration of Rhizosphere-Enhanced Treatment of Organics-Contaminated Soils on Native American Lands with Application to Northern FUD Sites.* ESTCP Final Report. CRREL LR-04-18. www.crrel.usace.army.mil/techput/CRREL_Reports/reports/ LR-04-18.pdf

Reynolds, C. M., Braley, W. A., Travis, M. D., Perry, L. B., and Iskandar, I. K. 1998. *Bioremediation of Hydrocarbon-Contaminated Soils and Groundwater in Northern Climates*. CRREL Special Report 98–5.

Reynolds, C. M., Travis, M., Braley, W. A., and Scholze, R. J. 1994. Applying field expedient bioreactors and landfarming in cold climates. In *Hydrocarbon Bioremediation*, Hinchee, R., Miller, R. N., and Hoeppel, R. E. (eds.), Chelsea, MI, Lewis Publishers, 100–6.

Rhodes, M., Wardell-Johnson, G. W., Rhodes, M. P., and Raymond, B. 2006. Applying network theory to the conservation of habitat tress in urban environments: a case study from Brisbane, Australia. *Conserv. Biol.* **20**: 861–70.

Rhykerd, R. L., Weaver, R. W., and McInnes, K. J. 1995. Influence of salinity on bioremediation of oil in soil. *Environ. Pollution* **90**: 127–30.

Richmond, S. A., Lindstrom, J. E., and Braddock, J. F. 2001. Assessment of natural attenuation of chlorinated aliphatics and BTEX in subarctic groundwater. *Environ. Sci. Technol.* **35(20)**: 4038–45.

Rike, A. G., Børresen, M., and Instances, A. 2002. Response of cold-adapted microbial populations in a permafrost profile to hydrocarbon contaminants. *Polar Rec.* **37(202)**: 239–48.

Rike, A. G., Haugen, K. B., Børresen, M., Engene, B., and Kolstad, P. 2003a. In situ biodegradation of petroleum hydrocarbons in frozen arctic soils. *Cold Reg. Sci. Technol.* **37(2)**: 97–120.

Rike, A. G., Haugen, K. B., Børresen, M., Kolstad, P., and Engene, B. 2003b. In-situ monitoring of hydrocarbon biodegradation in the winter months at Longyearbyen, Spitsbergen. *Proc. 3rd Assessment and Remediation of Contaminated Sites in Arctic and Cold Climates (ARCSACC) Conference*, Nahir, M., Biggar, K., and Cotta, G. (eds.), St. Joseph's Print Group, Edmonton, May 4–6, 268–78.

Rike, A. G., Haugen, K. B., and Engene, B. 2005. *In situ* biodegradation of hydrocarbons in arctic soil at sub-zero temperatures – field monitoring and theoretical simulation of the microbial activation temperature at a Spitsbergen contaminated site. *Cold Reg. Sci. Technol.* **41**: 189–209.

Riser-Roberts, E. 1998. *Remediation of Petroleum Contaminated Soils: Biological, Physical, and Chemical Processes*, Boca Raton, Lewis Publishers.

Ristinen, R. and Kraushaar, J. 1999. *Energy and Problems of a Technical Society*, New York, John Wiley and Sons.

Rivkina, E. M., Friedmann, E. I., McKay, C. P., and Gilichinsky, D. A. 2000. Metabolic activity of permafrost bacteria below the freezing point. *Appl. and Environ. Microbiol.* **66(8)**: 3230–3.

Rosenberg, E. 1992. Hydrocarbon-oxidising bacteria (Chapter 19). In *The Procaryotes: A Handbook on the Biology of Bacteria: Ecophysiology, Isolation, Identification, Applications. 2nd Edition*, Balows, A. *et al.* (eds.), New York, Springer-Verlag, 446–59.

Rothwell, D. R. and Davis, R. 1997. *Antarctic Environmental Protection: A Collection of Australian and International Instruments* NSW, The Federation Press.

Roura, R. 2004. Monitoring and remediation of hydrocarbon contamination at the former site of Greenpeace's World Park Base, Cape Evans, Ross Island, Antarctica. *Polar Rec.* **40**: 51–67.

Rowsell, S. 2003. A decision tree for selecting bioremediation in cold climates. *Proc. 3rd Assessment and Remediation of Contaminated Sites in Arctic and Cold Climates (ARCSACC) Conference*, Nahir, M., Biggar, K., and Cotta, G. (eds), St. Joseph's Print Group, Edmonton, May 4–6, 183–96.

Ruberto, L., Vazquez, S. C., and MacCormack, W. P. 2003. Effectiveness of the natural bacterial flora, biostimulation and bioaugmentation on the bioremediation of a hydrocarbon contaminated Antarctic soil. *Int'l. Biodeterioration & Biodegradation* **52**: 115–25.

Russell, N. J. 1990. Cold adaptation of microorganisms. *Phil. Trans. R. Soc. Lond.* **B 326**: 595–611.

Russell, N. J. 2000. Toward a molecular understanding of cold activity of enzymes from psychrophiles. *Extremophiles* **4**: 83–90.

Russell, N. J. 2002. Bacterial membranes: the effects of chill storage and food processing. An overview. *Int'l. J. Food Microbiol.* **79**: 27–34.

Ryden, B. E. and Kostov, L. 1977. Ground water and the water-frost cycle in a tundra mire, *Striae* **4**: 17–19.

Salanitro, J. P. 1993. The role of bioattenuation in the management of aromatic hydrocarbon plumes in aquifers. *Ground Water Monitor. & Remed.* **13**: 150–61.

Salminen, J. M., Tuomi, P. M., Suortti, A.-M., and Jørgensen, K. S. 2004. Potential for aerobic and anaerobic biodegradation of petroleum hydrocarbons in boreal subsurface. *Biodegradation* **15**: 29–39.

Sandvik, S., Lode, A., and Pedersen, T. A. 1986. Biodegradation of oily sludge in Norwegian soils. *Appl. Microbiol. and BioTechnol.* **23**: 297–301.

Saul, D. J., Aislabie, J., Brown, C. E., Harris, L., and Foght, J. M. 2005. Hydrocarbon contamination changes the bacterial diversity of soil from around Scott Base, Antarctica. *FEMs Microbiol. Ecol.* **53**: 141–55.

Schafer, A. N., Snape, I., and Siciliano, S. D. 2007. Soil biogeochemical toxicity endpoints for sub-Antarctic Islands contaminated with petroleum hydrocarbons. *Environ. Toxicol. Chem.* **26**: 890–7.

Schiewer, S. and Niemeyer, T. 2006. Soil heating and optimized nutrient addition for accelerating bioremediation in cold climates. *Polar Rec.* **42(1)**: 23–31.

Schinder-Keel, U., Bang Lejbølle, K., Baehler, E., Haas, D., and Keel, C. 2001. The sigma factor AlgU (AlgT) controls exopolysaccharide production and tolerance towards desiccation and osmotic stress in the biocontrol agent *Pseudomonas fluorescens* CHA0. *Appl. and Environ. Microbiol.* **67**: 5683–93.

Schmidtke, T., White, D., and Woolard, C. 1999. Oxygen release kinetics from solid phase oxygen in Arctic Alaska. *J. Haz. Mat'ls.* **B64**: 157–65.

Schnitzer, M. 1991. Soil organic matter – the next 75 years. *Soil Sci.* **151**: 41–58.

Schofield, R. K. 1935. The pH of the water in soil. *Proc. 3rd Int'l. Congress on Soil Sci.* **2**: 37–48; **3**: 182–6.

Scow, K. M. 1982. Rate of biodegradation. In *Handbook of Chemical Property Estimation Methods, Environmental Behavior of Organic Compounds*, Lyman, W. J., Reehl, W. F., and Rosenblatt, D. H. (eds.), New York, McGraw-Hill, Chapter 16.

Scow, K. M., Simkins, S., and Alexander, M. 1986. Kinetics of mineralisation of organic compounds at low concentrations in soils. *Appl. and Environ. Microbiol.* **51**: 1028–35.

Semple, K. T., Morris, A. W. J., and Paton, G. I. 2003. Bioavailability of hydrophobic organic contaminants in soils: fundamental concepts and techniques for analysis. *European J. Soil Sci.* **54**: 809–18.

SERDP. 2005. Final report. *SERDP and ESTCP Expert Panel Workshop on Research and Development Needs for the Environmental Remediation Application of Molecular Biological Tools.* http://docs.serdp-estcp.org/viewfile.cfm?Doc=SedimentsFinalReport.pdf.

Sexstone, A. J. and Atlas, R. M. 1977. Response of microbial populations in Arctic tundra soils to crude oil. *Can. J. of Microbiol.* **23**: 1327–33.

Seyfried, M. S. and Murdock, M. D. 1997. Use of air permeability to estimate infiltrability of frozen soils, *J. of Hydrol.* **202**: 95–107.

SFT. 1999. *Guidelines for the Risk Assessment of Contaminated Sites, Report 99:06.* Oslo, Norway, Norwegian Pollution Control Authority.

Shapley, D. 1974. Antarctica – world hunger for oil spurs Security Council review. *Science* **184**: 776–81.

Shields, D., Janzen, P., McCartney, D., and Man, A. 1997. In situ bioremediation in a sub-Arctic climate. *Proc. '97 In Situ and On-Site Bioremediation Symposium,* Columbus, OH, Battelle Press, 4(1): 319.

Shur, Y. L. 1988a. *Upper Permafrost Horizon and Thermokarst.* Novosibirsk, "Nauka".

Shur, Y. L. 1988b. The upper horizon of permafrost soil. *Proc. 5th Int'l. Permafrost Conf.,* Trondheim, Norway, 867–71.

Shur, Y. L., Hinkel, K. M., and Nelson, F. E. 2005. The transient layer: Implications for geocryology and climate-change science, *Permafrost Periglac.* **16**: 5–17.

Shur, Y. L. and Ping, C. L. 1994. Permafrost dynamics and soil formation. *Proc. of the Meeting on the Classification, Correlation, and Management of Permafrost-Affected Soils,* Soil Conservation Service, Lincoln, Nebraska, 112–17.

Sierra, J. and Renault, P. 1995. Oxygen consumption by soil microorganisms as affected by oxygen and carbon dioxide levels. *Appl. Soil Ecol.* **2**: 175–84.

Sims, J. L., Sims, R. C., and Matthews, J. E. 1989. *Bioremediation of Contaminated Surface Soils.* U.S. Environmental Protection Agency, EPA/600/9-89/073, Washington, D.C.

Smith, E. P., Lipkovich, I., and Ye, K. Y. 2002. Weight-of-evidence (WOE): Quantitative estimation of probability of impairment for individual and multiple lines of evidence. *Hum. Ecol. Risk Assess.* **8**: 1585–96.

Smith, M. W. and Burn, C. R. 1987. Outward flux of vapour from frozen soils at Mayo, Yukon, Canada: results and interpretation. *Cold Reg. Sci. Technol.* **13**: 143–54.

Snape, I., Ferguson, S. H., Harvey, P. M., and Riddle, M. J. 2006a. Investigation of evaporation and biodegradation of fuel spills in Antarctica: II – Extent of natural attenuation at Casey Station. *Chemosphere* **63**: 89–98.

Snape, I., Ferguson, S., and Revill, A. 2003. Constraints of rates of natural attenuation and in situ bioremediation of petroleum spills in Antarctica. *Proc.*

3rd Assessment and Remediation of Contaminated Sites in Arctic and Cold Climates (ARCSACC) Conference, Nahir, M., Biggar, K., and Cotta, G. (eds.), St. Joseph's Print Group, Edmonton, May 4–6, 257–61.

Snape, I., Gore, D. B., Cole, C. M., and Riddle, M. J. 2002. Contaminant dispersal and mitigation at Casey Station: an example of how applied geoscience research can reduce environmental risks in Antarctica. *Royal Soc. of New Zealand Bulletin* **35**: 641–8.

Snape, I., Harvey, P. M., Ferguson, S. H., Rayner, J. L., and Revill, A. T. 2005. Investigation of evaporation and biodegradation of fuel spills in Antarctica: I – a chemical approach using GC-FID. *Chemosphere* **61**: 1485–94.

Snape, I., Morris, C.E, and Cole, C. M. 2001. The use of permeable reactive barriers to control contaminant dispersal during site remediation in Antarctica. *Cold Reg. Sci. Technol.* **32**: 157–74.

Snape, I., Riddle, M. J., Gore, D. G., and Cole, C. M. 1998. *Interim Report on the Contaminated Sites of the 'Old' Casey Tip and Abandoned Wilkes Station.* A report to the Environmental Management Section, Australian Antarctic Division, Hobart, Australia.

Snape, I., Riddle, M. J., Stark, J. S., Cole, C. M., King, C. K., Duquesne, S., and Gore, D. B. 2001. Management and remediation of contaminated sites at Casey Station, Antarctica. *Polar Rec.* **37**: 199–214.

Snape, I., Siciliano, S., Schafer, A., Rayner, J. L., and Riddle, M. J. 2006b. Development of fuel spill remediation guidelines for Antarctica. *2nd SCAR Open Science Conference 'Antarctica in the Earth System'*, Hobart, 143.

Snape, I., Siciliano, S., Schafer, A., *et al.* 2006c. Development of petroleum remediation guidelines for polar regions. *Contaminants in Freezing Ground: Proc. 5th Int'l. Conf.*, Rike, A. G., Øvstedal, J., and Vethe, Ø. (eds.), Oslo, Norway, Norsk Geologisk Forening, 87.

Soehnlen, G. 1991. Cleansing Contaminated, Granular Soils by Controlled Freezing, Masters Report, Michigan State University.

Solé, R. V. and Montoya, J. M. 2001. Complexity and fragility in ecological networks. *Proc. Roy. Soc.* **268**: 2039–45.

Soloway, D. A., Nahir, M., Billowits, M. E., and Whyte, L. G. 2001. In situ bioremediation of diesel-contaminated soil in Canada's Arctic territory: A case study at the Whitehorse International Airport, Yukon Territory. *Polar Rec.* **37(202)**: 267–72.

Stallwood, B., Shears, J., Williams, P. A., and Hughes, K. A. 2005. Low temperature bioremediation of oil-contaminated soil using biostimulation and bioaugmentation with a *Pseudomonas* sp from maritime Antarctica. *J. of Appl. Microbiol.* **99**: 794–802.

Stark, S. C., Gardner, D., Snape, I., and Mclvor, E. 2003. Assessment of contamination by heavy metals and petroleum hydrocarbons at Atlas Cove Station, Heard Island. *Polar Rec.* **39**: 397–414.

Starr, R. C. and Cherry, J. A. 1994. In situ remediation of contaminated ground water: the funnel and gate system. *Ground Water* **32**: 465–76.

Stevenson, F. J. 1985. Geochemistry of soil humic substances. In *Humic Substances in Soil, Sediment, and Water: Geochemistry, Isolation, and Characterization*, Aiken, G. R., McKnight, D. M., Wershaw, R. L., and MacCarthy, P. (eds.), New York, Wiley-Interscience, 13–52.

Stevenson, F. J. and Cole, M. A. 1999. The nitrogen cycle in soil: Global and ecological aspects. In *Cycles of Soil: Carbon, Nitrogen, Phosphorus, Sulfur, Micronutrients*, New York, Wiley, 139–90.

Stow, J. P., Sova, J., and Reimer, K. J. 2005. The relative influence of distant and local (DEW-line) PCB sources in the Canadian Arctic. *Sci. Total Environ.* **342**: 107–18.

Stumm, W. and Morgan, J. J. 1996. *Aquatic Chemistry: Chemical Equilibria and Rates in Natural Waters, 3rd Edn*, New York, John Wiley & Sons.

Suarez, M. P. and Rifai, H. S. 1999. Biodegradation rates for fuel hydrocarbons and chlorinated solvents in groundwater. *Bioremed. J.* **3(4)**: 337–62.

Swedish EPA. 2002. Environmental quality criteria for contaminated sites. Swedish Environmental Protection Agency, viewed 21 December 2006, www.internat.naturvardsverket.se/index.php3?main = /documents/legal/assess/assedoc/cont.htm.

Thieringer, H. A., Jones, P. G., and Inouye, M. 1998. Cold shock and adaptation. *BioEssays* **20**: 49–57.

Thomas, H., Jensen, D., and Authier, B. 1995. Remediation of crude-oil-contaminated soils beneath a containment liner. *Proc. Geophysical Environmental 2000*, Reston, VA American Society of Civil Engineers, 52–8.

Thomassin-Lacroix, E. J. M., Eriksson, M., Reimer, K. J., and Mohn, W. W. 2002. Biostimulation and bioaugmentation for on-site treatment of weathered diesel fuel in Arctic soil. *Appl. Microbiol. Biotechnol.* **59**: 551–6.

Thomassin-Lacroix, E. J. M., Yu, Z., Eriksson, M., Reimer, K., and Mohn, W. W. 2001. DNA-based and culture-based characterization of a hydrocarbon-degrading consortium enriched from Arctic soil. *Can. J. of Microbiol.* **47**: 1107–15.

Thompson, B. A. W., Davies, N. W., Goldsworthy, P. M., *et al.* 2006. In situ lubricant degradation in Antarctic marine sediments. 1. Short-term changes. *Environ. Toxicol. Chem.* **25**: 356–66.

Tice, A. R., Anderson, D. M., and Banin, A. 1976. *The Prediction of Unfrozen Water Contents in Frozen Soils from Liquid Limit Determinations.* U.S. Army Cold Regions Research and Engineering Laboratory Report CRREL 76–8.

Tisdale, S. L., Nelson, W. L., Beaton, J. D., and Havlin, J. L. 1993. *Soil Fertility and Fertilizers*, New York, MacMillan.

Tishin, M. I. 1983. Thermal regime formation under large lakes in Central Yakutia. In *Thermal Physics Studies in Siberia Ktyolitozone*, Pavlov, A. (ed.), Novosibirsk "Nauka", 127–35 (in Russian).

Tolstikhin, N. I. and Tolstikhin, O. N. 1973. Underground and surface water of the permafrost region. In *General Permafrost Science*, Mel'nikov, P. I. and Tostikhin, N. I. (eds.), Novosibirsk, "Nauka", 192–229 (in Russian).

Torsvik, V. and Øvreås, L. 2002. Microbial diversity and function in soil: from genes to ecosystems. *Current Opinions in Microbiol.* **5**: 240–5.

TPHCWG. 1998a. *Analysis of Petroleum Hydrocarbons in Environmental Media*. Total Petroleum Hydrocarbon Criteria Working Group series; Volume 1. Amherst, MA, Amherst Scientific Publishers.

TPHCWG. 1998b. *Characterisation of C_6 to C_{35} Petroleum Hydrocarbons in Environmental Samples*. Total Petroleum Hydrocarbon Criteria Working Group series. Amherst, MA, Amherst Scientific Publishers.

Trefry, M. G. and Franzmann, P. D. 2003. An extended kinetic model accounting for non-ideal microbial substrate mineralisation in environmental samples. *Geomicrobiol. J.* **20**: 113–29.

Ulrich, A. C., Biggar, K. W., Armstrong, J., *et al.* 2006. Impact of cold temperatures on biodegradation rates. *Proc. Sea to Sky Geotechnique 2006, 59th Canadian Geotech. Conf., and 7th Joint CGS/IAH-CNC Groundwater Specialty Conf.*, Vancouver, Canada, paper no. 484.

UNEP-WCMC. 1994. Russian Arctic Oil Pipeline Spill. United Nations Environment Programme / World Conservation Monitoring Centre, Cambridge, UK, viewed 28 August 2006, www.unep-wcmc.org/latenews/emergency/usinsk_pipeline_1994/usinsk.htm.

United States Environmental Protection Agency. 1995. *How to Evaluate Alternative Cleanup Technologies for Underground Storage Tank Sites: A Guide for Corrective Action Plan Reviewers*. U.S. Environmental Protection Agency, EPA 510-B-95-007. Washington, DC.

USEPA. 1996. *Soil Screening Guidance: Technical Background Document. Second Edition.* Publication 9355.4-17A. Office of Emergency and Remedial Response, U.S. Environmental Protection Agency, Washington, DC.

UST. 2005. Contaminated Soil in Iceland. Environment and Food Agency of Iceland (UST), viewed 28 November 2006, http://english.ust.is/infobase/pollution-prevention/WasteManagementinIceland/ContaminatedsoilIniceland/nr/3064.

UST. 2006. *Waste Management in Iceland*. Environment and Food Agency of Iceland (UST).

van Everdingen, R. O. 1974. Groundwater in permafrost regions of Canada. *Proc. of Permafrost Hydrology* workshop, Can. National Committee for the International Hydrologic Decade, Ottawa, 83–93.

van Loon, W. K. P., van Haneghem, I. A., and Boshoven, H. P. A. 1988. Thermal and hydraulic conductivity of unsaturated sands. *5th Int'l. Symposium on Ground Freezing*, 81–90.

Van Stempvoort, D. R., Armstrong, J., and Mayer B. 2002. Bacterial sulfate reduction in biodegradation of hydrocarbons in low-temperature, high-sulfate groundwater, Western Canada. *Proc. '02 Petroleum Hydrocarbons Conf. and Organic Chemicals in Ground Water: Prevention, Detection, and Remediation*, Westerville, OH, National Ground Water Association (ed.), 244–59.

Van Stempvoort, D., Armstrong., J., and Mayer, B. 2007a. Microbial reduction of sulfate injected to gas condensate plumes in cold groundwater. *J. Contam. Hydrol.* (in press).

Van Stempvoort, D., Armstrong., J., and Mayer, B. 2007b. Seasonal recharge and replenishment of sulfate associated with biodegradation of a hydrocarbon plume, *Ground Water Monitor. Remed.* (in press).

Van Stempvoort, D. R., Bickerton, G., Lesage, S., and Millar, K. 2004. Cold-climate, in situ biodegradation of petroleum fuel in ground water, Moose Factory, Ontario, Canada. *Proc. '04 Petroleum Hydrocarbons and Organic Chemicals in Ground Water: Prevention, Assessment, and Remediation Conf.*, National Ground Water Association (ed.), Westerville, OH, 131–8.

Van Stempvoort, D. and Biggar, K. W. 2007. Potential for bioremediation of petroleum hydrocarbons in groundwater under cold climate conditions: A review. *Cold Reg. Sci. Technol.* (in press).

Van Stempvoort, D., Biggar, K. W., Iwakun, O., Bickerton, G., and Voralek, J. 2006. *Characterization of Fuel Spill Plumes in Fractured Rock at a Permafrost Site: Colomac Mine, NWT.* 2005/2006 Program Progress Report, April 2006, National Water Research Institute and University of Alberta.

Van Stempvoort, D., Maathuis, H., Jaworski, E., Mayer, B., and Rich, K. 2005. Oxidation of fugitive methane in ground water linked to bacterial sulfate reduction. *Ground Water* **43(2)**: 187–99.

Venosa, D. V., Haines, J. R., and Allen, D. M. 1992. Efficacy of commercial inocula in enhancing biodegradation of weathered crude oil contaminating a Prince William Sound beach. *J. Ind. Microbiol.* **10**: 1–11.

Vidali, M. 2001. Bioremediation. An overview. *Pure and Appl. Chem.* **73(7)**: 1163–72.

Vigil, M. F. and Kissel, D. E. 1991. Equations for estimating the amount of nitrogen mineralized from crop residues. *J. Soil Sci. Soc. of America* **55**: 757–61.

Virginia, R. A. and Wall, D. H. 1999. How soils structure communities in the Antarctic dry valleys. *Bioscience* **49**: 973–83.

Vorhees, D. J., Weisman, W. H., and Gustafson, J. B. 1999. *Human Health Risk-Based Evaluation of Petroleum Release Sites: Implementing the Working Group Approach.* Total Petroleum Hydrocarbon Criteria Working Group series; Volume 5. Amherst, MA, Amherst Scientific Publishers.

Waksman, S. A. 1924. Influence of microorganisms upon the carbon-nitrogen ratio in the soil. *J. Agricultural Sci.* **14**: 555–62.

Wall, D. H. and Virginia, R. A. 1999. Controls on soil biodiversity: insights from extreme environments. *Appl. Soil Ecol.* **13**: 137–50.

Walworth, J., Braddock, J., and Woolard, C. 2001. Nutrient and temperature interactions in bioremediation of cryic soils. *Cold Reg. Sci. Technol.* **32**: 85–91.

Walworth, J., Pond, A., Snape, I., Rayner, J. L., and Harvey, P. M. 2007. Nitrogen requirements for maximizing petroleum bioremediation in a sub-Antarctic soil. *Cold Reg. Sci. Technol.* (in press).

Walworth, J. L. and Reynolds, C. M. 1995. Bioremediation of a petroleum contaminated soil: Effects of phosphorus, nitrogen and temperature. *J. Soil Contam.* **4(3)**: 299–310.

Walworth, J. L., Woolard, C. R., Acomb, L., and Wallace, M. 1999. Nutrient and temperature interactions in bioremediation of petroleum-contaminated cryic soils. *In-Situ and On-Site Bioremediation* **5**(3): 505–10.

Walworth, J. L., Woolard, C. R., and Braddock, J. F. 1999. Nitrogen management in bioremediation. *Soil and Groundwater Cleanup* Feb/March: 12–15.

Walworth, J. L., Woolard, C. R., Braddock, J. F., and Reynolds, C. M. 1997a. Enhancement and inhibition of soil petroleum biodegradation through the use of fertilizer nitrogen: An approach to determining optimum levels. *J. Soil Contam.* **6(5)**: 465–80.

Walworth, J. L., Woolard, C. R., and Harris, K. C. 1997b. Bioremediation of petroleum-contaminated soil using fish bonemeal in cold climates. *AgroBorealis* **29**: 31–4.

Walworth, J. L., Woolard, C. R., and Harris, K. C. 2003. Nutrient amendments for contaminated peri-glacial soils: Use of cod bone meal as a controlled release nutrient source. *Cold Reg. Sci. Technol.* **43**: 1–8.

Wang, Z., Fingas, M., Blenkinsopp, S., *et al.* 1998. Comparison of oil composition changes due to biodegradation and physical weathering in different oils. *J. Chromatography A* **809**: 89–107.

Wang, Z. D. and Fingas, M. 2003. Fate and identification of spilled oils and petroleum products in the environment by GC-MS and GC-FID. *Energ. Sources* **25**: 491–508.

Wang, Z. D., Yang, C., Fingas, M., *et al.* 2005. Characterization, weathering, and application of sesquiterpanes to source identification of spilled lighter petroleum products. *Environ. Sci. Technol.* **39**: 8700–7.

Wartena, E. G. and Evenset, A. 1997. Effects of the Komi oil spill 1994 in the Nenets Okrug. Oil Components and Other Contaminants in Sediments and Fish from the Pechora River. 1995. Report APN514.789.1. Akvaplan-niva, Tromso.

Watanabe, K. and Hamamura, N. 2003. Molecular and physiological approaches to understanding the ecology of pollutant degradation. *Current Opinion in Biotechnol.* **14**: 289–95.

Waterhouse, E. J. and Roper-Gee, R. 2002. From dig and ship to watch and wait? Fuel spill management in the New Zealand Antarctic programme. *3rd Contaminants in Freezing Ground.* Snape, I. and Warren, R. (eds.), Hobart, Australian Antarctic Division.

Watson, S. W., Bock, E., Harms, H., Koops, H. P., and Hooper, A. B. 1989. Nitrifying bacteria. In *Bergey's Manual of Systematic Bacteriology*, Staley, J. T. *et al.* (eds.), Baltinpre, MD, William and Wilkins, 1808–43.

Watts, J. R., Corey, J. C., and McLeod, K. W. 1982. Land application studies of industrial waste oils. *Environ. Pollution* **28**: 165–75.

Westervelt, W. W., Lawson, P. W., Wallace, M. N., and Fosbrook, C. 1997. Intrinsic remediation of arctic diesel fuel near drinking water wells. *Proc. '97 In Situ and On-Site Biororemediation Symposium*, Columbus, OH Battelle Press, 4(1):61–6.

Westlake, D. W. S., Jobson, A. M., and Cook, F. D. 1977. *In situ* degradation of oil in a soil of the boreal region of the Northwest Territories. *Can. J. Microbiol.* **24**: 254–60.

Westlake, D. W. S., Jobson, A., Phillippe, R., and Cook, F. D. 1973. Biodegradability and crude oil composition. *Can. J. Microbiol.* **20**: 915–28.

White, D. M. 1995. Bioremediation of crude oil in the active layer overlying Alaska's North Slope Permafrost, Ph.D. Dissertation, Univ. of Notre Dame.

White, D. M., Collins, C. M., Barnes, D., and Byard, H. 2004. Effects of a crude oil spill on permafrost after 24 years in interior Alaska. *Proc. Cold Regions Engineering and Construction Conf.*, American Society of Civil Engineers, Edmonton, May 16–19.

White, D. M., Garland, D. S., Beyer, L., and Yoshikawa, K. 2004. Pyrolysis-GC/MS fingerprinting of environmental samples. *J. Analytical and Appl. Pyrolysis* **71**: 107–18.

White, D. M. and Irvine, R. L. 1996. The bituminous material in Arctic peat: implications for analyses of petroleum contamination. *J. Haz. Mat'ls.* **49**: 81–196.

White, D. M. and Irvine, R. L. 1998a. Analysis of bioremediation in organic soils. In: *Bioremediation: Principles and Practice, Volume 1, Fundamentals and Applications*, Sikdar, S. K. and Irvine, R. L. (eds.), Lancaster, PA, Technomic Publishing, 185–221.

White, D. M. and Irvine, R. L. 1998b. Potential applications for pyrolysis-GC/MS in bioremediation. *Environ. Monitoring and Assessment* **50**: 53–65.

White, D. M., Luong, H., and Irvine, R. L. 1998. Pyrolysis-GC/MS analysis of contaminated soils in Alaska. *J. Cold Regions Eng.* **12**: 1–10.

White, T. L. and Williams, P. J. 1994. Cryogenic alteration of frost-susceptible soils. *Proc. 7th Int'l. Symposium on Freezing Ground*, Nancy, France, 17–24.

White, T. L. and Williams, P. J. 1996. The role of microstructure – geotechnical properties of freezing soils. *Proc. 5th Int'l. Symposium on Thermal Eng. and Sci. for Cold Regions*, Ottawa, Canada, 415–26.

Whyte, L. G., Bourbonniere, L., Bellerose, C., and Greer, C. W. 1999a. Bioremediation assessment of hydrocarbon-contaminated soils from high arctic. *Bioremediation J.* **3(1)**: 69–79.

Whyte, L. G., Bourbonniere, L., and Greer, C. W. 1997. Biodegradation of petroleum hydrocarbons by psychrotrophic *Pseudomonas* strains possessing both alkane (*alk*) and naphthalene (*nah*) catabolic pathways. *Appl. and Environ. Microbiol.* **63**: 3719–23.

Whyte, L. G., Bourbonniere, L., Roy, R., and Greer, C. W. 1998. *Bioremediation Assessment of Whitehorse Airport Contaminated Aquifer*. Phase 1 – Final Report prepared for: Public Works and Government Services Canada. Environmental Microbiology, NRC – Biotechnology Research Institute, Montreal, Quebec, Canada.

Whyte, L. G., Goalen, B., Labbé, D., Greer, C. W., and Nahir, M. 2001. Bioremediation treatability assessment of hydrocarbon-contaminated soils from Eureka, Nunavut. *Cold Reg. Sci. Technol.* **32(2–3)**: 121–32.

Whyte, L. G., Greer, C. W., and Inniss, W. E. 1996. Assessment of the biodegradation potential of psychrotrophic microorganisms. *Can. J. Microbiol.* **42**: 99–106.

Whyte, L. G., Hawari, J., Zhou, E., *et al.* 1998. Biodegradation of variable chain length alkanes at low temperatures by a psychrotrophic *Rhodococcus* sp. *Appl. and Environ. Microbiol.* **64**: 2578–84.

Whyte, L. G. and Innis, W. E. 1992. Cold shock proteins and cold acclimation proteins in a psychrotrophic bacterium. *Can. J. Microbiol.* **38**: 1281–5.

Whyte, L. G., Labbé, D., Goalen, B., *et al.* 2003. In-situ bioremediation of hydrocarbon contaminated soils in the high arctic. *Proc. 3rd Assessment and Remediation of Contaminated Sites in Arctic and Cold Climates (ARCSACC) Conference*, Nahir, M., Biggar, K., and Cotta, G. (eds), St. Joseph's Print Group, Edmonton, May 4–6, 245–56.

Whyte, L. G., Schultz, A., van Beilen, J. B., *et al.* 2002a. Prevalence of alkane monooxygenase genes in Arctic and Antarctic hydrocarbon-contaminated and pristine soils. *FEMS Microbiol. Ecol.* **41**: 141–50.

Whyte, L. G., Slagman, S. J., Pietrantonio, F., *et al.* 1999b. Physiological adaptations involved in alkane assimilation at low temperatures by *Rhodococcus* sp. Strain Q15. *Appl. and Environ. Microbiol.* **65**: 2961–8.

Whyte, L. G., Smits, T. M. H., Labbe, D., *et al.* 2002b. Gene cloning and characterization of multiple alkane hydroxylases in *Rhodococcus* sp. strains Q15 and NRRL B-16531. *Appl. and Environ. Microbiol.* **68**: 5933–42.

Wiggert, D. C., Andersland, O. B., and Davies, S. H. 1997. Movement of liquid contaminants in partially saturated frozen granular soils. *Cold Reg. Sci. Technol.* **25**: 111–17.

Williams, P. J. 1968. *Unfrozen Water Content of Frozen Soils and Soil Moisture Suction.* Division of Building Research National Research Council of Canada, Research Paper no. **359**: 11–26.

Wilson, B. H., Bledsoe, B. E., Kampbell, D. H., *et al.* 1986. Biological fate of hydrocarbons at an aviation gasoline spill site. *Proc. Conf. on Petroleum Hydrocarbons and Organic Chemicals in Ground Water*, National Water Well Association (ed.), Columbus, OH, 78–90.

Wilson, J. L., Conrad, S. H., Mason, W. R., Peplinski, W., and Hagan, E. 1990. *Laboratory Investigation of Residual Liquid Organics From Spills, Leaks, and the Disposal of Hazardous Wastes.* Robert S. Kerr Environmental Research Laboratory Office of Research and Development, U.S. EPA, EPA/600/6-90/004.

Wilson, J., Rowsell, S., Chu, A., MacDonald, A, and Hetman, R. 2003. Biotreatability and pilot scale study for remediation of arctic diesel at 10 C. *Proc. 3rd Assessment and Remediation of Contaminated Sites in Arctic and Cold Climates (ARCSACC) Conference*, Nahir, M., Biggar, K., and Cotta, G. (eds), St. Joseph's Print Group, Edmonton, May 4–6, 279–89.

Wingrove, T. 1997. Diesel contamination remediation at a remote site in a cold climate. *Practice Periodical of Haz., Toxic, and Radioactive Waste Mgm't.* **1(1)**: 30–4.

Woinarski, A. Z., Snape, I., Stevens, G. W., and Morris, C. E. 2002. Development of a natural zeolite permeable reactive barrier for the treatment of contaminated water in Antarctica. *Proc. 3rd Int'l. Conf. on Contaminants in Freezing Ground*, Snape, I., and Warren, R. (eds.), Hobart, Australia, 14–18 April, 87–8.

Woinarski, A. Z., Snape, I., Stevens, G. W., and Stark, S. C. 2003. The effects of cold temperature on copper ion exchange by natural zeolite for use in a permeable reactive barrier in Antarctica. *Cold Reg. Sci. Technol.* **37(2)**: 159–68.

Woinarski, A. Z., Stevens, G. W., and Snape, I. 2006. A natural zeolite permeable reactive barrier to treat heavy-metal contaminated waters in Antarctica: kinetic and fixed-bed studies. *IChemE* **84(B2)**: 109–16.

Wong, R. C. K., Chu, A., Ng, R., and Duchscherer, T. M. 2003. An experimental study of biodegradation kinetics for distillated fractions of Alberta crude oil at 5 °C and 20 °C. *Proc. 3rd Assessment and Remediation of Contaminated Sites in Arctic and Cold Climates (ARCSACC) Conference*, Nahir, M., Biggar, K., and Cotta, G. (eds), St. Joseph's Print Group, Edmonton, May 4–6, 197–203.

Wood, J. A. and Williams, P. J. 1985. Further experimental investigation of regelation flow with an ice sandwich permeater. In *Freezing and Thawing of Soil-water Systems*, Anderson, D. M. and Williams, P. J. (eds.), Technical Council on Cold Regions Engineering Monograph, New York, American Society of Civil Engineers, 85–94.

Woolard, C. R., Walworth, J. L., and White, D. M. 2000. Contaminated soil bioremediation in cold climates: nutrient management strategies to enhance hydrocarbon biodegradation rates. *ISCORD 2000, Proc. 6th Int'l. Symposium on Cold Region Development*, Hobart, Tasmania, Australia, 48–51.

Woolard, C. R., White, D. M., Walworth, J. L., and Hannah, M. E. 1999a. The magnitude and variability of biogenic interference in cold regions soils. *J. Cold Regions Eng.* **13(3)**: 113–21.

Wrenn, B. A., Haines, J. R., Venosa, A. D., Kadkhodayan, M., and Suidan, M. T. 1994. Effects of nitrogen source on crude oil biodegradation. *J. Ind. Microbiol.* **13**: 279–86.

Yakimov, M. M., Giuliano, L., Bruni, V., Scarfi, S., and Golyshin, P. N. 1999. Characterization of Antarctic hydrocarbon-degrading bacteria capable of producing bioemulsifiers. *Microbiologica* **22**: 249–56.

Yen, Y. C., Cheng, K. C., and Fukusako, S. 1991. Review of intrinsic thermophysical properties of snow, ice, sea ice, and frost. *Proc. 3rd Int'l. Symposium on Cold Regions Heat Transfer*, 187–218.

Yu, Z., Stewart, G. R., and Mohn, W. W. 2000. Apparent contradiction: psychrotolerant bacteria from hydrocarbon-contaminated Arctic tundra soils that degrade diterpenoids synthesized by trees. *Appl. and Environ. Microbiol.* **66**: 5148–54.

Zarling, J. P. and Braley, W. A. 1988. Geotechnical thermal analysis. In *Embarkment Design and Construction in Cold Regions: Technical Council on Cold Regions Engineering Monograph*, Reston, VA, American Society of Civil Engineers, 35–44.

Zhou, J. 2003. Microarrays for bacterial detection and microbial community analysis. *Current Opinion in Microbiol.* **6**: 288–94.

Zhou, E. and Crawford, R. L. 1995. Effects of oxygen, nitrogen, and temperature on gasoline biodegradation in soil. *Biodegradation* **6**: 127–40.

Index

Made in the USA
Lexington, KY
31 August 2014